TICKS, CARROTS, AND A PARAKEET NAMED LIBERACE...
COULD ANY OF THEM BE DEADLY?

● When thirty-nine schoolchildren fell ill in northern Arkansas, health officials immediately suspected a variety of "parrot fever" because of a sick little bird that local children called Liberace. But was the bird to blame? The truth would prove as intriguing as it was unexpected.

● A forty-year-old dishwasher had a touch of dyspepsia. Within days he was dead. The only clue his physician had uncovered was a word the man uttered twice: *"Schlachtfest."* A *Schlachtfest* is a pork feast, and there was one at a German-American meeting-and-banquet hall. But why did only the dishwasher get sick?

● A woman doctor woke up feeling very sick with cramps, joint pain, and nausea. She drank a steaming cupful of milk and felt better. Later her symptoms worsened, especially when she and her husband visited their summer cottage. But he wasn't affected. Was her illness psychosomatic? Or was the doctor in jeopardy from a totally unexpected cause?

THREE OF THE FASCINATING CASES IN
THE MEDICAL DETECTIVES

BERTON ROUECHÉ joined the staff of *The New Yorker* in 1944. His "Annals of Medicine" department and his stature as a medical journalist have been recognized by numerous awards, including those from the Lasker Foundation and the American Medical Association. He lived and worked in Amagansett, Long Island, for many years, and died, at age 83, in April 1994.

THE
MEDICAL
DETECTIVES

Berton
Roueché

T·T

TRUMAN TALLEY BOOKS/PLUME
NEW YORK

PLUME
Published by the Penguin Group
Penguin Books USA Inc., 375 Hudson Street,
New York, New York 10014, U.S.A.
Penguin Books Ltd, 27 Wrights Lane,
London W8 5TZ, England
Penguin Books Australia Ltd, Ringwood,
Victoria, Australia
Penguin Books Canada Ltd, 10 Alcorn Avenue,
Toronto, Ontario, Canada M4V 3B2
Penguin Books (N.Z.) Ltd, 182-190 Wairau Road,
Auckland 10, New Zealand

Penguin Books Ltd, Registered Offices:
Harmondsworth, Middlesex, England

Published by Truman Talley Books/Plume, an imprint of Dutton Signet, a
division of Penguin Books USA Inc. Previously published in a different form
in a Truman Talley Books/Times Books hardcover edition.

First Plume Printing, March, 1991
30 29 28 27 26 25 24 23 22

 REGISTERED TRADEMARK—MARCA REGISTRADA

Library of Congress Cataloging-in-Publication Data
Roueché, Berton, 1911–1994
    The medical detectives/Berton Roueché.
      p. cm.
    "Truman Talley books."
    Includes index.
    1. Medicine—Case studies. I. Title
  RC66.R655   1988
  616'.09—dc20   90-14267
  ISBN: 0-452-26588-6   CIP
Printed in the United States of America

FOR

William Shawn

I hope that Lord Grey and you are well—no easy thing seeing that there are above 1500 diseases to which Man is subjected.

—LETTER TO LADY GREY
FROM SIDNEY SMITH

Contents

The Medical Detectives

CHAPTER **1**

Eleven Blue Men

AT ABOUT EIGHT O'CLOCK on Monday morning, September 25, 1944, a ragged, aimless old man of eighty-two collapsed on the sidewalk on Dey Street, near the Hudson Terminal. Innumerable people must have noticed him, but he lay there alone for several minutes, dazed, doubled up with abdominal cramps, and in an agony of retching. Then a policeman came along. Until the policeman bent over the old man, he may have supposed that he had just a sick drunk on his hands; wanderers dropped by drink are common in that part of town in the early morning. It was not an opinion that he could have held for long. The old man's nose, lips, ears, and fingers were sky-blue. The policeman went to a telephone and put in an ambulance call to Beekman-Downtown Hospital, half a dozen blocks away. The old man was carried into the emergency room there at eight-thirty. By that time, he was unconscious and the blueness had spread over a large part of his body. The examining physician attributed the old man's morbid color to cyanosis, a condition that usually results from an insufficient supply of oxygen in the blood, and also noted that he was diarrheic

3

and in a severe state of shock. The course of treatment prescribed by the doctor was conventional. It included an instant gastric lavage, heart stimulants, bed rest, and oxygen therapy. Presently, the old man recovered an encouraging, if painful, consciousness and demanded, irascibly and in the name of God, to know what had happened to him. It was a question that, at the moment, nobody could answer with much confidence.

For the immediate record, the doctor made a free-hand diagnosis of carbon-monoxide poisoning—from what source, whether an automobile or a gas pipe, it was, of course, pointless even to guess. Then, because an isolated instance of gas poisoning is something of a rarity in a section of the city as crammed with human beings as downtown Manhattan, he and his colleagues in the emergency room braced themselves for at least a couple more victims. Their foresight was promptly and generously rewarded. A second man was rolled in at ten-twenty-five. Forty minutes later, an ambulance drove up with three more men. At eleven-twenty, two others were brought in. An additional two arrived during the next fifteen minutes. Around noon, still another was admitted. All of these nine men were also elderly and dilapidated, all had been in misery for at least an hour, and all were rigid, cyanotic, and in a state of shock. The entire body of one, a bony, seventy-three-year-old consumptive named John Mitchell, was blue. Five of the nine, including Mitchell, had been stricken in the Globe Hotel, a sunless, upstairs flophouse at 190 Park Row, and two in a similar place, called the Star Hotel, at 3 James Street. Another had been found slumped in the doorway of a condemned building on Park Row, not far from City Hall Park, by a policeman. The ninth had keeled over in front of the Eclipse Cafeteria, at 6 Chatham Square. At a quarter to seven that evening, one more aged blue man was brought in. He had been lying, too sick to ask for help, on his cot in a cubicle in the Lion Hotel, another flophouse, at 26 Bowery, since ten o'clock that morning. A clerk had finally looked in and seen him.

By the time this last blue man arrived at the hospital, an investigation of the case by the Department of Health, to which all outbreaks of an epidemiological nature must be reported, had been under way for five hours. Its findings thus far had not been illuminating. The investigation was conducted by two men. One was the Health Department's chief epidemiologist, Dr. Morris Greenberg, a small, fragile, reflective man of fifty-seven, who is now acting director of the Bureau of Preventable Diseases; the other was Dr. Ottavio Pellitteri, a field epidemiologist, who, since 1946, has been administrative medical inspector for the Bureau. He is thirty-six years old, pale, and stocky, and has a bristling black mustache. One day, when I was in Dr. Greenberg's office, he and Dr. Pellitteri told me about the case. Their recollection of it is, understandably, vivid. The derelicts were the victims of a type of poisoning so rare that only ten previous outbreaks of it had been recorded in medical literature. Of these, two were in the United States and two in Germany; the others had been reported in France, England, Switzerland, Algeria, Australia, and India. Up to September 25, 1944, the largest number of people stricken in a single outbreak was four. That was in Algeria, in 1926.

The Beekman-Downtown Hospital telephoned a report of the occurrence to the Health Department just before noon. As is customary, copies of the report were sent to all the Department's administrative officers. "Mine was on my desk when I got back from lunch," Dr. Greenberg said to me. "It didn't sound like much. Nine persons believed to be suffering from carbon-monoxide poisoning had been admitted during the morning, and all of them said that they had eaten breakfast at the Eclipse Cafeteria, at 6 Chatham Square. Still, it was a job for us. I checked with the clerk who handles assignments and found that Pellitteri had gone out on it. That was all I wanted to know. If it amounted to anything, I knew he'd phone me before making a written report. That's an arrangement we have here. Well, a couple of hours later I got a call from him. My interest perked right up."

"I was at the hospital," Dr. Pellitteri told me, "and I'd talked to the staff and most of the men. There were ten of them by then, of course. They were sick as dogs, but only one was in really bad shape."

"That was John Mitchell," Dr. Greenberg put in. "He died the next night. I understand his condition was hopeless from the start. The others, including the old boy who came in last, pulled through all right. Excuse me, Ottavio, but I just thought I'd get that out of the way. Go on."

Dr. Pellitteri nodded. "I wasn't at all convinced that it was gas poisoning," he continued. "The staff was beginning to doubt it, too. The symptoms weren't quite right. There didn't seem to be any of the headache and general dopiness that you get with gas. What really made me suspicious was this: Only two or three of the men had eaten breakfast in the cafeteria at the same time. They had straggled in all the way from seven o'clock to ten. That meant that the place would have had to be full of gas for at least three hours, which is preposterous. It also indicated that we ought to have had a lot more sick people than we did. Those Chatham Square eating places have a big turnover. Well, to make sure, I checked with Bellevue, Gouverneur, St. Vincent's, and the other downtown hospitals. None of them had seen a trace of cyanosis. Then I talked to the sick men some more. I learned two interesting things. One was that they had all got sick right after eating. Within thirty minutes. The other was that all but one had eaten oatmeal, rolls, and coffee. He ate just oatmeal. When ten men eat the same thing in the same place on the same day and then all come down with the same illness . . . I told Greenberg that my hunch was food poisoning."

"I was willing to rule out gas," Dr. Greenberg said. A folder containing data on the case lay on the desk before him. He lifted the cover thoughtfully, then let it drop. "And I agreed that the oatmeal sounded pretty suspicious. That was as far as I was willing to go. Common, ordinary, everyday food poisoning—I gathered that was what Pellitteri had in mind—wasn't a very satisfying answer. For one thing, cyanosis is hardly symptomatic of that. On the other hand, diarrhea and severe vomiting are, almost invariably. But they weren't in the clinical picture, I found, except in two or three of the cases. Moreover, the incubation periods—the time lapse between eating and illness—were extremely short. As you probably know, most food poisoning is caused by eating something that has been contaminated by bacteria. The usual offenders are the staphylococci—they're mostly responsible for boils and skin infections and so on—and the salmonella. The latter are related to the typhoid organism. In a staphylococcus case, the first symptoms rarely develop in under two hours. Often, it's closer to five. The incubation period in the other ranges from twelve to thirty-six hours. But here we were with something that hit in thirty minutes or less. Why, one of the men had got only as far as the sidewalk in front of the cafeteria before he was knocked out. Another fact that Pellitteri had dug up struck me as very significant. All of the men told him that the illness had come on with extraordinary suddenness. One minute they were feeling fine, and the next minute they were practically helpless. That was another point against the ordinary food-poisoning theory. Its onset is never that fast. Well, that suddenness began to look like a lead. It led me to suspect that some drug might be to blame. A quick and sudden reaction is characteristic of a great many drugs. So is the combination of cyanosis and shock."

"None of the men were on dope," Dr. Pellitteri said. "I told Greenberg I was sure of that. Their pleasure was booze."

"That was O.K.," Dr. Greenberg said. "They could have got a toxic dose of some drug by accident. In the oatmeal, most likely. I couldn't help thinking that the oatmeal was relevant to our problem. At any rate, the drug idea was very persuasive."

"So was Greenberg," Dr. Pellitteri remarked with a smile. "Actually, it was the only explanation in sight that seemed to account for everything we knew about the clinical and environmental picture."

"All we had to do now was prove it," Dr. Greenberg went on mildly. "I asked Pellitteri to get a blood sample from each of the men before leaving the hospital for a look at the cafeteria. We agreed he would send the specimens to the city toxicologist, Dr. Alexander O. Gettler, for an overnight analysis. I wanted to know if the blood contained methemoglobin. Methemogiobin is a compound that's formed only when any one of several drugs enters the blood. Gettler's report would tell us if we were at least on the right track. That is, it would give us a yes-or-no answer on drugs. If the answer was yes, then we could go on from there to identify the particular drug. How we would go about that would depend on what Pellitteri was able to turn up at the cafeteria. In the meantime, there was nothing for me to do but wait for their reports. I'd theorized myself hoarse."

Dr. Pellitteri, having attended to his bloodletting with reasonable dispatch, reached the Eclipse Cafeteria at around five o'clock. "It was about what I'd expected," he told me. "Strictly a horse market, and dirtier than most. The sort of place where you can get a full meal for fifteen cents. There was a grind house on one side, a cigar store on the other, and the 'L' overhead. Incidentally, the Eclipse went out of business a year or so after I was there, but that had nothing to do with us. It was just a coincidence. Well, the place looked deserted and the door was locked. I knocked, and a man came out of the back and let me in. He was one of our people, a health inspector for the Bureau of Food and Drugs, named Weinberg. His bureau had stepped into the case as a matter of routine, because of the reference to a restaurant in the notification report. I was glad to see him and to have his help. For one thing, he had put a temporary embargo on everything in the cafeteria. That's why it was closed up. His main job, though, was to check the place for violations of the sanitation code. He was finding plenty."

"Let me read you a few of Weinberg's findings," Dr. Greenberg said, extracting a paper from the folder on his desk. "None of them had any direct bearing on our problem, but I think they'll give you a good idea of what the Eclipse was like—what too many restaurants are like. This copy of his report lists fifteen specific violations. Here they are: 'Premises heavily infested with roaches. Fly infestation throughout premises. Floor defective in rear part of dining room. Kitchen walls and ceiling encrusted with grease and soot. Kitchen floor encrusted with dirt. Refuse under kitchen fixtures. Sterilizing facilities inadequate. Sink defective. Floor and walls at serving tables and coffee urns encrusted with dirt. Kitchen utensils encrusted with dirt and grease. Storage-cellar walls, ceiling, and floor encrusted with dirt. Floor and shelves in cellar covered with refuse and useless material. Cellar ceiling defective. Sewer pipe leaking. Open sewer line in cellar.' Well . . ." He gave me a squeamish smile and stuck the paper back in the folder.

"I can see it now," Dr. Pellitteri said. "And smell it. Especially the kitchen, where I spent most of my time. Weinberg had the proprietor and the cook out there, and I talked to them while he prowled around. They were very cooperative. Naturally. They were scared to death. They knew nothing about gas in the place and there was no sign of any, so I went to work on the food. None of what had been prepared for breakfast that morning was left. That, of course, would have been too much to hope for. But I was able to get together some of the kind of stuff that had gone into the men's breakfast, so that we could make a chemical determination at the Department. What I took was ground coffee, sugar, a mixture of evaporated milk and water that passed for cream, some bakery rolls, a five-pound carton of dry oatmeal, and some salt. The salt had been used in preparing the oatmeal. That morning, like every morning, the cook told me, he had prepared six gallons of oatmeal, enough to serve around a hundred and twenty-five people. To make it, he used five pounds of dry cereal, four gallons of water—regular city water—and a handful of salt. That was his term—a handful. There was an open gallon can of salt standing on the stove. He said the handful he'd put in that morning's oatmeal had come from that. He refilled the can on the stove every morning from a big supply can. He pointed out the big can—it was up on a shelf—and as I was getting it down to take with me, I saw another can, just like it, nearby. I took that one down, too. It was also full of salt, or, rather, something that looked like salt. The proprietor said it wasn't salt. He said it was saltpetre—sodium nitrate—that he used in corning beef and in making pastrami. Well, there isn't any harm in saltpetre; it doesn't even act as an antiaphrodisiac, as a lot of people seem to think. But I wrapped it up with the other loot and took it along, just for fun. The fact is, I guess, everything in that damn place looked like poison."

After Dr. Pellitteri had deposited his loot with a Health Department chemist, Andrew J. Pensa, who promised to have a report ready by the following afternoon, he dined hurriedly at a restaurant in which he had confidence and returned to Chatham Square. There he spent the evening making the rounds of the lodging houses in the neighborhood. He had heard at Mr. Pensa's office that an eleventh blue man had been admitted to the hospital, and before going home he wanted to make sure that no other victims had been overlooked. By midnight, having covered all the likely places and having rechecked the downtown hospitals, he was satisfied. He repaired to his office and composed a formal progress report for Dr. Greenberg. Then he went home and to bed.

The next morning, Tuesday, Dr. Pellitteri dropped by the Eclipse, which was still closed but whose proprietor and staff he had told to return for questioning. Dr. Pellitteri had another talk with the proprietor and the cook. He also had a few inconclusive words with the rest of the cafeteria's employees—two dishwashers, a busboy, and a counterman. As he was leaving, the cook, who had apparently passed an uneasy night with his conscience, remarked that it was possible that he had absent-mindedly refilled the salt can on the stove from the one that contained saltpetre. "That was interesting," Dr. Pellitteri told me, "even though such a possibility had already occurred to me, and even though I didn't know whether it was important or not. I assured him that he had nothing to worry about. We had been certain all along that nobody had deliberately poisoned the old men." From the Eclipse, Dr. Pellitteri went on to Dr. Greenberg's office, where Dr. Gettler's report was waiting.

"Gettler's test for methemoglobin was positive," Dr. Greenberg said. "It had to be a drug now. Well, so far so good. Then we heard from Pensa."

"Greenberg almost fell out of his chair when he read Pensa's report," Dr. Pellitteri observed cheerfully.

"That's an exaggeration," Dr. Greenberg said. "I'm not easily dumfounded. We're inured to the incredible around here. Why, a few years ago we had a case involving some numskull who stuck a fistful of potassium-thiocyanate crystals, a very nasty poison, in the coils of an office water cooler, just for a practical joke. However, I can't deny that Pensa rather taxed our credulity. What he had found was that the small salt can and the one that was supposed to be full of sodium nitrate both contained sodium *nitrite*. The other food samples, incidentally, were O.K."

"That also taxed my credulity," Dr. Pellitteri said.

Dr. Greenberg smiled. "There's a great deal of difference between nitrate and nitrite," he continued. "Their only similarity, which is an unfortunate one, is that they both look and taste more or less like ordinary table salt. Sodium nitrite isn't the most powerful poison in the world, but a little of it will do a lot of harm. If you remember, I said before that this case was almost without precedent—only ten outbreaks like it on record. Ten is practically none. In fact, sodium-nitrite poisoning is so unusual that some of the standard texts on toxicology don't even mention it. So Pensa's report was pretty startling. But we accepted it, of course, without question or hesitation. Facts are facts. And we were glad to. It seemed to explain everything very nicely. What I've been saying about sodium-nitrite poisoning doesn't mean that sodium nitrite itself is rare. Actually, it's fairly common. It's used in the manufacture of dyes and as a medical drug. We use it in treating certain heart conditions and for high blood pressure. But it also has another important use, one that made its presence at the Eclipse sound plausible. In recent years, and particularly during the war, sodium nitrite has been used as a substitute for sodium nitrate in preserving meat. The government permits it but stipulates that the finished meat must not contain more than one part of sodium nitrite per five thousand parts of meat. Cooking will safely destroy enough of that small quantity of the drug." Dr. Greenberg shrugged. "Well, Pellitteri had had the cook pick up a handful of salt—the same amount, as nearly as possible, as went into the oatmeal—and then had taken this to his office and found that it weighed approximately a hundred grams. So we didn't have to think twice to realize that the proportion of nitrite in that batch of cereal was considerably higher than one to five thousand. Roughly, it must have been around one to about eighty before cooking destroyed part of the nitrite. It certainly looked as though Gettler, Pensa, and the cafeteria cook between them had given us our answer. I called up Gettler and told him what Pensa had discovered and asked him to run a specific test for nitrites on his blood samples. He had, as a matter of course, held some blood back for later examination. His confirmation came through in a couple of hours. I went home that night feeling pretty good."

Dr. Greenberg's serenity was a fugitive one. He awoke on Wednesday morning troubled in mind. A question had occurred to him that he was unable to ignore. "Something like a hundred and twenty-five people ate oatmeal at the Eclipse that morning," he said to me, "but only eleven of them got sick. Why? The undeniable fact that those eleven old men were made sick by the ingestion of a toxic dose of sodium nitrite wasn't enough to rest on. I wanted to know exactly how much sodium nitrite each portion of that cooked oatmeal had contained. With Pensa's help again, I found out. We prepared a batch just like the one the cook had made on Monday. Then Pensa measured out six ounces, the size of the average portion served at the Eclipse, and analyzed it. It contained two and a half grains of sodium nitrite. That explained why the hundred and fourteen other people did not become ill. The toxic dose of sodium nitrite is three grains. But it didn't explain how each of our eleven old men had received an additional half grain. It seemed extremely unlikely that the extra touch of nitrite had been in the oatmeal when it was served. It had to come in later. Then I began to get a glimmer. Some people sprinkle a little salt, instead of sugar, on hot cereal. Suppose, I thought, that the bus-boy, or whoever had the job of keeping the table salt shakers filled, had made the same mistake that the cook had. It seemed plausible. Pellitteri was out of the office—I've forgotten where—so I got Food and Drugs to step over to the Eclipse, which was still under embargo, and bring back the shakers for Pensa to work on. There were seventeen of them, all good-sized, one for each table. Sixteen contained either pure sodium chloride or just a few inconsequential traces of sodium nitrite mixed in with the real salt, but the other was point thirty-seven per cent nitrite. That one was enough. A spoonful of that salt contained a bit more than half a grain."

"I went over to the hospital Thursday morning," Dr. Pellitteri said. "Greenberg wanted me to check the table-salt angle with the men. They could tie the case up neatly for us. I drew a blank. They'd been discharged the night before, and God only knew where they were."

"Naturally," Dr. Greenberg said, "it would have been nice to know for a fact that the old boys all sat at a certain table and that all of them put about a spoonful of salt from that particular shaker on their oatmeal, but it wasn't essential. I was morally certain that they had. There just wasn't any other explanation. There was one other question, however. Why did they use so *much* salt? For my own peace of mind, I wanted to know. All of a sudden, I remembered Pellitteri had said they were all heavy drinkers. Well, several recent clinical studies have demonstrated that there is usually a subnormal concentration of sodium chloride in the blood of alcoholics. Either they don't eat enough to get sufficient salt or they lose it more rapidly than other people do, or both. Whatever the reasons are, the conclusion was all I needed. Any animal, you know, whether a mouse or a man, tends to try to obtain a necessary substance that his body lacks. The final question had been answered."

[*1947*]

CHAPTER 2

A Pig from Jersey

———————————◆————————————

AMONG THOSE WHO PASSED through the general clinic of Lenox
Hill Hospital, at Seventy-sixth Street and Park Avenue, on Monday morning, April 6, 1942, was a forty-year-old Yorkville dishwasher whom I will call Herman Sauer. His complaint, like his
occupation, was an undistinguished one. He had a stomach ache.
The pain had seized him early Sunday evening, he told the examining physician, and although it was not unendurably severe, its
persistence worried him. He added that he was diarrheic and
somewhat nauseated. Also, his head hurt. The doctor took his
temperature and the usual soundings. Neither disclosed any cause
for alarm. Then he turned his attention to the manifest symptoms.
The course of treatment he chose for their alleviation was unex-
ceptionable. It consisted of a dose of bismuth subcarbonate, a
word of dietetic advice, and an invitation to come back the next
day if the trouble continued. Sauer went home under the comforting impression that he was suffering from nothing more serious
than a touch of dyspepsia.

Sauer was worse in the morning. The pain had spread to his

chest, and when he stood up, he felt dazed and dizzy. He did not, however, return to Lenox Hill. Instead, with the inconstancy of the ailing, he made his way to Metropolitan Hospital, on Welfare Island. He arrived there, shortly before noon, in such a state of confusion and collapse that a nurse had to assist him into the examining room. Half an hour later, having submitted to another potion of bismuth and what turned out to be an uninstructive blood count, he was admitted to a general ward for observation. During the afternoon, his temperature, which earlier had been, equivocally, normal, began to rise. When the resident physician reached him on his evening round, it was a trifle over a hundred and three. As is customary in all but the most crystalline cases, the doctor avoided a flat-footed diagnosis. In his record of the case, he suggested three compatible possibilities. One was aortitis, a heart condition caused by an inflammation of the great trunk artery. The others, both of which were inspired by an admission of intemperance that had been wrung from Sauer in the examining room, were cirrhosis of the liver and gastritis due to alcoholism. At the moment, the doctor indicated, the last appeared to be the most likely.

Gastritis, aortitis, and cirrhosis of the liver, like innumerable other ailments, can seldom be repulsed by specific medication, but time is frequently effective. Sauer responded to neither. His fever held and his symptoms multiplied. He itched all over, an edema sealed his eyes, his voice faded and failed, and the seething pains in his chest and abdomen advanced to his arms and legs. Toward the end of the week, he sank into a stony, comalike apathy. Confronted by this disturbing decline, the house physician reopened his mind and reconsidered the evidence. His adaptability was soon rewarded. He concluded that he was up against an acute and, to judge from his patient's progressive dilapidation, a peculiarly rapacious infection. It was an insinuating notion, but it had one awkward flaw. The white-blood-cell count is a reliable barometer of infection, and Sauer's count had been entirely normal. On Wednesday, April 15, the doctor requested that another count be made. He did not question the accuracy of the original test, but the thought had occurred to him that it might have been made prematurely. The report from the laboratory was on his desk when he reached the hospital the following day. It more than confirmed his hunch. It also relieved him simultaneously of both uncertainty and hope. Sauer's white count was morbidly elevated by a preponderance of eosinophiles, a variety of cell that is produced by several potentially epidemic diseases but just one as formidably dishevelling as the case in question. The doctor put down the report and called the hospital superintendent's office. He asked the clerk who answered the phone to inform the Department of Health, to which the appearance of any disease of an epidemiological nature must be promptly communicated, that he had just uncovered a case of trichinosis.

The cause of trichinosis is a voracious endoparasitic worm, *Trichinella spiralis,* commonly called trichina, that lodges in the muscle fibres of an animal host. It enters the host by way of the alimentary canal, and in the intestine produces larvae that penetrate the intestinal walls to enter the blood stream. The worm is staggeringly prolific, and it has been known to remain alive, though quiescent, in the body of a surviving victim for thirty-one years. In general, the number of trichinae that succeed in reaching the muscle determines the severity of an attack. As such parasitic organisms go, adult trichinae are relatively large, the males averaging one-twentieth of an inch in length and the females about twice that. The larvae are less statuesque. Pathologists have found as many as twelve hundred of them encysted in a single gram of tissue. Numerous animals, ranging in size from the mole to the hippopotamus, are hospitable to the trichina, but it has a strong predilection for swine and man. Man's only important source of infection is pork. The disease is perpetuated in swine by the practice common among hog raisers of using garbage, some of which inevitably contains trichinous meat, for feed. Swine have a high degree of tolerance for the trichina, but man's resistive powers are feeble. In 1931, in Detroit, a man suffered a violent seizure of trichinosis as a result of merely eating a piece of bread buttered with a knife that had been used to slice an infested sausage. The hog from which the sausage was made had appeared to be in excellent health. Few acute afflictions are more painful than trichinosis, or more prolonged and debilitating. Its victims are occasionally prostrated for many months, and relapses after apparent recoveries are not uncommon. Its mortality rate is disconcertingly variable. It is usually around six percent, but in some outbreaks nearly a third of those stricken have died, and the recovery of a patient from a full-scale attack is almost unheard of. Nobody is, or can be rendered, immune to trichinosis. Also, there is no specific cure. In the opinion of most investigators, it is far from likely that one will ever be found. They are persuaded that any therapeutic agent potent enough to kill a multitude of embedded trichinae would probably kill the patient, too.

Although medical science is unable to terminate, or even lessen the severity of, an assault of trichinosis, no disease is easier to dodge. There are several dependable means of evasion. Abstention from pork is, of course, one. It is also the most venerable, having been known, vigorously recommended, and widely practiced for at least three thousand years. Some authorities, in fact, regard the Mosaic proscription of pork as the pioneering step in the development of preventive medicine. However, since the middle of the nineteenth century, when the cause and nature of trichinosis were illuminated by Sir James Paget, Rudolf Virchow, Friedrich Albert von Zenker, and others, less ascetic safeguards have become available. The trichinae are rugged but not indestructible. It has been amply demonstrated that thorough cooking (until the meat is bone-white) will make even the wormiest pork harmless. So will refrigeration at a maximum temperature of five degrees for a minimum of twenty days. So, just as effectively, will certain scrupulous methods of salting, smoking, and pickling.

Despite this abundance of easily applied defensive techniques, the incidence of trichinosis has not greatly diminished over the globe in the past fifty or sixty years. In some countries, it has even increased. The United States is one of them. Many epidemiologists are convinced that this country now leads the world in trichinosis.

It is, at any rate, a major health problem here. According to a compendium of recent autopsy studies, approximately one American in five has at some time or another had trichinosis, and it is probable that well over a million are afflicted with it every year. As a considerable source of misery, it ranks with tuberculosis, syphilis, and undulant fever. It will probably continue to be one for some time to come. Its spread is almost unimpeded. A few states, New York among them, have statutes prohibiting the feeding of uncooked garbage to swine, but nowhere is a very determined effort made at enforcement, and the Bureau of Animal Industry of the United States Department of Agriculture, although it assumes all pork to be trichinous until proved otherwise, requires packing houses to administer a prophylactic freeze to only those varieties of the meat—frankfurters, salami, prosciutto, and the like—that are often eaten raw. Moreover, not all processed pork comes under the jurisdiction of the Department. At least a third of it is processed under local ordinances in small, neighborhood abattoirs beyond the reach of the Bureau, or on farms. Nearly two per cent of the hogs slaughtered in the United States are trichinous.

Except for a brief period around the beginning of this century, when several European countries refused, because of its dubious nature, to import American pork, the adoption of a less porous system of control has never been seriously contemplated here. One reason is that it would run into money. Another is that, except by a few informed authorities, it has always been considered unnecessary. Trichinosis is generally believed to be a rarity. This view, though hallucinated, is not altogether without explanation. Outbreaks of trichinosis are seldom widely publicized. They are seldom even recognized. Trichinosis is the chameleon of diseases. Nearly all diseases are anonymous at onset, and many tend to resist identification until their grip is well established, but most can eventually be identified by patient scrutiny. Trichinosis is occasionally impervious to bedside detection at any stage. Even blood counts sometimes inexplicably fail to reveal its presence at any stage in its development. As a diagnostic deadfall, it is practically unique. The number and variety of ailments with which it is more or less commonly confused approach the encyclopedic. They include arthritis, acute alcoholism, conjunctivitis, food poisoning, lead poisoning, heart disease, laryngitis, mumps, asthma, rheumatism, rheumatic fever, rheumatic myocarditis, gout, tuberculosis, angioneurotic edema, dermatomyositis, frontal sinusitis, influenza, nephritis, peptic ulcer, appendicitis, cholecystitis, malaria, scarlet fever, typhoid fever, paratyphoid fever, undulant fever, encephalitis, gastroenteritis, intercostal neuritis, tetanus, pleurisy, colitis, meningitis, syphilis, typhus, and cholera. It has even been mistaken for beriberi. With all the rich inducements to error, a sound diagnosis of trichinosis is rarely made, and the diagnostician cannot always take much credit for it. Often, as at Metropolitan Hospital that April day in 1942, it is forced upon him.

The report of the arresting discovery at Metropolitan reached the Health Department on the morning of Friday, April 17. Its form was conventional—a postcard bearing a scribbled name, address, and diagnosis—and it was handled with conventional dispatch. Within an hour, Dr. Morris Greenberg, who was then chief epidemiologist of the Bureau of Preventable Diseases and is now its director, had put one of his fleetest agents on the case, a field epidemiologist named Lawrence Levy. Ten minutes after receiving the assignment, Dr. Levy was on his way to the hospital, intent on tracking down the source of the infection, with the idea of alerting the physicians of other persons who might have contracted the disease along with Sauer. At eleven o'clock, Dr. Levy walked into the office of the medical superintendent at Metropolitan. His immediate objective was to satisfy himself that Sauer was indeed suffering from trichinosis. He was quickly convinced. The evidence of the eosinophile count was now supported in the record by more graphic proof. Sauer, the night before, had undergone a biopsy. A sliver of muscle had been taken from one of his legs and examined under a microscope. It teemed with *Trichinella spiralis*. On the basis of the sample, the record noted, the pathologist who made the test estimated the total infestation of trichinae at upward of twelve million. A count of over five million is almost invariably lethal. Dr. Levy returned the dossier to the file. Then, moving on to his more general objective, he had a word with the patient. He found him bemused but conscious. Sauer appeared at times to distantly comprehend what was said to him, but his replies were faint and rambling and mostly incoherent. At the end of five minutes, Dr. Levy gave up. He hadn't learned much, but he had learned something, and he didn't have the heart to go on with his questioning. It was just possible, he let himself hope, that he had the lead he needed. Sauer had mentioned the New York Labor Temple, a German-American meeting-and-banquet hall on East Eighty-fourth Street, and he had twice uttered the word *"Schlachtfest."* A *Schlachtfest,* in Yorkville, the Doctor knew, is a pork feast.

Before leaving the hospital, Dr. Levy telephoned Dr. Greenberg and dutifully related what he had found out. It didn't take him long. Then he had a sandwich and a cup of coffee and headed for the Labor Temple, getting there at a little past one. It was, and is, a shabby yellow-brick building of six stories, a few doors west of Second Avenue, with a high, ornately balustraded stoop and a double basement. Engraved on the façade, just above the entrance, is a maxim: "Knowledge Is Power." In 1942, the Temple was owned and operated, on a non-profit basis, by the Workmen's Educational Association; it has since been acquired by private interests and is now given over to business and light manufacturing. A porter directed Dr. Levy to the manager's office, a cubicle at the end of a dim corridor flanked by meeting rooms. The manager was in, and, after a spasm of bewilderment, keenly cooperative. He brought out his records and gave Dr. Levy all the information he had. Sauer was known at the Temple. He had been employed there off and on for a year or more as a dishwasher and general kitchen helper, the manager related. He was one of a large group of lightly skilled wanderers from which the cook was accustomed to recruit a staff whenever the need arose. Sauer had last worked at the Temple on the nights of March 27 and March 28.

On the latter, as it happened, the occasion was a *Schlachtfest.*
Dr. Levy, aware that the incubation period of trichinosis is
usually from seven to fourteen days and that Sauer had presented
himself at Lenox Hill on April 6, motioned to the manager to
continue. The *Schlachtfest* had been given by the Hindenburg
Pleasure Society, an informal organization whose members and
their wives gathered periodically at the Temple for an evening of
singing and dancing and overeating. The arrangements for the
party had been made by the secretary of the society—Felix Lindenhauser, a name which, like those of Sauer and the others I shall
mention in connection with the *Schlachtfest,* is a fictitious one.
Lindenhauser lived in St. George, on Staten Island. The manager's
records did not indicate where the pork had been obtained. Probably, he said, it had been supplied by the society. That was frequently the case. The cook would know, but it was not yet time
for him to come on duty. The implication of this statement was
not lost on Dr. Levy. Then the cook, he asked, was well? The
manager said that he appeared to be. Having absorbed this awkward piece of information, Dr. Levy inquired about the health of
the others who had been employed in the kitchen on the night of
March 28. The manager didn't know. His records showed, however, that, like Sauer, none of them had worked at the Temple
since that night. He pointed out that it was quite possible, of
course, that they hadn't been asked to. Dr. Levy noted down their
names—Rudolf Nath, Henry Kuhn, Frederick Kreisler, and William Ritter—and their addresses. Nath lived in Queens, Kreisler
in Brooklyn, and Kuhn and Ritter in the Bronx. Then Dr. Levy
settled back to await the arrival of the cook. The cook turned up
at three, and he, too, was very cooperative. He was feeling fine,
he said. He remembered the *Schlachtfest.* The pig, he recalled, had
been provided by the society. Some of it had been ground up into
sausage and baked. The rest had been roasted. All of it had been
thoroughly cooked. He was certain of that. The sausage, for example, had been boiled for two hours before it was baked. He had
eaten his share of both. He supposed that the rest of the help had,
too, but there was no knowing. He had neither seen nor talked to

any of them since the night of the feast. There had been no occasion to, he said.

Dr. Levy returned to his office, and sat there for a while in meditation. Presently, he put in a call to Felix Lindenhauser, the secretary of the society, at his home on Staten Island. Lindenhauser answered the telephone. Dr. Levy introduced himself and stated his problem. Lindenhauser was plainly flabbergasted. He said he was in excellent health, and had been for months. His wife, who had accompanied him to the *Schlachtfest,* was also in good health. He had heard of no illness in the society. He couldn't believe that there had been anything wrong with that pork. It had been delicious. The pig had been obtained by two members of the society, George Muller and Hans Breit, both of whom lived in the Bronx. They had bought it from a farmer of their acquaintance in New Jersey. Lindenhauser went on to say that there had been twenty-seven people at the feast, including himself and his wife. The names and addresses of the company were in his minute book. He fetched it to the phone and patiently read them off as Dr. Levy wrote them down. If he could be of any further help, he added as he prepared to hang up, just let him know, but he was convinced that Dr. Levy was wasting his time. At the moment, Dr. Levy was almost inclined to agree with him.

Dr. Levy spent an increasingly uneasy weekend. He was of two antagonistic minds. He refused to believe that Sauer's illness was not in some way related to the *Schlachtfest* of the Hindenburg Pleasure Society. On the other hand, it didn't seem possible that it was. Late Saturday afternoon, at his home, he received a call that increased his discouragement, if not his perplexity. It was from his office. Metropolitan Hospital had called to report that Herman Sauer was dead. Dr. Levy put down the receiver with the leaden realization that, good or bad, the *Schlachtfest* was now the only lead he would ever have.

On Monday, Dr. Levy buckled heavily down to the essential but unexhilarating task of determining the health of the twenty-seven men and women who had attended the *Schlachtfest.* Although his attitude was half-hearted, his procedure was methodical, unhurried, and objective. He called on and closely examined each of the guests, including the Lindenhausers, and from each procured a sample of blood for analysis in the Health Department laboratories. The job, necessarily involving a good deal of leg work and many evening visits, took him the better part of two weeks. He ended up, on April 30th, about equally reassured and stumped. His findings were provocative but contradictory. Of the twenty-seven who had feasted together on the night of March 28, twenty-five were in what undeniably was their normal state of health. Two, just as surely, were not. The exceptions were George Muller and Hans Breit, the men who had provided the pig. Muller was at home and in bed, suffering sorely from what his family physician had uncertainly diagnosed as some sort of intestinal upheaval. Breit was in as bad a way, or worse, in Fordham Hospital. He had been admitted there for observation on April 10. Several diagnoses had been suggested, including rheumatic myocarditis, pleurisy, and grippe, but none had been formally retained. The nature of the two men's trouble was no mystery to Dr. Levy. Both, as he was subsequently able to demonstrate, had trichinosis.

On Friday morning, May 1, Dr. Levy returned to the Bronx for a more searching word with Muller. Owing to Muller's debilitated condition on the occasion of Dr. Levy's first visit, their talk had been brief and clinical in character. Muller, who was now up and shakily about, received him warmly. Since their meeting several days before, he said, he had been enlivening the tedious hours of illness with reflection. A question had occurred to him. Would it be possible, he inquired, to contract trichinosis from just a few nibbles of raw pork? It would, Dr. Levy told him. He also urged him to be more explicit. Thus encouraged, Muller displayed an unexpected gift for what appeared to be total recall. He leisurely recounted to Dr. Levy that he and Breit had bought the pig from a farmer who owned a place near Midvale, New Jersey. The farmer had killed and dressed the animal, and they had delivered the carcass to the Labor Temple kitchen on the evening of

March 27. That, however, had been only part of their job. Not wishing to trouble the cook and his helpers, who were otherwise occupied, Muller and Breit had then set about preparing the sausage for the feast. They were both experienced amateur sausage makers, he said, and explained the process—grinding, maccrating, and seasoning—in laborious detail. Dr. Levy began to fidget. Naturally, Muller presently went on, they had been obliged to sample their work. There was no other way to make sure that the meat was properly seasoned. He had taken perhaps two or three little nibbles. Breit, who had a heartier taste for raw pork, had probably eaten a trifle more. It was hard to believe, Muller said, that so little —just a pinch or two—could cause such misery. He had thought his head would split, and the pain in his legs had been almost beyond endurance. Dr. Levy returned him sympathetically to the night of March 27. They had finished with the sausage around midnight, Muller remembered. The cook had departed by then, but his helpers were still at work. There had been five of them. He didn't know their names, but he had seen all or most of them again the next night, during the feast. Neither he nor Breit had given them any of the sausage before they left. But it was possible, of course, since the refrigerator in which he and Breit had stored the meat was not, like some, equipped with a lock . . . Dr. Levy thanked him, and moved rapidly to the door.

Dr. Levy spent the rest of the morning in the Bronx. After lunch, he hopped over to Queens. From there, he made his way to Brooklyn. It was past four by the time he got back to his office. He was hot and gritty from a dozen subway journeys, and his legs ached from pounding pavements and stairs and hospital corridors, but he had tracked down and had a revealing chat with each of Sauer's kitchen colleagues, and his heart was light. Three of them —William Ritter, Rudolf Nath, and Frederick Kreisler—were in hospitals. Ritter was at Fordham, Nath at Queens General, and Kreisler at the Coney Island Hospital, not far from his home in Brooklyn. The fourth member of the group, Henry Kuhn, was sick in bed at home. All were veterans of numerous reasonable but incorrect diagnoses, all were in more discomfort than danger, and all, it was obvious to Dr. Levy's unclouded eye, were suffering from trichinosis. Its source was equally obvious. They had prowled the icebox after the departure of Muller and Breit, come upon the sausage meat, and cheerfully helped themselves. They thought it was hamburger.

Before settling down at his desk to compose the final installment of his report, Dr. Levy looked in on Dr. Greenberg. He wanted, among other things, to relieve him of the agony of suspense. Dr. Greenberg gave him a chair, a cigarette, and an attentive ear. At the end of the travelogue, he groaned. "Didn't they even bother to cook it?" he asked.

"Yes, most of them did," Dr. Levy said. "They made it up into patties and fried them. Kuhn cooked his fairly well. A few minutes, at least. The others liked theirs rare. All except Sauer. He ate his raw."

"Oh," Dr. Greenberg said.

"Also," Dr. Levy added, "he ate two."

[*1950*]

CHAPTER 3

A Game of Wild Indians

DURING THE SECOND WEEK in August, 1946, an elderly man, a middle-aged woman, and a boy of ten dragged themselves, singly and painfully, into the Presbyterian Hospital, in the Washington Heights section of Manhattan, where their trouble was unhesitatingly identified as typhoid fever. This diagnosis was soon confirmed by laboratory analysis, and on Thursday morning, August 15th, a report of the outbreak was dutifully telephoned to the Department of Health. It was received and recorded there, in accordance with the routine in all alarms of an epidemiological nature, by a clerk in the Bureau of Preventable Diseases named Beatrice Gamso. Miss Gamso is a low-strung woman and she has spent some thirty callousing years in the Health Department, but the news gave her a turn. She sat for an instant with her eyes on her notes. Then, steadying herself with a practiced hand, she swung around to her typewriter and set briskly about dispatching copies of the report to all administrative officers of the Department. Within an hour, a reliable investigator from the Bureau was on his way to Washington Heights. He was presently followed by one of his colleagues, a Department public-health nurse, several agents from the Bureau of Food and Drugs, and an inspector from the Bureau of Sanitary Engineering.

Typhoid fever was among the last of the massive pestilential fevers to yield to the probings of medical science, but its capitulation has been complete. It is wholly transparent now. Its clinical manifestations (a distinctive rash and a tender spleen, a fiery fever and a languid pulse, and nausea, diarrhea, and nosebleed), its cause (a bacillus known as *Eberthella typhosa*), and its means of transmission have all been clearly established. Typhoid is invariably conveyed by food or drink contaminated with the excreta of its victims. Ordinarily, it is spread by someone who is ignorant, at least momentarily, of his morbid condition. One reason for such unawareness is that for the first several days typhoid fever tends to be disarmingly mild and indistinguishable from the countless fleeting malaises that dog the human race. Another is that nearly five per cent of the cases become typhoid carriers, continuing indefinitely to harbor a lively colony of typhoid bacilli in their systems. The existence of typhoid carriers was discovered by a group of German hygienists in 1907. Typhoid Mary Mallon, a housemaid and cook who was the stubborn cause of a total of fifty-three cases in and around New York City a generation ago, is, of course, the most celebrated of these hapless menaces. About seventy per cent, by some unexplained physiological fortuity, are women. The names of three hundred and eighty local carriers are currently on active file in the Bureau of Preventable Diseases. They are called on regularly by public-health nurses and are permanently enjoined from any employment that involves the handling of food. More than a third of all the cases that occur here are traced to local carriers but, because of the vigilance of the Health Department, rarely to recorded carriers; new ones keep turning up. Most of the rest of the cases are of unknown or out-of-town origin. A few are attributable to the products of polluted waters (clams and oysters and various greens).

The surveillance of carriers is one of several innovations that in little more than a generation have forced typhoid fever into an abrupt tractability throughout most of the Western world. The others include certain refinements in diagnostic technique, the institution of public-health measures requiring the chlorination of city-supplied water and proscribing the sale of unpasteurized milk, and the development of an immunizing vaccine. Since late in the nineteenth century, the local incidence of typhoid fever has dropped from five or six thousand cases a year to fewer than fifty, and it is very possible that it may soon be as rare as smallpox. Banishment has not, however, materially impaired the vigor of *Eberthella typhosa.* Typhoid fever is still a cruel and withering affliction. It is always rambunctious, generally prolonged, and often fatal. It is also one of the most explosive of communicable diseases. The month in which it is most volcanic is August.

The investigator who led the sprint to Washington Heights that August morning in 1946 was Dr. Harold T. Fuerst, an epidemiologist, and he and Dr. Ottavio J. Pellitteri, another epidemiologist, handled most of the medical inquiry. One afternoon, when I was down at the Bureau, they told me about the case. Miss Gamso sat at a desk nearby, and I noticed after a moment that she was following the conversation with rapt attention. Her interest, it turned out, was entirely understandable. Typhoid-fever investigations are frequently tedious, but they are seldom protracted. It is not unusual for a team of experienced operatives to descry the source of an outbreak in a couple of days. Some cases have been riddled in an afternoon. The root of the trouble on Washington Heights eluded detection for almost two weeks, and it is probable that but for Miss Gamso it would never have been detected at all.

"I got to Presbyterian around eleven," Dr. Fuerst told me. "I found a staff man I knew, and he led me up to the patients. It was typhoid, all right. Not that I'd doubted it, but it's routine to take a look. And they were in bad shape—too miserable to talk. One —the woman—was barely conscious. I decided to let the questioning go for the time being. At least until I'd seen their histories. A clerk in the office of the medical superintendent dug them out for me. Pretty skimpy—name, age, sex, occupation, and address, and a few clinical notations. About all I got at a glance was that they weren't members of the same family. I'd hoped, naturally, that they would be. That would have nicely limited the scope of the investigation. Then I noticed something interesting. They weren't a family, but they had a little more in common than just typhoid. For one thing, they were by way of being neighbors. One of them lived at 502 West 180th Street, another at 501 West 178th Street, and the third at 285 Audubon Avenue, just around the corner from where it runs through the five-hundred block of West 179th Street. Another thing was their surnames. They were different, but they weren't dissimilar. All three were of Armenian origin. Well, Washington Heights has an Armenian colony—very small and very clannish. I began to feel pretty good. I didn't doubt for a minute that the three of them knew each other. Quite possibly they were friends. If so, it was reasonable to suppose that they might recently have shared a meal. It wasn't very likely, of course, that they had been the only ones to share it. Ten-year-old boys don't usually go out to meals without their parents. Maybe there had been a dozen in on it. It could even have been some sort of national feast. Or a church picnic. Picnic food is an ideal breeding ground for the typhoid organism. It can't stand cooking, but it thrives in raw stuff—ice cream and mayonnaise and so on. And if a carrier had happened to have a hand in the arrangements . . . I decided we'd do well to check and see if there was an Armenian carrier on our list."

"We found one, all right," Dr. Pellitteri said. "A widow named Christos—she died a year or two ago—who lived on West 178th Street."

"To be sure, we had only three cases," Dr. Fuerst went on. "But I didn't let that bother me. I've never known an outbreak of typhoid in which everybody who was exposed got sick. There are always a certain number who escape. They either don't eat whatever it is that's contaminated or they have a natural or an acquired immunity. Moreover, the incubation period in typhoid—the time it takes for the bug to catch hold—varies with the individual. Ten days is about the average, but it can run anywhere from three to thirty. In other words, maybe we had seen only the vanguard. There might be more to come. So in the absence of anything better, the Armenian link looked pretty good. I called the Bureau and told Bill Birnkrant—he was acting director at the time—what I thought, and he seemed to think the same. He said he'd start somebody checking. I went back upstairs for another try at the patients."

"That's when the rest of us began to come into the picture," Dr. Pellitteri said. "My job was the recent social life of the Armenian colony. Ida Matthews, a public-health nurse, took the carrier angle. Neither of us had much luck. The file listed twelve carriers in Washington Heights. As I remember, the only Armenian was Mrs. Christos. At any rate, the nurse picked her first. I remember running into Miss Matthews somewhere on Audubon toward the end of that first afternoon. She told me what progress she had made. None. Mrs. Christos was old and sick, and hadn't been out of her apartment for a month. Miss Matthews said there was no reason to doubt the woman's word, as she had a good reputation at the Department—very cooperative, obeyed all the rules. Miss Matthews was feeling pretty gloomy. She'd had high hopes. Well, I knew how she felt. I'd hit nothing but dead ends myself. Our patients didn't seem to be friends. Apparently, they just knew each other. The priest at the Gregorian church in the neighborhood— Holy Cross Armenian Apostolic, on West 187th Street—knew of no recent feasts or festivals. He hadn't heard of any unusual amount of illness in the parish, either. No mysterious chills and fevers. And the Armenian doctors in the neighborhood said the same. They had seen nothing that resembled typhoid except the cases we already had. Before I gave up for the day, I even got in touch with an Armenian girl who used to work at the Department. The only thing I could think of at the moment was a check of the Armenian restaurants. When I mentioned that, she burst out laughing. It seems Armenians don't frequent Armenian restaurants. They prefer home cooking."

"I got Pellitteri's report the next morning," Dr. Fuerst said.

"And Miss Matthews'. I was back at the hospital, and when I called Birnkrant, he gave me the gist of them. I can't say I was greatly surprised. To tell the truth, I was relieved. The Armenian picnic I'd hypothesized the day before would have created a real mess. Because the hospital had reported two new cases. Two women. They lived at 500 West 178th Street and 611 West 180th Street, but they weren't Armenians. One was Italian. The other was plain American. So we were right back where we started. Only, now we had five cases instead of three, and nothing to tie them together but the fact that they all lived in the same neighborhood. And had the same brand of typhoid. There are around a dozen different strains, you know, which sometimes complicates matters. About the only thing Birnkrant and I could be sure of was that the feast theory—any kind of common gathering—was out. I'd had a word with the new patients. They had never even heard of each other. So the link had to be indirect. That gave us a number of possibilities. The source of infection could be water— either drinking water or a swimming pool. Or it could be commercial ice. Or milk. Or food. Drinking water was a job for Sanitary Engineering. The others, at the moment, were up to us—meaning Pellitteri and me. They were all four conceivable. Even ice. You can find a precedent for anything and everything in the literature on typhoid. But just one was probable. That was food. Some food that is sold already prepared—like potato salad or frozen custard —or one that is usually eaten raw. All we had to do was find out what it was, and where they got it, and how it got that way. Birnkrant and I figured out the area involved. It came to roughly four square blocks. I don't know if you know that part of Washington Heights. It's no prairie. Every building is a big apartment house, and the ground floors of most are stores. At least a fourth have something to do with food."

"I was in the office when Fuerst called," Dr. Pellitteri said. "Before he hung up, I got on the phone and we made the necessary arrangements about questioning the patients and their families— who was to see who. Then I took off. I wasn't too pessimistic. The odds were against a quick answer, but you never know. It was just possible that they all bought from the same store. Well, as it happened, they did. In a way. The trouble was it wasn't one store. It was practically all of them. Fuerst had the same experience. We ended up at the office that evening with a list as long as my arm —half a dozen fruit-and-vegetable stands, four or five groceries, a market that sold clams, and an assortment of ice-cream parlors and confectioneries and delicatessens. Moreover, we couldn't even be sure the list included the right store. Most people have very strange memories. They forget and they imagine. You've got to assume that most of the information they give you may be either incomplete or inaccurate, or both. But there was a right store— we knew that. Sanitary Engineering had eliminated drinking water, and we had been able to rule out swimming and milk and ice. Only one of the group ever went swimming, all but one family had electric refrigerators, and none of them had drunk unpasteurized milk. It had to be contaminated food from a store. That much was certain."

"It was also certain that we had to have some help," Dr. Fuerst said. "Pellitteri and I could have handled a couple of stores. Or even, at a pinch, three or four. But a dozen or more—it would take us weeks. Let me give you an idea what an investigation like that involves. You don't just walk in the store and gaze around. You more or less take it apart. Every item of food that could conceivably cause trouble is examined, the physical setup is inspected for possible violations of the Sanitary Code, and all employees and their families are interviewed and specimens taken for laboratory analysis. So we needed help, and, of course, we got it. Birnkrant had a conference with the Commissioner the next morning and they talked it over, and the result was an engineer and another nurse and a fine big team from Food and Drugs. Very gratifying."

"And Miss Matthews," Dr. Pellitteri said. "We had her back again. She had finally finished with her carriers. They were all like the first. None had violated any of the rules."

"As expected," Dr. Fuerst said. "The average carrier is pretty cooperative. Well, that was Saturday. By Monday, we had made a certain amount of progress. We hadn't found anything yet, but the field was narrowing down. And all of a sudden we got a little nibble. It came from a confectionery called Pop's, on 178th Street, around noon. Pop's had been well up on our list. They sold ice cream made on the premises, and the place was a neighborhood favorite. Which meant it got a very thorough going over. But we were about ready to cross it off—everything was in good shape, including the help—when it developed that the place had just changed hands. Pop had sold out a week before, and he and his wife, who'd helped him run it, were on the way to California. Needless to say, Pop's went back on the list, and at the top. Also, somebody did some quick checking. Pop and his wife were driving, and their plan was to spend a few days with friends in Indianapolis. That gave us a chance. We called Birnkrant and he called Indianapolis—the State Health Department. They were extremely interested. Naturally. They said they'd let us know."

Dr. Fuerst lighted a cigarette. "Then we got a jolt," he said. "Several, in fact. The first was a call from the hospital. Four new cases. That brought the total up to nine. But it didn't stay there long. Tuesday night, it went to ten. I don't mind saying that set us back on our heels. Ten cases of typhoid fever in less than a week in one little corner of the city is almost unheard of in this day and age. The average annual incidence for the whole of Washington Heights is hardly half a case. That wasn't the worst of it, though. The real blow was that tenth case. I'll call him Jones. Jones didn't fit in. The four Monday cases, like the three Armenians and the Italian and the American, all lived in that one four-block area. Jones didn't. He lived on 176th Street, but way over west, almost on Riverside Drive. An entirely different neighborhood. I had a word with Jones the first thing Wednesday morning. I remember he worked for the post office. That's about all I learned. He hardly knew where he was. When I left the hospital, I called on his wife. She wasn't much help, either. She did all the family marketing, she told me and she did it all within a block or two of home. That was that. She was very definite. On the other hand, there was Mr. Jones. He had typhoid, which doesn't just happen, and it was the same strain as all the rest. So either it was a very strange coincidence or she was too upset to think. My preference, until proved otherwise, was the latter. I found a phone, and called Birnkrant and gave him the latest news. He had some news for me. Indianapolis had called. They had located Pop and his wife and made the usual tests. The results were negative."

"I don't know which was the most discouraging," Dr. Pellitteri said. "Jones, I guess. He meant more work—a whole new string of stores to check. Pop had been ninety per cent hope. He merely aroused suspicion. He ran a popular place, he sold homemade ice cream, and when the epidemic broke, he pulled out. Or so it appeared from where we stood. It hurt to lose him. Unlikely or not, he had been a possibility—the first specific lead of any kind that we had been able to find in a week of mighty hard work. During the next few days, it began to look more and more like the last. Until Friday evening. Friday evening we got a very excited call from the laboratory. It was about a batch of specimens we had submitted that morning for analysis. One of them was positive for *E. typhosa.* The man's name doesn't matter. It didn't even then. What did matter was his occupation. He was the proprietor of a little frozen-custard shop—now extinct—that I'll call the Jupiter. The location was interesting, too. It was a trifle outside our area, but still accessible, and a nice, easy walk from the Joneses'. Food and Drugs put an embargo on the Jupiter that night. The next morning, we began to take it apart."

"I missed that," Dr. Fuerst said. "I spent Saturday at the hospital. It was quite a day. We averaged a case an hour. I'm not exaggerating. When I finally left, the count was nine. Nine brand new cases. A couple of hours later, one more turned up. That made twenty, all told. Fortunately, that was the end. Twenty was the grand total. But, of course, we didn't know that then. There was no reason to believe they wouldn't just keep coming."

"The rest of us had the same kind of day," Dr. Pellitteri said. "Very disagreeable. There was the owner of the Jupiter—poor devil. You can imagine the state he was in. All of a sudden, he was out of business and a public menace. He didn't even know what a typhoid carrier was. He had to be calmed down and instructed. That was the beginning. It got worse. First of all, the Jupiter was as clean as a whistle. We closed it up—had to, under the circumstances—and embargoed the stock, but we didn't find anything. That was peculiar. I can't explain it even now. He was either just naturally careful or lucky. While that was going on, we went back to the patients and questioned them again. Did they know the Jupiter? Were they customers? Did they ever buy anything there? We got one yes. The rest said no. Emphatically. If there had been a few more yeses—even three or four—we might have wondered. But they couldn't all be mistaken. So the Jupiter lead began to look pretty wobbly. Then the laboratory finished it off. They had a type report on the Jupiter organism. It wasn't the *E. typhosa* we were looking for. It was one of the other strains. That may have been some consolation to Mr. Jupiter. At least, he didn't have an epidemic on his conscience. But it left us uncomfortably close to the end of our rope. We had only a handful of stores still to check. If we didn't find the answer there, we were stumped. We didn't. We crossed off the last possibility on Tuesday morning, August 27. It was Number Eighty. We'd examined eighty stores and something like a thousand people, and all we had to show for it was a new carrier."

"Well, that was something," Dr. Fuerst said. "Even if it was beside the point. But we also had another consolation. None of the patients had died. None was going to. They were all making excellent progress."

"That's true enough," Dr. Pellitteri said. "But we couldn't claim much credit for that." He paused, and shifted around in his chair. "About all we can take any credit for is Miss Gamso, here," He smiled. "Miss Gamso saved the day. She got inspired."

Miss Gamso gave me a placid look. "I don't know about inspired," she said. "It was more like annoyed. I heard them talking—Dr. Birnkrant, and these two, and all the rest of them—and I read the reports, and the days went by and they didn't seem to be getting anywhere. That's unusual. So it was irritating. It's hard to explain, but I got to thinking about that carrier Mrs. Christos.

There were two things about her. She lived with a son-in-law who was a known food handler. He was a baker by trade. Also, where she lived was right in the middle of everything—519 West 178th Street. That's just off Audubon. And Audubon is the street where practically all our cases did most of their shopping. Well, there was one store in particular—a fruit-and-vegetable market called Tony's—on almost everybody's list. The address was 261 Audubon Avenue. Then I really got a brainstorm. It was right after lunch on Tuesday, August 27. I picked up the telephone and called the bureau that registers house numbers at the Borough President's office, and I asked them one question. Did 519 West 178th Street and 261 Audubon Avenue happen by any chance to be the same building? They asked me why I wanted to know. I wasn't talking, though. I just said was it, in a nice way, and the man finally said he'd see. When he came back, I was right. They were one and the same. I was so excited I thought I'd burst. Dr. Pellitteri was sitting right where he is now. He was the first person I saw, so I marched straight over and told him. He kind of stared at me. He had the funniest expression." Miss Gamso smiled a gentle smile. "I think he thought I'd gone crazy."

"I wouldn't say that," Dr. Pellitteri said. "I'll admit, however, that I didn't quite see the connection. We'd been all over Tony's —it was almost our first stop—and there was no earthly reason to question Miss Matthews' report on Mrs. Christos. The fact that they occupied the same building was news to me. To all of us, as I recall. But what if they did? Miss Gamso thought it was significant or suspicious or something. The point escaped me. When she mentioned the son-in-law, though, I began to get a little more interested. We knew him, of course—anybody who lives with a carrier is a potential cause of trouble—and checked on him regularly. But it was just possible that since our last checkup he had become infected. That happens. And although we hadn't found him working in any of the stores, he could have come and gone a couple of weeks before we started our investigation. At any rate, it was worth looking into. Almost anything was, by then. I went up that afternoon. I walked past Tony's on the way to 519. There wasn't any doubt about their being in the same building. Tony's is gone now, like Mrs. Christos, but the way it was then, his front door was about three steps from the corner, and around the corner about three more steps was the entrance to the apartments above. The Christos flat was on the fifth floor—Apartment 53. Mrs. Christos and her son-in-law were both at home. They let me in and that's about all. I can't say they were either one delighted to see me. Or very helpful. She couldn't add anything to what she had already told Miss Matthews. The son-in-law hardly opened his mouth. His last regular job, he said, had been in January, in a cafeteria over in Astoria. Since then, he'd done nothing but odd jobs. He wouldn't say what, when, or where. I couldn't completely blame him. He was afraid that if we got to questioning any of his former employers, they'd never take him on again. When I saw how it was, I arranged for a specimen and, for the moment, let it go at that. There was no point in getting rough until we knew for sure. I told him to sit tight. If he was positive, I'd be back in a hurry. I got the report the next day. He wasn't. He was as harmless as I am. But by then it didn't matter. By that time, it was all over. To tell the truth, I had the answer before I ever left the building."

Dr. Pellitteri shook his head. "I walked right into it," he said. "It was mostly pure luck. What happened was this. On the way out, I ran into the superintendent—an elderly woman. I was feeling two ways about the son-in-law—half sympathetic and half suspicious. It occurred to me that the superintendent might have some idea where he'd been working the past few weeks. So I stopped and asked. She was a sour old girl. She didn't know and didn't care. She had her own troubles. They were the tenants, mainly. She backed me into a corner and proceeded to unload. The children were the worst, she said—especially the boys. Always thinking up some new devilment. For example, she said, just a few weeks ago, toward the end of July, there was a gang of them up on the roof playing wild Indians. Before she could chase them off, they'd stuffed some sticks down one of the plumbing vent pipes. The result was a stoppage. The soil pipe serving one whole tier of apartments blocked and sprang a leak, and the bathroom of the bottom apartment was a nice mess. I hadn't been paying much attention until then. But at that point— Well, to put it mildly, I was fascinated. Also, I began to ask some questions. I wanted to know just what bathroom had flooded. The answer was Apartment 23. What were the other apartments in that tier? They were 33, 43, and 53. What was underneath Apartment 23? A store— Tony's Market, on the corner. Then I asked for a telephone. Birnkrant's reaction was about what you'd expect. Pretty soon, a team from Sanitary Engineering arrived. They supplied the details and the proof. Tony stored his fruits and vegetables in a big wooden walk-in refrigerator at the rear of his store. When Sanitary Engineering pulled off the top, they found the soil pipe straight overhead. The leak had been repaired almost a month before, but the sawdust insulation in the refrigerator roof was still damp from the waste that had soaked through. It wasn't Tony's fault. He hadn't known. It wasn't anybody's fault. It was just one of those things. So that was that."

"Not entirely," Dr. Fuerst said. "There was still Jones to account for. It wasn't necessary. The thing was settled. But I was curious. I had a talk with him the next day. We talked and talked. And in the end, he remembered. He was a night walker. Every evening after dinner, he went out for a walk. He walked all over Washington Heights, and usually, somewhere along the line, he stopped and bought something to eat. It was generally a piece of fruit. As I say, he finally remembered. One night, near the end of July, he was walking down Audubon and he came to a fruit stand and he bought an apple. On the way home, he ate it."

[1952]

CHAPTER 4

The Incurable Wound

—————————◆·◆—————————

ON OCTOBER 30, 1951, a woman I'll call Mabel Tate, the wife of a West Texas cotton planter, was admitted to the Parkland City–County Hospital, in Dallas, with a tentative diagnosis of bulbar poliomyelitis. The record also noted, as is usual in ambiguous cases, two possible variant readings. They were epidemic encephalitis and, at the suggestion of the Tate family doctor, influenza. The general nature of her trouble, however, was somewhat less uncertain. All major signs and symptoms reflected a virus invasion, and one of massive, if not overwhelming, proportions. Mrs. Tate was blazing with fever, she was wildly agitated, and she was unable to speak, unable to swallow, and unable to move her left arm. Four days later, she sank into a coma, and died. Something about the manner of her death prompted the attending physician to request a clarifying post-mortem examination. The autopsy was done, with Mr. Tate's consent, early the following day. When the attending physician reached the hospital that morning, a report of the laboratory findings was on his desk. It began, "Encephalomyelitis with demonstrable Negri bodies in central motor neurons . . ." There was no need to read any further. That emphatically answered his question. Negri bodies are distinctive clusters of cellular substance whose presence in the brain has just one denotation. Mrs. Tate was a victim of rabies.

The attending physician once more sought out Mr. Tate. He told him what the pathologist had found and what the finding meant. That being the case, he went on to explain, two corollary conclusions were obvious. One was that Mrs. Tate had been attacked and bitten by a rabid animal. The other related to the approximate time of the attack. In view of the usual incubation period of rabies, he felt, it had probably taken place between two and six weeks earlier. The doctor spread his hands. All that remained was to establish the specific source of infection. It could have been a dog. It could have been a cat or a fox. It might even have been a skunk. There were numerous possibilities. Mr. Tate nodded. He appreciated the doctor's position. He doubted, though, if he could be of much help. It depended on what the doctor meant by an animal. His wife had been bitten, all right, and fairly recently, too. On October 9, to be exact. But it wasn't a dog or a cat or any of those. It was a bat. His wife had come across it lying in the road near their house. She had thought it was dead, and stooped down to take a look. The next thing she knew, it had jumped up and given her a nasty nip on the left arm. Then it had flown away.

The doctor hesitated. Very curious, he said. And certainly a most curious coincidence. He shrugged, and rose. But, of course, that was all it could be. The only species of bat in which rabies had ever been demonstrated was the vampire, and its range was limited to tropical Latin America. He was forced to conclude that Mr. Tate was mistaken. There must have been another animal episode. It could have happened as much as a year before. Such cases were uncommon but possible. Either Mr. Tate had forgotten or his wife had neglected to tell him. The doctor returned to his office and took out the record of the case. He closed it with the notation "Rabies, source unknown."

Officially, the animal responsible for the death of Mrs. Tate is still not known. There is little possibility now that its identity will ever be irrefutably established. The rules of scientific evidence are too rigid for that. Nevertheless, in the opinion of most interested epidemiologists, the case no longer presents much of a riddle. Several subsequent events, they feel, have rendered it to all practical purposes clear. The first of these occurred on a cattle ranch some thirty miles southeast of Tampa, Florida, on June 23, 1953. Around ten o'clock that morning, the stockman's son, a boy of seven whom I'll name David Bonner, was playing in the back yard when a bat burst out of a nearby clump of trees. He called to his father, who was at work a short distance away, and pointed. Mr. Bonner glanced up, and stared. It was odd enough to see a bat abroad in the full light of day, but the creature's behavior was even stranger. The bat, when Mr. Bonner first caught sight of it, was circling the house. An instant later, it turned and streaked straight for the woods. Then it was back again—flying high, low, and every which way. Suddenly, from almost directly overhead, it swooped. David screamed, and tried to run. But it was too late. The bat was already upon him. Mr. Bonner crossed the yard in a bound. He caught his son and swung him about. The bat was clinging to the boy's chest, its teeth sunk deep in his flesh, and blood was staining his shirt. Mr. Bonner broke its grip with a backhand swipe. It dropped, with a strangled hiss, to the ground. He gave it a kick, for good measure. Then he picked up his son and carried him into the house.

David was more frightened than hurt. While Mrs. Bonner held and comforted him, his father examined the bite. It was an ugly wound but a small one, and not, Mr. Bonner decided, in any sense serious. There seemed no need to call a doctor. He cleaned the bite with soap and water, dusted it with sulfanilamide, and covered it with a gauze dressing. That—for the moment, at least—appeared to be sufficient. It didn't, however, put his mind altogether at rest.

The circumstances of the assault, he had to admit, were, if nothing else, uncomfortably queer. Mrs. Bonner agreed. They held a hurried conference and reached a prompt decision. Mr. Bonner fetched his jacket and a paper bag, and returned to the back yard. The bat was lying where he had kicked it. Its fur was sandy brown, with yellow overtones, and except for its saucer ears and its long, web-fingered forearms, it might have been a field mouse. It was also, he was relieved to find, dead. He scooped it into the bag, and went on to the garage and his car. Forty minutes later, he was in the Tampa office of the Florida State Board of Health, closeted with a staff epidemiologist.

Mr. Bonner began the interview with a brief account of the incident. He then produced the bat and stated the reason for his visit. He wanted to have the creature examined. It was his understanding, he said, that bats were capable of transmitting rabies. He remembered reading in a livestock journal that they had been linked to an outbreak of the disease among cattle somewhere in South America. That was true, the doctor replied. There had, in fact, been many such cases, and not only among cattle. Several human cases were also on record. Bat rabies, as it was awkwardly called, was endemic in several Latin-American countries. They included Brazil (where the phenomenon was first reported), Honduras, Mexico, Colombia, Venezuela, Surinam, and the island of Trinidad. But, he pointed out, the bats involved were not ordinary bats. They were bats of a kind unknown outside the tropics. They were true, or bloodsucking, vampires. The bat that Mr. Bonner had brought with him was a harmless Florida yellow, a member of the species *Dasypterus floridanus.* It subsisted, like all other bats in the United States, exclusively on insects. Those were the facts. They didn't, of course, explain the attack. He had no theory about that. It was his opinion, though, that the facts held no cause for alarm. The doctor paused. However, he added, it was impossible to deny that the bat had behaved very strangely, and he quite understood how Mr. Bonner felt. A certain amount of uneasiness was only natural. Consequently, in order to settle the matter, he would send the bat along to the board's local laboratory for a routine brain examination. The result, he was confident, would be completely reassuring.

Mr. Bonner left the Board of Health at a little past one. By the time he reached home, it was almost two. At three, he was called to the telephone. It was the epidemiologist in Tampa, and he sounded stunned. He was calling, he said, from the laboratory. A bacteriologist there had just finished a microscopic examination of the bat's brain, and Mr. Bonner, incredibly, was right. The findings were positive for rabies. Arrangements were now being made for the usual confirmatory tests. They involved the inoculation of laboratory mice with bat-brain material, and would be done at the main State Board of Health laboratory, in Jacksonville. But that was largely a formality. The microscopic evidence was in every essential conclusive. Mr. Bonner's son had been bitten by a rabid animal, and it was imperative that preventive measures be taken at once. Could he bring the boy in to the Tampa office that afternoon? Mr. Bonner could, and did. The Pasteur treatment, as the immunizing procedure against rabies is called (in commemoration of its creator), requires a subcutaneous injection of antirabies vaccine every day for two weeks. David completed the course, apparently with success, on July 7, but because of the variable length of the incubation period in rabies, the summer was well over before it could be said that he was in all probability out of danger. That he had been in danger was beyond dispute by then. The Jacksonville tests had confirmed the fact, and so had an even more elaborate investigation, conducted at the request of the Florida authorities by the United States Public Health Service, at its Virus and Rickettsia Laboratory, in Montgomery, Alabama. It was also certain by then that David's experience could not be dismissed as an isolated freak of misfortune. Late in September, while he was still under regular observation, a woman I'll call Frances Roberts suffered an almost identical attack, and that was closely followed by a third. The scene of both was eastern Pennsylvania.

The second Pennsylvania episode, though the least unequivocal of the three cases, was by far the most unsettling in its implications. Unlike the others, it happened in a city, and indoors—in a tavern in the central business section of Harrisburg. Its victim was a used-car salesman I'll identify as Carl Dayton. Shortly after midnight on Saturday, November 28, 1953, Mr. Dayton was standing with a group of friends at the tavern bar. Something brushed his face. He stumbled back, looked up, and saw a bat. It was dodging from wall to wall, just below the ceiling, and was heading toward the rear of the room. There was an open window there, but the bat made no attempt to escape. Instead, it circled back to the bar, lower now and moving fast. The bartender tried to whip it down with a towel, and one of the customers swung his hat at it. Both of them missed. Another struck out with a rolled newspaper, and caught it a staggering blow. It fell to the floor at Mr. Dayton's feet. He squatted down for a look, then sprang up with a yell and began to pound on the bar. His friends stood frozen, and stared. The bat was fixed to the back of his hand, and before he could shake it off, it had bitten his thumb to the bone. The bartender was the first to recover. He slammed the bat across the room, and this released the others. They charged the bat and stomped it to death and threw it into the street. That, to the impairment of the subsequent investigation, was the last of the bat. Then, more sensibly, they inspected Mr. Dayton's wound. It was obvious that he needed medical attention, and after bandaging his thumb with a handkerchief, they fetched a cab and sent him off to Harrisburg Polyclinic Hospital. From the hospital, where an interne (either unimpressed or unconvinced by his explanation of the accident) was satisfied to merely clean, close, and properly dress the wound, Mr. Dayton went home to bed.

Mr. Dayton, like David Bonner, is still alive, and for much the same reason. In his case, too, chance decisively intervened. Within an hour after the accident, a reporter on the Harrisburg *Patriot,* the city's morning newspaper, emerged from his office, hailed a cab, and headed home. In the course of the trip, the driver began to talk. There was one thing about hacking, he said—anything could happen. Take tonight, for example. He had just come back from hauling a man to Polyclinic Hospital, and guess what was the matter with him. He'd been bitten on the thumb—by a bat! It was a mean-looking wound, too. His whole hand was covered with blood. But who ever heard of a bat attacking a man? He didn't know whether to believe it or not. Neither did the reporter, but it struck him that, if true, it was a possible story. The following day, on the way to work, he stopped by the hospital. A glance at the outpatient record established the facts of the matter. He then, with providential thoroughness, dropped around to the office of Ernest J. Witte, chief of the Division of Veterinary Public Health of the Pennsylvania Department of Health, and asked him what they meant. The case they described was news to Dr. Witte, but he answered without hesitation. They meant, he said, reaching for the telephone, that his division would investigate the incident at once. One phase of the inquiry would involve a search for the bat. Another, infinitely more urgent, would be concerned with Mr. Dayton. He must be found and returned to the hospital for immediate prophylactic treatment. The bat, in all probability, had been rabid.

Dr. Witte's hunch, though spontaneous, was anything but blind. He had good reason to associate belligerent bats with rabies. The relationship, indeed, was one with which he happened to be peculiarly familiar. His knowledge derived not only from the alerting example of the Bonner episode, a bulletin on which the Public Health Service had promptly dispatched to all state health officers, but also, more recently, from direct professional experience. That had been provided by the case of Frances Roberts. On the afternoon of September 29, as Dr. Witte later reported to the American Public Health Association, Mrs. Roberts, the wife of an amateur ornithologist of Boiling Springs, an upland resort about twenty miles west of Harrisburg, had accompanied her husband on a canoeing jaunt across a lake near their home. Toward six o'clock, deciding to stretch their legs before turning back, they beached their craft on a wooded shore, and Mr. Roberts wandered off to observe a flock of waterfowl.

Mrs. Roberts stayed by the canoe, and she was standing there,

Dr. Witte noted in his report, when "a bat suddenly landed on [her] upper arm, and bit her without warning or provocation. The woman . . . was startled by the attack and could not immediately identify the object clinging to her arm. Because of her fright, she does not recall distinct biting sensations, although she was conscious of the creature's scratching. Still not knowing what the object was, she finally grabbed the bat with her other hand and threw it against a nearby fence, where it remained stunned by the blow. [Mr. Roberts] was attracted by the commotion and quickly identified the animal as a [hoary, or *Lasiurus cinereus*] bat. Being a naturalist, with considerable knowledge and background in wildlife, he quickly recognized the behavior of the bat to be abnormal. He had the presence of mind to act with swiftness and cleverly trapped the creature in a pail, which he had nearby, and covered it with a newspaper. In a matter of minutes, the party headed back by canoe . . . to their home. [Mrs. Roberts] proceeded immediately to the doctor for treatment. . . . The physician reported that the patient received attention within one hour after the biting episode. He scrubbed the wound thoroughly with surgical soap and cauterized the wound, using an electric cautery. There were three distinct tooth marks on the upper arm, between the elbow and shoulder. He then called the [State] Health Department for advice on the handling of this case. Motivated to a large degree by the reports of the Florida experience, we immediately recommended antirabies prophylaxis." At the same time, Mr. Roberts was asked to deliver the bat to the Harrisburg laboratory of the department's Bureau of Animal Industry. He did so the following morning. "Touch preparations and, later, sections of the bat's brain revealed typical Negri bodies," Dr. Witte continued. "These were confirmed by the Director of Laboratories, Pennsylvania Department of Health, and by the Virus and Rickettsia Laboratory of the U. S. Public Health Service. Two rabbits were injected intracranially with the bat-brain material. Both animals developed clinical symptoms of rabies and died [within] twenty-seven days. [Meanwhile], starting October 1, [Mrs. Roberts] received fourteen injections of vaccine. [She] suffered no adverse reactions during the entire course of treatment. . . . As of this date [November 10, 1953], she remains in good health, but is still under her physician's care."

Although it was Dr. Witte who brought the attack on Mrs. Roberts to general medical attention, his report was not the first account of her misadventure. The first was a newspaper story, less comprehensive but equally stirring, that was widely published throughout the East within a day or two of the incident. Among those whom it particularly stirred was Frederick R. Taylor, an internist and professor of medical literature at the Bowman Gray School of Medicine of Wake Forest College, in Wake Forest, North Carolina. The news did not merely startle Dr. Taylor. It also inspired him to reflection, the nature of which he presently communicated to a colleague in Georgia. His letter, which has been preserved, began with a forceful summary of the Roberts case. This was followed by some lines to the effect that he had long been imperturbably aware of the existence of rabies in Latin-American vampires. "But," he then exclaimed, "an ordinary, insectivorous bat!!! What would happen if the Western bats that live literally by the millions in Carlsbad Caverns, New Mexico, got an epidemic started there? I have seen a high cloud of countless hordes of bats come out of the caverns' mouth at dusk. Too horrible to contemplate!"

Dr. Taylor's letter was dated October 19, 1953. Little more than two years later, on February 1, 1956, a news story authorized by the New Mexico State Department of Public Health appeared in the Santa Fe *New Mexican* under a six-column headline reading, "CARLSBAD CAVE BATS INFECTED WITH RABIES." "Rabies," it began, "has been discovered among the millions of bats at Carlsbad Caverns. It was a rabies epidemic which caused the death of hundreds of the Caverns bats in August and September of last year." The account continued:

Last Aug. 20, officials of the National Park Service at Carlsbad noticed dead and dying bats in increasing numbers. They were found on the floor of the caverns and in its entrance. Ranchers in the area also found dead bats. At that time it was thought that extensive insecticide spraying might have caused the deaths during the 10-day epidemic. But tests by the U. S. Public Health Service found no evidence of this. Instead, tests were begun to see if rabies had caused the deaths.

Lt. Col. Kennet Burns, chief of the veterinary virus laboratory at Ft. Sam Houston, Tex., collected specimens of dead and dying bats for examination while the epidemic was going on. Virus examinations by Burns revealed the presence of rabies in more than 50 per cent of the specimens, the department said. In addition, blood samples from a large number of live bats collected in flight at the caverns after the epidemic showed the presence of antibodies against rabies, indicating that many of the bats had been exposed to the disease some time in the past. . . .

The story also stated that although no human being had ever been known to be bitten by a bat while visiting the caverns, the health authorities had warned people against touching any of the creatures they might find dead or dying there.

Rabies is one of around sixty human diseases now known, or confidently supposed, to be of viral origin. Its causative agent is thus a member of the most mysterious form of life on earth. About all that can be said of the viruses is that they are supremely small (some are only just within the reach of an electron microscope), infinitely numerous (not even the bacteria are more ubiquitous), and almost incomparably specialized. All viruses are obligate intracellular parasites. They share with the rickettsiae the otherwise unique distinction of being unable to grow or reproduce outside the protoplasmic tissue of a living host. In general, the severity of a viral invasion reflects the functional importance of the particular cells to which the invaders are drawn. The virus of rabies is a neurotropic virus. Like the viruses of poliomyelitis and the several encephalitides, it has a special affinity for the cells of the central nervous system. It has, however, little else in common with any other virus. Its range, for one thing, is extraordinarily wide. Unlike the great majority of viruses (including the agents of smallpox, measles, yellow fever, poliomyelitis, infectious hepatitis, and the common cold), which can find in nature fewer than half a dozen satisfactory habitats, it is able to exist comfortably and abundantly proliferate in any warm-blooded animal. Its means of transmission is also peculiarly its own. Most viruses insinuate themselves into a host through either the respiratory passage or the gastrointestinal tract. A few are conveyed by bloodsucking insects. The rabies virus enters by way of a bite contaminated with the saliva of one of its victims. In this respect, it might seem to resemble the various mosquito-borne viruses, but the resemblance is merely apparent. The latter are transmitted in the natural course of the carrier's search for food. There is nothing natural about the transfer of the rabies virus. It wrings collaboration from its carrier hosts by torturing them into a homicidal fury. The incubation period of rabies (or the interval between the implantation of the organism and its establishment in the brain) is largely determined by the depth of the wound, its proximity to the brain, and the size of the original viral colony. This period, though disconcertingly variable, is seldom shorter than fifteen days and almost never longer than a year. But whether the virus reaches its destination in days or weeks or months, the result is inevitably the same. Rabies, in man, is a fatal disease.

The symptomatology of rabies is essentially the same in all susceptible animals. There are only superficial differences. The onset of the disease is generally mild and always indistinct. In man (and, insofar as can be determined, most comparably complicated animals), its earliest manifestations are those of any infection—a little fever, a dull headache, a scratchy throat, occasional nausea. This phase frequently lasts for two or three days, and sometimes even four, and is followed by a tingling pain at the site of the wound—the first diagnostically significant indication of rabies. Its grip, already fixed beyond release, then suddenly tightens. The muscles stiffen, the nerves tense, and the mind begins to fray with temper and apprehension. Anxiety quickens into fear. There is a vivid sense of approaching doom, a certainty of death. "A [rabid] patient weighed down with terror often becomes maniacal," D. L. Harris, medical director of the Pasteur Clinic in St. Louis, noted in a recent clinical study. "An excessive flow of thick tenacious saliva pours over his face and neck and becomes smeared on his hands and clothes and over the bedding and floor. These periods of rage are followed by moments of calm in which [he] usually shows anxiety for the safety of those around him and warns them of the approach of another crisis. Hyperesthesia of the skin to changes of temperature, and especially to currents of air, and increased sensitiveness to sound and light mark the progress of cerebral irritation. Convulsions are brought on by the least irritation and by the slightest current of air . . . the breath comes in spasms, dyspnea is extreme, and there are epileptiform seizures or tetanic rigidity. Hydrophobia is rarely absent. . . . When the patient [attempts to drink], there is an immediate viselike contraction of the muscles of deglutition with an excruciatingly painful spasm of the glottis and the pharynx. The body trembles with convulsive movements, the jaws are clenched, respiration is impossible. . . . After several attempts to drink, the pain is so terrible that despite intense thirst [the patient] cannot be induced to try to swallow liquids, and the sight of water or mention of the word brings on an attack. As a rule, death occurs after two or three days from cardiac or respiratory failure."

Although all highly developed animals are equally responsive to its gothic embrace, the rabies virus has its favored circle of hosts. It is naturally most inclined to frequent those best equipped to further its spread. This largely confines its normal range to the more prolific and short-tempered carnivores, a group that includes the fox, the wolf, the coyote, the jackal, the skunk, the mongoose, the cat, and the dog. Of these, the last, for reasons still obscure, has always been its most consistently conspicuous victim. The dog is also the animal in which its depredations were first recognized as those of a specific disease. Just when that occurred is uncertain. An illusion in the *Iliad* to "canine madness" has persuaded many medical historians that rabies may have been known to the Mediterranean world as early as the tenth century before Christ, and most believe, on the basis of rather stronger internal evidence, that the fifth-century Greek philosopher Democritus, who is chiefly remembered as a pioneer atomic theoretician and the teacher of Hippocrates, was probably conscious of its existence. The first explicit reference to rabies of which there is any record was set down by Aristotle, around 335 B.C., in his *Historia Animalium.* "Dogs suffer from three diseases: lyssa, quinsy, and sore feet," he noted. "Lyssa drives the animal mad, and any animal whatever, excepting man, will take the disease if bitten by a dog so afflicted; the disease is fatal to the dog itself, and to any animal it may bite, man excepted." Lyssa, a transliteration of λύσσα, means "frenzy," and is the name by which rabies was originally known. The Romans gave the disease its modern name, which derives from *rabere,* the Latin for "to rage," and has been in common usage since the first Christian century.

The Romans also modernized the Greek conception of rabies. A gifted encyclopedist of the early empire named Aulus Cornelius Celsus was among the first to raise his eyes from the pages of *Historia Animalium* and look squarely at the world around him. Having done so, he proceeded to challenge the first of Aristotle's comfortable exceptions as myopically veterinarian. All animals, he decided in his classic *De Medicina,* were equally susceptible to rabies. Celsus was willing, however, to concede Aristotle's second exception. It was possible, his studies informed him, that the disease could be mastered in man. He then went on to propose a still valid preventive technique ("the wound . . . must be cauterized") and, less acutely, an antidote and a course of treatment. This consisted of thirty herbal ingredients (including poppy tears, Illyrian iris, Gallic nard, white pepper, male frankincense, and turpentine) mixed with honey and dissolved in a tumbler of wine. Its omission, he added, was risky. "When too little has been done for such a wound it usually gives rise to a fear of water," he wrote. "In these cases there is very little hope for the sufferer. But still there is just one remedy, to throw the patient unawares into a water tank which he has not seen beforehand. If he cannot swim, let him sink under and drink, then lift him out. If he can swim, push him under at intervals so that he drinks his fill of water even against his will. For so his thirst and dread of water are removed at the same time. Yet this procedure incurs a further danger, that a spasm of sinews, provoked by the cold water, may carry off a weakened body. Lest this should happen, he must be taken straight from the tank and plunged into a bath of hot oil."

Celsus's uneasy concession that rabies need not be fatal to man was accepted without recorded dispute for fifteen hundred years. So, except for certain pharmacological refinements, were his methods of breaking its hold. Pedanius Dioscorides, whose *De Materia Medica* was the standard pharmacopoeia throughout the Roman era, contented himself with offering two alternative antidotes. One was a draught of hippocampus, or sea-horse, ashes. The other had as its active principle the leaves of the bladder campion. "This, being beaten when it is green, with old swine's grease, is good for the mad-dog-bitten," he wrote. Rufus of Ephesus, a second-century physiologist, preferred a draught of "wormwood, aristolochia, Lycian thorn, decoction of river-crayfish, water germander, rock-parsley, and the root called gentian." Even Galen, the most observant, as well as the most imaginative, medical investigator in the millennia between Hippocrates and the Renaissance, had nothing to add to Celsus but a polished definition: "[Rabies] is a disease that follows the bite of a mad dog and is accompanied by an aversion to drinking liquids, convulsions, and hiccups. Sometimes maniacal attacks supervene."

After Galen, and the subsequent canonization of Greco-Roman medicine, the illumination of rabies, like that of all disease, was considered complete, and the subject complacently closed. The first attempt to reopen it was made in the sixteenth century. A Veronese savant named Hieronymus Fracastorius is usually celebrated for this act of desecration. Rabies, he announced in 1546, in his precocious *Contagions, and Contagious Diseases and Their Treatment,* was an infectious disease, always communicated by the injection of saliva into the blood, and, notwithstanding the protestations of pharmacy, always irremediably fatal. He also emphasized this novel conception of the disease in a dissertation on hunting dogs. "What particularly calls for the care of the skilled mind," he wrote, "is when, inflamed with rabies, [the dog] attacks now these, now those and, turning against the master himself, he inflicts the incurable wound." Fracastorius was a man of towering intellectual stature. In addition to being a notable physician, he was a poet (the term "syphilis" derives from his *A Poetical History of the French Disease*), a botanist, a geographer, a musician, a mathematician, and an astronomer, and his morbid view of rabies received a respectful hearing. It even, for a time, attracted a few admirers. But hope and habit were too strong, and within a generation the more congenial classic conception resumed its interrupted vogue.

Celsus's hydrotherapeutic regimen, adapted to the ducking stool, was commonly prescribed in cases of rabies throughout the sixteenth and seventeenth centuries, and at least on occasion (an essay by Oliver Goldsmith, written around 1765, refers to "a little boy bit in the leg, and gone down to be dipped in the salt water") during much of the eighteenth century. His pharmacological influence continued even longer. In 1806, the New York State Legislature passed, without recorded opposition, a bill entitled "An Act for Granting a Compensation to John M. Crous, for Discovering and Publishing a Cure for the Canine Madness." Crous's cure, for which he was granted a thousand dollars, was a tablet to be swallowed with water. Its components included the pulverized jawbone of a dog, the dried false tongue of a newly foaled colt, and a pinch of corroded copper taken from an English penny minted in the reign of George I. Other American physicians of that time, perhaps less impressed by royalty, favored a remedy composed of bole armeniac, alum, chalk, elecampane, and black pepper. They also had confidence, as did many European doctors, in the curative powers of concretions, similar to kidney stones, that are sometimes found in the intestines of deer, goats, and other herbivorous animals and that, because they were used as a specific in the treatment of rabies, became known as madstones. Such concretions, being formed of mineral salts, are porous and somewhat absorbent. These qualities helped to support the belief that a madstone applied to a rabic wound would promptly extract the venom. "This afternoon called on by a man in Jeffersonville to apply the madstone to a little son bitten a day or two previous," an Indiana judge, soldier, and statesman named John McCoy noted in his journal on June 9, 1848. "Rode thro' the rain and reached there about sunset. Induced to think the dog mad." In 1879, at an auction in Texas, a madstone brought two hundred and fifty dollars. That would be the equivalent of about a thousand dollars today.

The supposition that rabies could be cured by some curious pill or poultice expired with the nineteenth century. The absolute lethality of the disease is now universally accepted. One reason for this abrupt resignation to reality is that the evidence assembled by modern medical science leaves no room for doubt. Another is that the truth is no longer unbearable. Since the eighteen-eighties, when Pasteur was inspired to adapt to rabies his epochal discovery that the pathogenic properties of a microorganism can be attenuated (by drying, or treatment with certain chemicals, and passage through a succession of laboratory animals) without affecting its capacity to generate protective antibodies, a reliable means of hobbling the disease has been everywhere at hand.

Pasteur conceived the idea of rabies prophylaxis in 1880. By the end of 1883, he and his associates at the Ecole Normale, in Paris, were able to produce a stable strain of suitably domesticated virus. This was followed by two series of experiments establishing beyond dispute that the strain was immunologically effective in dogs. The first of these was brought to a brilliant close in June, 1884, with a formal trial before a committee of scientists appointed by the French Government. For this definitive test, Pasteur chose two previously vaccinated dogs, two untreated dogs, and two untreated rabbits. After being examined by the committee, the six animals were anesthetized and trephined. Each animal was then identically inoculated with a quantity of material drawn from the brain of a demonstrably rabid dog. When the operation was completed, the animals were separately confined, and all received the same postoperative care. Two weeks later, the four controls, or untreated animals, developed rabies, and died. The vaccinated dogs remained in normal health. The second series of experiments, though begun at about the same time, continued into the following year, and the results were equally emphatic. They showed that it was possible to immunize a dog against rabies not only before but, if the step was undertaken promptly, after exposure to the disease. Pasteur emerged from this revolutionary triumph with a vision of one even more revolutionary. "What I aspire to [now] is the possibility of treating a man after a bite with no fear of accidents," he wrote in the spring of 1885. ". . . I have not yet dared to treat human beings after bites from rabid dogs. But the time is not far off."

The time, as it turned out, was only a few weeks off. Pasteur treated his first human patient on July 6, 1885. This now famous pioneer was a nine-year-old Alsatian boy named Joseph Meister. Two days before, while walking on a country road near his home, he had been attacked by a plainly rabid dog, knocked down, and bitten fourteen times. When Pasteur saw him, at the request of a family doctor, the boy was more dead than alive. In fact, Pasteur later recalled, it was only the apparent hopelessness of the case that induced him to attempt its treatment. The procedure he used was a freehand adaptation of the one he had developed in his most recent experiments with dogs, and it took ten days. During that time, the boy received thirteen inoculations, of increasingly potent vaccine. His immediate reaction was encouraging, and it continued satisfactory throughout the treatment. At the end of a month, his wounds having healed, he seemed to be fully recovered. He was. Joseph Meister lived to be sixty-four. He died in 1940, a suicide.

The rehabilitation of Joseph Meister, which Pasteur described in a paper, entitled "Méthode pour Prévenir la Rage après Morsure" and presented at a meeting of the Académie des Sciences on October 26, 1885, created an instant and appreciative stir throughout the medical world. "[Rabies], that dread disease against which all therapeutic measures had hitherto failed, has at last found a remedy," the formidable neuropathologist Edmé-Félix-Alfred Vulpian proclaimed. Assisted by this and other resounding testimonials, the Pasteur treatment, as the procedure came to be called, was in international use within a decade, and it has since been administered many thousands of times, with sufficient success to establish its worth as a reliable defensive tool. Or so it is generally assumed. To what extent the Pasteur treatment protects human beings against the development of rabies, however, is not known, and probably (in view of the natural scarcity of volunteers available for a series of controlled experiments) never will be. Its powers, in any event, are somewhat less than total. In a recent monograph, Harald N. Johnson, a staff member of the Rockefeller Foundation, observes, "On the basis of clinical evidence, there seems to be no doubt that rabies vaccine is effective in preventing the disease in the majority of the instances in which there is an expected incubation period of more than one month." But such an incubation period can only be expected in cases involving bites on the arms, legs, or torso. The chances that the Pasteur treatment will prevent the development of the disease when the victim is bitten severely on the head or neck are slight.

Only one more or less controlled test of rabies immunization in human beings has ever been made. That was conducted by a World Health Organization team in 1954, in Iran. Its purpose was to evaluate an antirabies serum developed that year by Hilary Koprowski, assistant director of viral and rickettsial research at the Lederle Laboratories of the American Cyanamid Company. Serum differs from vaccine in that it contains—rather than merely stimulates the body to produce—the immunizing agents known as antibodies. A rabid wolf had burst into a mountain village, not far from the W. H. O. team's station, and bitten twenty-nine men, women, and children. As a matter of course, the Pasteur treatment was prescribed at once for all the victims. In addition, seventeen of the group, whose wounds included bites on the head or neck, were given immediate injections of serum. Eleven of them received one injection, the others two or more. The results were unmistakably clear. Twenty-five of the victims, including all who had received at least two injections of serum, survived. Of the four who died, three had been given only the Pasteur treatment, and the other a single serum inoculation. The limited efficacy of the Pasteur treatment is not, unfortunately, its only flaw. It has others. It is unpleasantly long (the present regimen, even when supplemented by serum, requires from fourteen to twenty-one days), it is usually expensive (the average injection costs about five dollars), and, above all, it is disturbingly dangerous. Reactions to antirabies treatment range from those common in allergic conditions—erythematous or urticarial rashes, edema, syncope—to one known as neuroparalytic accident. Neuroparalytic accident varies in degree from a polyneuritis to ascending encephalomyelitis. The latter, in an uncomfortable number of cases, is permanently incapacitating, and sometimes fatal.

The imperfections of the Pasteur treatment are not, of course, sufficient to deny it a place in the modern medical kit. There is, after all, nothing with which to replace it. In the opinion of most investigators, however, the imperfections are pronounced enough to discourage its use in any but cases of certain—or suspicious but unverifiable—exposure. It is also their urgent conviction that postexposure prophylaxis is, at best, an indirect defense against the menace of rabies. "There can be no question that the ultimate solution to the rabies problem is predicated on the control and eventual elimination of the disease from animal populations," the *American Journal of Public Health* commented editorially in May, 1955. "This may be accomplished by the setting up of transmission barriers, such as animal immunization, elimination of stray dogs, and the reduction of excessive numbers of wildlife vectors." It has been accomplished in a considerable number of countries. Britain, where a system of controls, rigidly enforced by the Ministry of Agriculture and Fisheries, was established around 1900, is perhaps the most notable of these. The last human exposure to rabies in England occurred nearly fifty years ago, and except for a handful of cases among imported dogs held in quarantine, there have been no outbreaks among animals there since shortly after the First World War. The Scandinavian countries—Denmark, Sweden, and Norway—have, by similar exertions, achieved almost as admirable a record, and so, among others, have Australia, New Zealand, and Malaya.

The record of the United States, despite the existence of an elaborate apparatus of legislative controls, is less imposing. Except for Hawaii, where rabies has somehow never gained a foothold, few parts of this country are wholly free of the disease. Last year (1953), around half a million Americans were treated for bites inflicted by animals. Of these, sixty thousand were judged to have been exposed to rabies and received the Pasteur treatment. Three of them died. There were nine additional fatalities among persons who received incomplete or no treatment. The lowest incidence of human rabies in recent years was ten cases, in 1949. The highest was fifty-six cases, in 1944. Among domestic animals the average annual mortality is between seven and eight thousand. The persistence of rabies in man and beast throughout the United States has been variously explained, but two factors are considered decisive. One of these is indifference. Although many states have laws that specify a certificate of vaccination as a prerequisite for obtaining a dog license, and although all make some provision for the disposal of strays, such measures are seldom enforced, and then only in moments of epidemic panic. The other is the still enormous number of wild animals among which the rabies organism is endlessly perpetuated. This indigenous reservoir includes not only such conspicuous vectors as the fox and the skunk but badgers, raccoons, beavers, squirrels, and, since the early nineteen-fifties, the insectivorous bat.

The full significance of the bat attacks on Frances Roberts, Carl Dayton, David Bonner, and Mabel Tate has yet to be determined. One thing, however, seems certain. These four people were not the victims of a fleeting freak of nature. Their experiences have since been duplicated elsewhere in the country. Three more attacks by rabid bats were reported in 1954. All occurred in Texas—the first, early in April, in San Antonio, and the second and third, in May and July, near Austin. The victims were a youth of twenty and two small children. Another was reported in October, 1955, in Madera, California, and involved a middle-aged man. Two more attacks—one certain and the other probable—were added to the record in 1956, and in May, 1957, an eleven-year-old boy was attacked near his home in Jamesville, Wisconsin, and bitten on the arm and chin. The 1956 victims were a soldier on maneuvers in Louisiana and a Texas State Health Department field epidemiologist named George C. Menzies. Dr. Menzies, at the time of his exposure, had been collecting specimens of cave-dwelling bats in central Texas to be examined for evidence of rabies infection. How and when he was exposed is not known. It is only known that he returned to his home, in Austin, on January 1, and the following morning developed symptoms of rabies. Two days later he was dead. The six other victims received the Pasteur treatment, and survived.

Dr. Menzies's last assignment was one of a number of similar studies that have been undertaken in collaboration with the United States Public Health Service since the Bonner episode in 1953. The investigation, which was understandably intensified by the harrowing discovery at Carlsbad Caverns two years later, is expected to continue for at least another year. Some months will then be required to accurately assess its results. The preliminary findings, however, have been tentatively correlated by Ernest S. Tierkel, chief of Rabies Control Activities at the service's Communicable Disease Center (now the Center for Disease Control) in Atlanta, and they are hardly reassuring. "During the last eighteen months or so, various field units have bagged in the neighborhood of ten thousand bats, in sixteen different states," Dr. Tierkel says. "About a hundred and fifty of them were positive for rabies. The group included four species of tree-living, or solitary, bats and eight species of cave-dwellers, or colonials. All, of course, insectivorous. Every state in which we've made a thorough study has yielded its quota of positives. The list, at the moment, is Alabama, California, Florida, Georgia, Louisiana, Michigan, Minnesota, Montana, New Mexico, New York, Ohio, Oklahoma, Pennsylvania, Texas, Utah, and Wisconsin. Those are the facts that we have to work with. What they mean—their epidemiological significance—is what we hope to find out.

"In the early phases of our investigation, one possible conjecture was that what we were turning up wasn't really rabies. We thought it might be a new virus disease of bats so closely related antigenically to the rabies virus that the two couldn't be differentiated by the usual laboratory tests. But a little more laboratory work disposed of that possibility. The disease is definitely rabies. Another basic question is whether the disease has always been present in the insectivorous bats of the United States and we have only just discovered it, or whether it represents a recent northward invasion into this country from the vampire-bat-rabies areas in Latin America. I'm inclined to suspect that the latter is the answer. We know, at any rate, that the Mexican free-tails of our Southwest migrate deep into the vampire country of Mexico. According to some authorities, the vampires and the free-tails even share the same winter caves. We hope that's all they share. If it turns out, as some preliminary findings have suggested, that our bats also share the vampires' resistance to rabies, we're up against an extremely difficult problem. Vampires—some of them, at least —are known to be capable of transmitting the disease for long periods of time without showing any signs of illness themselves. In other words, they're like Typhoid Mary. They're true carriers. If our bats have that capacity, if we find that they sometimes attack simply because they're frightened and not because they've been driven into a frenzy by the disease, and if we also find that the bat represents an important reservoir of rabies in the United States . . . Well, those are only possibilities, of course. We don't have the data yet to even hazard an answer. But what if they're shown to be facts? I think it would be a very good idea to tighten up our system of rabies controls."

[*1956*]

CHAPTER 5

$CH_3CO_2C_6H_4CO_2H$
(Aspirin)

———————————◆◆———————————

AROUND FIVE O'CLOCK on the afternoon of Wednesday, May 4, 1955, an office boy walked into the office of Dr. Harold Jacobziner, Assistant Commissioner for Maternal and Child Health of the New York City Health Department and chief of the department's newly established Poison Control Center, with a memorandum initialed by a clerk in the report room of the center. Dr. Jacobziner had his hat on his head and his briefcase in his hand, but the message turned him back to his desk. He sat down, glanced again at the paper, and paused for a moment of geographical calculation. Then he picked up the telephone and dialed the department's Bureau of Public Health Nursing. An official of his acquaintance answered the call, and after a brisk exchange of civilities Dr. Jacobziner got briskly down to business. He had a job for one of the nurses on the staff of the municipal health center that served the Corona section of Queens. The assignment was an investigatory visit to the home of a couple named (I'll say) Mr. and Mrs. Francis R. Poole. They lived on Alburtis Avenue, in Corona. According to a formal notification just received by the Poison

Control Center, their son, a boy of three named Richard, had been admitted to Whitestone Memorial Hospital a few hours earlier. It was another case of acute acetylsalicylic-acid intoxication.

Acetylsalicylic acid is the universal comforter known in the Esperanto of the laboratory as $CH_3CO_2C_6H_4CO_2H$ and almost everywhere else as aspirin. The latter is a term of proprietary origin that has achieved an all but total nomenclatural ascendancy over science. This triumph, though unique in medical history, was in no way uniquely accomplished. In common with all such fabrications that have outgrown the stigma of trade, the widespread acceptance of the name is generally attributable to its distinctive sound, to its attractive brevity, and to many years of pertinacious advertising. Nevertheless, unlike much commercial coinage, "aspirin" is not altogether an etymological freak. Its lineage is acceptably legitimate. It derives, roughly but rationally, from *Spiraea,* a botanical genus whose more prominent members (bridal wreath, meadowsweet, hardhack) are natural sources of salicylic acid, the active principle in aspirin. As it happens, however, the vegetable secretion of the acid is far from confined to plants of the *Spiraea* group. It also occurs in many other shrubs (jasmine, madder, partridgeberry), and in many legumes (peas, beans, clover), grasses (wheat, rye, sugar cane), and trees (beech, birch, olive, poplar, willow). The *Spiraea* were merely among the first such plants to be identified by modern pharmacology.

That the bark, fruit, and leaves of these plants contain some revitalizing agent has long been common, if largely rustic, knowledge. Potions rich in salicylic acid are as old as herbal therapy, and almost as ubiquitous. Practically all races seem to have early grasped their usefulness. A draught whose ingredients included the juice of willow bark was esteemed by many North American Indian tribes as an antipyretic, or fever reducer. Another aboriginal people—the Hottentots of South Africa—made good use of an essentially similar decoction to ease the agony of rheumatism. The willow, among other salicylate plants, was also held in high regard throughout the early Mediterranean world. Hippocrates, in the fourth century before Christ, recognized its capacity for relieving both pain and fever. He also perspicaciously recommended topical applications of willow leaves, presumably as an antiseptic, in the *post-partum* care of maternity cases, and, less perspicaciously, the juice of poplar bark for various eye diseases. A generation later, Theophrastus, a pioneer botanist who succeeded Aristotle as head of the Lyceum at Athens, proclaimed the analgesic excellence of madder bark. He was, in addition, commendably inspired to urge the inclusion of madder in the manual of mild but effective diuretics. Theophrastus's success in thus broadening the therapeutic range of the salicylates was presently matched by other investigators. Around A.D. 75, Dioscorides, a Greek surgeon in Roman military service, rose from a series of experiments with the discovery that a paste composed of willow ash would safely remove such callosites as corns. In the course of this study, he had stumbled upon the further fact that his corn cure was almost as useful in combating the torments of gout. Pliny the Elder, the celebrated Roman encyclopedist, placed willow juice on the diuretic list. He then went on to propose an infusion of poplar bark as a specific for sciatica, and, amending, if not improving upon, Hippocrates, suggested poplar gum to druggists in search of a satisfactory eyewash. To these innovations, more imaginative minds rapidly added an unguent of willow bark as a cure for earache, one of willow leaves as a dressing for bloody wounds, and a willow poultice to dissipate fistulas and erysipelatous lesions. By the end of the second century, when Galen completed his valiant thirty-volume pharmacopoeia, all the remedial powers—great, small, illusory—of the salicylates had been noted and defined. Galen was the mightiest of the Greco-Roman empirical pharmacists. He was also the last pharmacist of any stature anywhere until almost modern times. With the foundering of Rome, the whole of rational herbal therapy was lost to most of the Western world in the mists of Christian mysticism for well over a thousand years. ("All diseases of Christians are to be ascribed to demons," St. Augustine announced in the fifth century.) But for a scattering of practical folk-physicians, too simple to comprehend the complexities of fashionable piety, it might well have been lost forever.

The classic grasp of botanical medicine managed to survive both Imperial Rome and the Holy Roman Empire only in the country kitchen. Among salicylate plants, the willow was the first to be recalled to the attention of science. An eighteenth-century English clergyman and naturalist named Edward Stone is usually considered its rediscoverer. His find, though unexpected, was not entirely adventitious. He knew, in a way, just what he was looking for. Stone, like many others of his time and inquisitive bent, had long diverted himself with the hope of finding in some common plant an inexpensive antipyretic substitute for Peruvian bark (or cinchona), as quinine was then called. One afternoon in 1763, while enjoying a rural ramble, he got wind of the fact (possibly during a pause in a country kitchen) that willow bark was locally much admired as a household remedy for the feverish chills of ague. A hint was all he needed. He sampled a piece of bark, found that it shared with cinchona an "extraordinary bitterness," and buckled down to work. Some months later, in a letter to the president of the Royal Society of London for Improving Natural Knowledge, he summarized his findings. His paper began with a word on the willow. "As this tree," he noted, "delights in a moist or wet soil, where agues chiefly abound, the general maxim, that many natural maladies carry their cures along with them, or that their remedies lie not far from their causes, was so very apposite to this particular case, that I could not help applying it; and that this might be the intention of Providence here, I must own, had some little weight with me." Then, having delivered himself of this rather shaky ratiocination, Stone got firmly down to cases. Providence had provided him with fifty of them. All were victims of "agues, and intermitting disorders," and although their seizures varied in severity, all were placed on the same regimen—twenty grains of powdered willow bark dissolved in a dram of water, administered every four hours. The results, he was happy to record, had been uniformly excellent.

Stone's triumphant report, which was presently published in the

Philosophical Transactions of the Royal Society, had two triumphant results. The first, and by far the more direct and immediate, was a burst of corroborative clinical tests that promptly returned the willow to a reputable place in the professional medicine chest. The other was the introduction of the willow into the laboratories of scientific chemistry, which were then just opening. There, in the first decades of the nineteenth century, it happened to catch the eye of a cosmopolitan company of investigators whose particular interest was the chemical composition of medically useful plants. Their attention, once focused, was almost at once productive. The effective essence of willow bark was identified by the Italian chemists Fontana and Brugnatelli in 1826. Three years later, a French experimental pharmacist, Henri Leroux, accomplished its isolation. In 1838, the extraction of salicylic acid was accomplished by the Italian chemist Rafelle Piria. At about the same time, in Switzerland, an apothecary named Pagenstecher found that it could also be obtained from meadowsweet and bridal wreath, and his provocative find was closely followed by a succession of explorations, in England, Germany, and elsewhere, that quickly established its presence in a multitude of other plants. The inquiry then came to a halt. For it was almost as quickly established that salicylic acid, despite its abundance of sources, was too costly to extract to be of much practical use. Also, for the moment, the investigators had satisfied their curiosity.

Their curiosity remained quiescent for close to a generation. It revived, like the whole of organic chemistry, with the stentorian enunciation, by the German chemists Friedrich Wöhler and Justus von Liebig, of the theory of synthesis, or the interrelation of organic and inorganic substances. Among those who heard their call was a professor of chemistry at the University of Leipzig named Hermann Kolbe. To Kolbe, the sound was a summons. It summoned him to work, and to fame. In 1874, after a period of patient stalking, he emerged from his laboratory and held up to view in the pages of the *Journal für praktische Chemie* an economical procedure for the production of synthetic salicylic acid. The importance of Kolbe's achievement, which posterity has recorded as a milestone in the history of synthetics, was readily grasped by all but the most myopic, and the halls of progressive chemistry rang with appreciative applause. So, understandably, did those of contemporary medicine, but only for a time. The hopes it excited in medical science soon and suddenly withered. Although Kolbe had solved one practical problem, his solution brought another to light, which revealed itself at the first round of clinical trials. Pure salicylic acid, though a model of strength in dulling pain and banishing fever, was by no means a model drug. Its salubrious powers were more than offset by its high degree of toxicity. Large or frequent doses, the reporting clinicians agreed, almost always resulted in nausea, often in spasms of vomiting, and sometimes even in coma. The inquiry once more shifted its course. What was needed now, its participants perceived, was a neutralizing agent, or buffer. That meant searching out some substance that would combine with salicylic acid (not all elements are physiologically compatible), render it sufficiently bland for general clinical use, and yet leave unimpaired its therapeutic vigor. The chemists' resilience, for once, was fully and finally rewarded. There are, to the best of scientific knowledge, just three compounds that meet these requirements with anything approaching precision, and all were known and in service before the end of the century. One of them—sodium salicylate—turned up at the very first cast, in 1876. In 1885, the Polish pharmacologist Marcellus von Nencki identified phenyl salicylate as another. The third is acetylsalicylic acid. It was introduced in 1899.

Acetylsalicylic acid had its pharmacological origin in the laboratories of Friedrich Bayer & Company, later a subsidiary of I. G. Farbenindustrie, near Düsseldorf. It was first developed there, early in 1899, by a young research chemist named Felix Hoffmann. Hoffmann was not, nor did he claim to be, the actual discoverer of acetylsalicylic acid. That distinction had been achieved some forty years before by an Alsatian chemist named Charles Frédéric von Gerhardt. What Hoffmann discovered was its usefulness. Von Gerhardt had encountered acetylsalicylic acid in 1853 while exploring the interaction of various salts and acids, but he considered it merely a novelty. Like Paul Gelmo, the precocious Viennese student who in 1908 casually created and indifferently abandoned the coal-tar derivative known since its re-creation (by Gerhard Domagk, in 1933) as sulfanilamide, von Gerhardt was born too soon to comprehend what he had wrought. Hoffmann, like Domagk, was more favorably placed. His accomplishment, however, can be laid only partly to the instructive passage of time. He had other, less common advantages. Of these, two were decisive. One was a uniquely compelling source of inspiration in the person of his father. The elder Hoffmann was an invalid, a racked and crippled victim of rheumatoid arthritis. Of the remedies then at hand, including both sodium and phenyl salicylate, none could give him any adequate relief, and his life had become an agony. This wrenched the son from his workaday studies and drove him in search of some effective medicament. Hoffmann's other great gift was good luck. This guided his steps past the standard works in the scientific library to a compendium of chemical freaks and drones, and enabled him to find what he was looking for. His next move was conventional. Having satisfied himself, by laboratory test and domestic trial, of the worth of acetylsalicylic acid, he assembled his notes and carried them dutifully to the Bayer Company's director of pharmacological research. That was Heinrich Dreser, the architect of, among other things, diacetylmorphine, or heroin, and then an imposing figure in European science. Dreser reacted to the data set before him in a manner befitting his rank. He took one look, and took over. It was he who piloted acetylsalicylic acid through its first full-dress clinical evaluation. It was he who discarded its natural name as being hard to pronounce, hard to remember, and impossible to patent, and replaced it with the commercially seemlier "Aspirin." And it was he who wrote, and cheerfully signed, the pioneering report, confidently entitled "Pharmakologisches über Aspirin (Acetylsalicylsäure)," whose publication in *Pflügers Archiv für die gesamte Physiologie des Menschen und der Tiere* first brought the compound to the attention of medicine. That was toward the end of 1899. Its merits took it from there.

Acetylsalicylic acid (as most clinicians still choose to call it), or aspirin (the name practically everyone else prefers), or Aspirin (as its original proprietors were profitably privileged to insist on its being called until the general expiration of their patents around 1917), is an extraordinary drug. Almost everything about it is unusual. It comes, for one thing, impressively close to being the oldest therapeutic agent in continuing active demand. Only such venerable specifics as quinine (for malaria), colchicum (for gout), digitalis (for cardiac weakness), and, perhaps, the bromides (for relief of nervous tension) are of greater age and durability. Unlike these tottering centenarians, however, it has more than merely held its own against the massive ingenuities of mid-twentieth-century pharmacology. In 1935, the annual consumption of acetylsalicylic acid in the United States was estimated at four million pounds. By 1944, it had risen to nearly eight million. According to the least galvanic estimates, it now exceeds eleven million pounds, or, in terms of the standard dosage, approximately sixteen billion five-grain tablets. The true figure is probably well in excess of that, for, in general, such estimates refer only to the acetylsalicylic acid that is marketed under the name of aspirin. They seldom include that contained in the several more recently developed proprietary preparations—Bufferin, Anacin, Empirin, and Alka-Seltzer, among others—of which it is a major ingredient. The popularity of acetylsalicylic acid is almost equally immoderate elsewhere in the world. In some countries, notably Canada and those of Western Europe, it has, if anything, inspired an even fiercer veneration (England, for example, with hardly a fourth the population of the United States, consumes at least three million pounds of it a year), and there are no countries in which it is unknown, unappreciated, or unavailable. It is, in fact, the most widely used drug on earth. It is also the cheapest (a five-grain tablet costs about half a cent), one of the safest (many aspirant suicides have survived doses of more than two hundred grains), and, among drugs of a comparable nature, including its companion salicylates, without much doubt the best. Some authorities go further than that. They consider it the most generally useful drug in the entire armamentarium of medicine.

It is not hard to understand why. The great majority of drugs, as the heft of the modern pharmacopoeia is enough to suggest, are more conspicuous for the refinement than for the variety of their therapeutic feats. Hundreds, like insulin (the specific for *diabetes melitus*) and heparin (an anticoagulant), are limited in usefulness to one particular function, and there are many hundreds more whose present scope is only a trifle less narrow. Even the widest-ranging antibiotics, for all the multiplicity of horrors—anthrax, brucellosis, endocarditis, meningitis, tularemia, and syphilis, among others—over which they have an almost sovereign domination, are anything but widely effective. Their range is wholly confined to diseases of infectious, or microbial, origin. Acetylsalicylic acid is differently endowed. Although it is incapable of mastering any disease, it possesses a masterful sweep. Few drugs embody more, or more various, powers. As an analgesic, an antipyretic, an antirheumatic, and—or at least so a recent study at the Boston University School of Medicine, in which it appeared to show a marked capacity for preventing the recurrence of kidney stones, would suggest—an anticalculus, its palliative reach embraces practically all the smaller miseries, as well as some of those of considerable stature, that try the human race.

Just how acetylsalicylic acid performs this symphony of services has yet to be fully fathomed. Fifty years of investigation have illuminated little more than the general route it follows through the body. Salicylic acid, the active principle of the drug, is absorbed directly from the small intestine into the blood stream. It remains there, moving with the current of the circulation, for, in most cases, ten to fourteen hours. It then passes into the kidneys and is excreted in the urine. It is also known to form in the kidneys a compound called glucuronide, whose presence increases the solubility of calcium—a phenomenon that would seem to explain its apparent anticalculous action. The essential physiology of the acid is otherwise very largely opaque. About the only further probability is that each of its distinctive effects is achieved by a different mechanism. There is no possible functional connection between its capacity for generating glucuronide and any of its other attributes, and it seems almost as certain that the processes by which it blocks pain, reduces fever, and suppresses joint inflammations are also completely unrelated. Of these three established powers, the first is the least mysterious. Most physiologists believe that the site of salicylic acid's analgesic action is the thalamus, the chief sensory reception center, located in the forebrain. There, in some obscure manner, the acid (or impulses set in motion by it) impedes the normal transmission of stimuli conveying the sensation of pain. The best evidence suggests that acetylsalicylic acid has the power to raise the threshold against pain perception by about thirty-five per cent. "A very small dose of aspirin has a measurable effect in raising the pain threshold," Stewart Wolf, head of the Department of Medicine of the University of Oklahoma School of Medicine, and Harold G. Wolff, professor of medicine at the Cornell University Medical College, noted in a recent monograph. "Increasing the dose has a progressively greater effect up to about ten grains, beyond which no additional pain-killing effect is manifest, although larger doses do somewhat prolong the duration of effect. With ten grains of aspirin, maximum effect is achieved in about one and a half hours. Thereafter its action in raising the pain threshold begins to dwindle. The effect can be kept at a peak most satisfactorily not by giving larger doses but by frequently repeating the ten-grain dose, say at two-hour intervals. . . . Combinations of analgesic drugs are no more effective than is the single strongest drug in the combination. The most that can be achieved in terms of deadening pain is the effect of ten grains of aspirin." This, of course, is not to say that acetylsalicylic acid is the strongest of all analgesics. It is merely the strongest that can be safely used without medical supervision. Morphine, for example, is capable of raising the threshold against pain perception as high as seventy per cent, but it also affects pain reaction; that is, it simultaneously blocks the passage of stimuli conveying such sensations as tension and anxiety. The artificial avoidance of tension and anxiety is the basis of addiction. Acetylsalicylic acid has no effect whatever on pain reaction.

Almost nothing is known about the way acetylsalicylic acid functions as a febrifuge. There are no absolute data on the dynamics involved, and only a few reasonably firm hypotheses. The firmest of these concern the general procedure by which a reduction of fever is accomplished. It was once held that the phenomenon could be ascribed largely to a braking action on the machinery of heat production. Recent research has substantially altered this view. The explanation now in vogue suggests that antipyresis results primarily from a simple quickening of the body's natural facilities for releasing excessive heat—the peripheral blood vessels and the sweat glands. Its ultimate cause, as currently envisioned, is more complex. Animal experiments have demonstrated that, contrary to snap assumption, this hyperactivity is induced not by direct stimulation of the superficial thermostatic apparatus but indirectly, through the central nervous system. In the opinion of most investigators, the specific focus of impact is probably one or another of the numerous regulatory nuclei that are clustered near the midbrain. Beyond this point, however, all suppositions falter. The nature of the power that acetylsalicylic acid has over these nuclei eludes even the nimblest imaginations, except for one twilit hypothesis: The control it exerts would seem to be somewhat less than total. For, as a mountain of empirical evidence makes clear, its influence is limited to remission of fever. It is peculiarly, and providentially, incapable, even in massive doses, of forcing the temperature of the body below 98.6 degrees, or normal.

Everything about the antirheumatic powers of acetylsalicylic acid is elusive. No one has any understanding of their fundamental character. It can be said only that they very definitely exist. They are, in fact, so plainly pronounced that acetylsalicylic acid has stood unchallenged for almost half a century as the drug of majority choice in the symptomatic treatment of rheumatic fever, rheumatoid arthritis, and other inflammations of connective tissue. Its position, though violently shaken by the cyclonic appearance of cortisone and ACTH in 1949, is still far from insecure. Many clinicians—especially in Britain, where the adrenocortical hormones were welcomed with a rather less millennial euphoria than in their native America—have never ceased to prefer it. "Modern hormone therapy notwithstanding, salicylate, either as the sodium salt or as acetylsalicylic acid—remains the safest and most satisfactory drug in the routine treatment of rheumatism," Stanley Graham, professor of child health at the University of Glasgow, wrote in a recent report to the *British Medical Journal.* Much the same opinion has been expressed by William McK. Jefferies, assistant professor of medicine at the Western Reserve University School of Medicine, in the *New England Journal of Medicine.* "Although ACTH, cortisone, or hydrocortisone in adequate doses can at least temporarily control the symptoms of rheumatoid arthritis in practically every case," Dr. Jefferies observed, "the chronic nature of this disease and the hazards of prolonged maintenance of steroid therapy make it generally advisable to try other accepted methods of treatment, such as salicylates and physiotherapy, first." To these views, Sir Henry Cohen and his fellow members of the Joint Committee of the (British) Medical Research Council and Nuffield Foundation on Clinical Trials of Cortisone, ACTH, and Other Therapeutic Measures in Chronic Rheumatic Diseases, which recently completed a definitive two-year study, have added the opinion that "for practical purposes there has been remarkably little to choose between cortisone and aspirin in the management of [rheumatoid arthritis]." Except, of course, as the committee must have felt it unnecessary to point out, in terms of price (the average effective dose of cortisone costs almost a dollar) and prudence.

Although, because of its low toxicity, its inability to dull the pain-reaction sense, and its lack of any effect on normal body temperature, acetylsalicylic acid ranks high among the safest of chemotherapeutic agents, it is by no means entirely innocuous. It has its savage side. Like most substances, including many foods,

it can produce a variety of allergic reactions—dizziness, ear ringing, nausea, vomiting, skin eruptions, asthmatic seizures, muscular spasms—in susceptible individuals. Some of these idiosyncrat-
ics are so exquisitely responsive that a single grain can have a
shattering, or even a fatal, impact. Cases have been reported of
people's being thrown into convulsions upon receiving a blood
transfusion from a donor whose gift contained a just perceptible
trace of the salicylate radical. Its touch can also dishevel constitu-
tions that are not congenitally vulnerable. The cruel acidity that
seriously flaws most salicylates is not altogether neutralized in the
acetylsalicylic compounds. Chemistry has merely blunted its bite.
There are certain diseases—peptic ulcer, for example, and coro-
nary thrombosis—in which the use of acetylsalicylic acid may
precipitate such complications as acute dyspepsia, prolonged
retching, and hemorrhage. A general debility, such as that
brought on by many essentially trivial ailments, may also, in a
manner not yet fully understood, magnify its pernicious powers.
It is, in addition, corrosive enough to sometimes enfeeble the most
robustly well. No one can long consume immoderate amounts of
aspirin (fifty or more grains a day) without suffering some degree
of intoxication, and even a tablet or two, if swallowed whole on
an empty stomach, often will cause a twinge of sour, heartburning
discomfort. For all their abundance, however, the reprehensibly
rash are not particularly prominent among those whom acetylsalicylic acid most commonly shakes or sickens. Nor, since they are
generally made wary by early experience or experienced counsel,
are the sensitive or the unwell. Its usual victims—in this country,
at least—are children. According to the United States Food and
Drug Administration, thousands of cases of salicylate poisoning
of sufficient seriousness to require the services of a physician occur
in the United States every year. In 1952, for instance, the total,
including a hundred and thirteen cases that ended in death,
reached nearly seventeen thousand. Of that number, nearly three
fourths, or more than thirteen thousand, were children—young
children. Hundreds were mere infants, a year old or less, and none
was more than five. Most of them, like the boy whose case was

brought to the attention of Dr. Jacobziner that afternoon last May, were around three.

The investigation into the case of Richard Poole that Dr. Jacobziner had requested fell to a visiting nurse named Veronica Flynn. Miss Flynn received the assignment when she reported for duty at the Corona District Health Center on Thursday morning, May 5, and she set about discharging it without excitement or delay. A chronicle of her findings, which she wrote and posted late that afternoon, reached Dr. Jacobziner the following day. It made, he found, sad but familiar reading.

"I had fifteen others almost exactly like it in my files," Dr. Jacobziner said the other day. "All children, all accidents, all totally inexcusable. Only the details were different. Fifteen cases may not sound like very many. It wouldn't be today, of course. But it was then. In May, 1955, my file on acetylsalicylic-acid intoxication didn't go back very far. It only went back a couple of months—to the middle of March, in fact. That was when we established our Poison Control Center, with Harry Raybin, an extremely competent chemist, as its technical director. There wasn't any file on aspirin or any other kind of poisoning until then. The accumulation of epidemiological data is merely one phase of our work at the center. The function of the center is primarily educational. That takes two forms. One is to provide doctors and hospitals in the city with a source of accurate emergency information when they come up against some unfamiliar kind of poison. There are hundreds of potentially toxic chemical substances in daily household use. Every now and then, some child takes it into his head to sample one of them. In many instances, the doctor who gets the case has only the most general notion of what to do. He may know what it was that the child ate. That usually isn't much of a mystery. It was Johnson's Cream Furniture Wax, say, or Noxon Metal Polish, or Shinola. But he doesn't know what's in it. He doesn't know its chemistry. Our people do. When we get a call for help, we have the answer ready, and also the antidote, if any. In return for that service, however, we require something of the doctors and hospitals. They must notify us promptly of every case of chemical poisoning that comes to their attention. That's essential to the other aspect of our educational function. It gives us a chance to study the particular how and why of such outbreaks, and, often enough, to do something to prevent their recurrence. Well, the Poole case gave us something to study, all right. Not because it was unusual in any of its particulars. In that respect, as I say, Miss Flynn's report was only too familiar. It didn't tell us much of anything that we didn't already know about the causes of acetylsalicylic-acid poisoning. What made it a study was this: There was almost nothing that is known about the problem that wasn't present. It was practically a model of carelessness, bad luck, and ignorance.

"Miss Flynn's report was also something of a model. It couldn't have been more explicit. She began with a note on the boy. He was still at Whitestone Memorial, and his condition, at the moment of writing, was listed as serious. The report then took up the results of her investigation. She had paid a visit to the Poole home and talked to Mrs. Poole. The Pooles had a flat in a remodeled house —living room, kitchen, bathroom, bedroom. Richard was their only child. Mrs. Poole, however, was pregnant. She was in her fifth month, often felt unwell, and spent much of her time lying down. Mr. Poole was a subway guard, and worked an early shift. He was not at home at the time of the accident. That was somewhere between eight-thirty and nine on Wednesday morning. He had been gone since six. Mrs. Poole was at home, but in bed and asleep. She didn't know when Richard had got up. He had had a slight cold for a couple of days, and she had supposed that he would sleep late, too. 'Hoped' might be a better word. Assuming, that is, that she thought about it at all. At any rate, she was mistaken. He didn't sleep late. He was up by at least half past eight. Up and about, running loose in the apartment, for all practical purposes alone. And only three years old. That was the setting. It makes a nice picture, doesn't it? Absolutely custom-built for trouble, and absolutely standard. They all begin that way.

"It was almost nine o'clock when Mrs. Poole woke up. When

she finally woke up, I should say. A little earlier, around eight-thirty, she had opened her eyes for a minute, and noticed that Richard's bed was empty. Or so she decided later. At the moment, it didn't make much impression on her. She was too comfortable, too sleepy. The second time was different. She didn't just wake up —she was awakened. By a noise. It sounded like breaking glass, and came from somewhere in the flat. She rolled over and sat up, and this time there wasn't any doubt about it. Richard wasn't in his bed. She called to him, and he answered her at once and in his normal voice. He was playing in the living room. That relieved her mind, but it didn't explain the crash. She got out of bed and went to investigate. Her first stop was the bathroom, and that was it. The medicine-cabinet door was open, there was an aspirin bottle on the floor, and the basin was a litter of broken glass and vitamin pills. She stared at the mess, and then walked into the living room and asked the child what had happened. Candy, he said. He was hungry. He'd been looking for candy. In the bathroom? she said. They didn't keep candy in the bathroom. He knew that perfectly well. She told him he ought to be ashamed of himself. And so on, until he began to cry. She gave up in exasperation, went back to the bathroom, and cleaned it up. Then she put on some coffee, and returned to the bedroom to dress. She was almost dressed when it suddenly dawned on her that the aspirin bottle had been empty. Not only empty but uncapped—and no tablets on the floor or in the basin or anywhere. But it shouldn't have been empty. They had only bought it on Sunday—the day Richard caught his cold. She dug it out of the wastebasket and looked at it. It was a small bottle, but there should have been at least thirty tablets left in it.

"Miss Flynn later saw the bottle. Mrs. Poole had kept it, and showed it to her during their talk. The facts—or, rather, some of them—were on the label. It was, as expected, baby aspirin. That means colored pink for eye appeal and flavored orange for palatability. One and a half grains per tablet. There had been forty tablets in all. But, of course, no warning. Nothing about potential toxicity. Only some mumbo jumbo in fine print advising adult supervision. Nothing to frighten the customer. Nothing that frightened Mrs. Poole, anyway. At least, not at first. I don't mean that she wasn't upset. All her reactions were normal. But mostly she was simply bewildered. It was quite obvious what had happened to the rest of the aspirin pills. She realized that at once. Richard had as good as told her. He'd eaten them. What she couldn't understand was why. It seemed fantastic. What had got into him? Why had he wanted to eat a bottle of pills—a bottle of medicine? It wasn't fantastic, however. Not in the least. As she learned the minute she asked him. He didn't know it was medicine. He thought it was candy. Why? Because it tasted like candy. That was one reason. There was also another reason—just as good, or better. Mrs. Poole was like all the other mothers in our files. She had told him it was candy. She had told him so on Sunday and Monday and again on Tuesday, and probably many other times as well. It had seemed, she thought, the easiest way of getting him to take it. What Mrs. Poole did next was also painfully typical. She looked at Richard, and he looked all right. Then she asked him how he felt. He said he felt fine. I suppose she was still uneasy, but that seemed to be that. So she relaxed. It would be easy enough to blame her, but you can't. It wouldn't be fair. She didn't know that anything serious had happened. She had the universal attitude that aspirin isn't really a drug. It's something else. Because drugs are something they sell only in drugstores, and you can get a dose of aspirin anywhere—at a lunchroom or a soda fountain or on a train. It's just aspirin.

"I'm speaking for myself, of course. That's only my interpretation. Miss Flynn didn't interpret. She simply gave the facts. And one of the facts was that Mrs. Poole, understandably or not, wasn't greatly alarmed. Aspirin poisoning doesn't manifest itself in a minute. It takes a little time to catch hold. So when Richard went on playing in a perfectly normal manner, she was reassured. She calmed down. She finished dressing and helped him into his clothes, and they had breakfast. Richard didn't eat much, but there was nothing strange about that. He hadn't eaten much since Sunday. He never had much appetite with a cold. After breakfast, she got him settled with his toys in the living room, and went to work. There were the dishes to wash and the beds to make and the rest of the household chores. That must have been about ten o'clock. Well, around ten-thirty she stopped for a rest and a cigarette. Richard seemed still absorbed in his toys. She asked him how he was feeling. He didn't answer. Instead, he began to cry. She dropped down on the floor and put her arms around him, but he twisted away. That wasn't like him. He was almost never cranky. She pulled him back and felt his forehead. It was cool but wet. He was streaming with perspiration. Even his clothes were damp. After a moment, he stopped crying, and began to whine and whimper. His stomach hurt, he said. So did his head. He hurt all over. Then, all of a sudden, he vomited.

"I don't know what Mrs. Poole thought. The chances are she was too frightened to actually think. But she did what any mother would do. She picked him up and carried him in to his bed and tried to make him comfortable. Then she took his temperature. It was normal. As it always is, of course, in such cases. But it was reassuring to her, and by eleven o'clock she thought he seemed a little better. The sweating had stopped and he was resting quietly. That usually happens at about that point. Sweating and irritability are early symptoms. They soon give way to just the reverse—dryness and lethargy. Now he wasn't really resting. He was doped. In any event, he didn't seem better for long. Half an hour later, he vomited again. Then he started breathing very hard and fast—another classic symptom. That, I gathered, was when the truth began to penetrate. Hyperventilation can be a frightening sight, and it must have terrified Mrs. Poole. She couldn't call a doctor. She knew no doctor to call. The Pooles didn't have a doctor—not even an obstetrician. But I must say she kept her head. She didn't collapse or run screaming out to the neighbors. She wrapped the boy up in a blanket and took him straight to Whitestone Memorial. Also, apparently, she told them exactly what had happened. Aspirin intoxication is an extremely difficult condition to diagnose. Unless there's a clue in the history, you have to work it out by test and elimination, and even then it's very often missed. Its actual incidence is probably many times greater than any of us realize. The treatment is less complicated. It includes a prompt gastric lavage, parenteral administration of some dextrose and saline solution to restore the fluid balance, and the usual supportive measures. Richard was admitted to the hospital, according to the record, at one o'clock. That meant he would have had his lavage in reasonably good time—around twelve-thirty—since they attended to that in the emergency room, on arrival. Very satisfactory.

"Well, that was the report. It told us what we needed to know, and I could assume that Miss Flynn had given Mrs. Poole some necessary instruction in the care of drugs and children. I marked it for filing and put it aside, and went on to whatever came next. As you must. Even the saddest is only one among many. By the time I got back from lunch that day, I had practically forgotten the case. And then Miss Flynn called. She had just been talking to Whitestone. Richard Poole was dead. He had died around noon. Respiratory failure. She thought I'd want to know for the record. I thanked her and hung up. I won't pretend I was more than conventionally shocked. It happens too often for that. I was, however, very considerably startled—it didn't seem in line with the facts. My impression had been that they added up to a very good chance for recovery. Thirty tablets of the size he had taken made a total of forty-five grains. Forty-five grains of aspirin is enough to cause a lot of trouble in a child of three. Anything over ten can be dangerous. As a general thing, though, it isn't fatal. I got out the report for another look. But a glance was all I needed. It was all there. It had simply slipped my mind. One of the anomalies of acetylsalicylic acid is that illness enormously intensifies its action, especially in small children, and Richard had been more or less sick since Sunday. Also, he swallowed it on an empty stomach."

[*1956*]

* * *

AUTHOR'S NOTE: The physiology of aspirin is somewhat clearer now than it was in 1955 when the above depiction was written. It is now believed, on the basis of the investigations of several researchers (including, especially, John R. Vane at the Royal College of Surgeons, in London, 1971), that aspirin achieves its incongruous collection of efficacies—the reduction of fever, the reduction of inflammation, the easing of aches and pains—by inhibiting the body's production of the hormonelike substances known as prostaglandins. Prostaglandin production is also linked to the natural formation of blood clots. It is the ability of aspirin to interfere with clotting that causes aspirin's chief drawback—its tendency to cause internal bleeding. But this very flaw has been recently turned to an advantage, and aspirin is now widely used to protect against heart attacks and strokes, by blocking clot formation.

CHAPTER 6

The Liberace Room

―――――――――――◆―――――――――――

WHEN TOM D. Y. CHIN, assistant chief of the Kansas City, Kansas, field station of the Epidemiology Branch of the United States Public Health Service, and his three associates—George B. Paxton, an epidemiologist; George W. Beran, a veterinarian; and Mrs. Jennie H. Rakich, a nurse—pulled out of the station parking lot and headed south on the afternoon of Thursday, March 3, 1955, they knew next to nothing about the job ahead of them. They knew where they were going. Their destination was a town of around two thousand population called Mountain Home, the seat of Baxter County, in northern Arkansas. They knew that Mountain Home was in the grip of an epidemic. During the previous two or three weeks, twenty-nine pupils at the Mountain Home elementary school had been stricken with what had been gingerly described as an acute febrile disease; in addition to fever, which ranged in some cases as high as 106 degrees, the symptoms were headache, cough, and prostration. They also knew that parakeets were kept as classroom pets in several of the lower grades. But that was all. It was all that anyone knew, including the Baxter County health officer, a Mountain Home physician named Benjamin Saltzman, who had requested the assistance of the Public Health Service; Alexander D. Langmuir, chief of the Epidemiology Branch, at Atlanta, who had approved the request and forwarded it to Kansas City; and Michael L. Furcolow, the Kansas City station chief, who had passed the assignment on to Dr. Chin and his team. Everything else was conjecture.

"I won't say the diagnostic possibilities were endless," Dr. Chin recalls. "They were merely numerous. The clinical picture —even the little we had of it—did narrow the field a bit. It seemed to suggest a respiratory infection. Still, quite a few diseases answer to that general description, so there was plenty of room for speculation. The trouble could have been one of the rickettsial diseases—Q fever, for example. It could have been influenza. It could have been a fungus infection—coccidioidomycosis or histoplasmosis or blastomycosis. It could have been acute miliary tuberculosis. It could have been psittacosis. It could even—at a stretch—have been typhoid fever. Those were some of the choices. They were the possibilities that the signs and symptoms brought immediately to mind. They all came in for a certain amount of discussion at the briefing conference we had with Furcolow just before we left, and, under the circumstances, one looked no more likely than the next. Except, perhaps, psittacosis. Psittacosis is a virus pneumonia transmitted to human beings by birds. Until about twenty years ago, it was generally accepted that the only birds susceptible to that particular virus were psittacines—members of the parrot family. Hence the name of the disease. Since then, we've learned the truth—all birds are vulnerable to psittacosis. Nevertheless, most cases of psittacosis in human beings are traceable to parrots. Particularly, these days, to parakeets. Well, a parakeet in a classroom is hardly evidence of psittacosis, but, taken with the clinical data we had, it was something to think about. It was certainly enough to make us very curious about the health of those birds. That, of course, was largely why we had a veterinarian on the team. It was also why Furcolow gave the assignment to me, instead of going himself. Not that I'm an authority on psittacosis, but my special field is virology.

"Mountain Home is about three hundred miles from Kansas City. As I remember, we made the trip in seven hours. Our only stop was Joplin, in southwestern Missouri, for dinner and gas. We reached Mountain Home around ten o'clock, and spent the night in a motel on the outskirts of town. As it turned out, we made it our headquarters. The next morning, as soon as we'd had some breakfast, we headed for the office of the county health department. Mountain Home, like most little Southern county seats, is built around a courthouse square, so we found it without any trouble. It was barely eight o'clock, but Dr. Saltzman and his county nurse, a woman named Margaret Whitmore, were already there and waiting. They gave us a hero's welcome. The town was in an uproar, Saltzman said. Parents, teachers, school board, businessmen—everybody was clamoring for action. They wanted results. He and two other local doctors were about at the end of their rope. They had tried every drug in the wonder book. None of them —old or new—seemed to help. Some of the children were back at school, but no one could take much credit for that. They had simply got well. And there were a good many others still as sick as ever. Thirteen were flat on their backs in bed. But now—he actually heaved a sigh of relief—his troubles were over. The Marines had landed, and the situation would soon be well in hand. I knew what he meant—that we had the staff, the epidemiological experience, and the technical facilities. But that was the way he put it.

"I'm afraid I didn't exactly share his optimism. Quite the reverse, as a matter of fact. And for several reasons. One of them was that Dr. Saltzman had nothing to add to what we already knew. Nothing helpful, I mean. Except for what he said was a rather equivocal set of chest X-rays, his epidemic-aid request had fully covered the physical findings. I went over his pictures later, and he was right. Most of them showed some cloudiness, but nothing that wasn't generally characteristic of any respiratory infection. It looked a little like psittacosis. But it also looked like miliary tuberculosis. Or histoplasmosis. The only other information he had was a change in the number of cases. Our total of twenty-nine was out of date. Since turning in his call for help, he had checked again with the local doctors and found some cases that they hadn't got around to reporting. Also, while we were on the road, a brand-new case had turned up. The total number of cases now was thirty-six—out of a school enrollment of three hundred and eighty-six. What depressed me most of all, however, was Dr. Saltzman himself. He wasn't at all what I had hoped to find. I had pictured him as somebody dim and doddering. That had made it possible to hope that he was simply out of his depth —that the outbreak might not be as tricky as it sounded. Well, I dropped that notion in a hurry. Ten minutes' conversation was enough to convince me that I'd met a really good man. Far too good to be easily baffled. If he was stumped, there were no two ways about it. We were in for a lot of hard work.

"All that, of course, is largely by the way. Whatever Dr. Saltzman might have been or said wouldn't have made much difference. I might have settled down to work in somewhat higher spirits, but our procedure would have been the same. It isn't often that a public-health investigator can take anything for granted. The epidemiological discipline requires him to see and judge for himself. And the only place to start is at the bottom—with the victim and his environment. In this case, that meant thirty-six children, a school, and an unknown number of parakeets. The division of labor was no immediate problem. Beran's job was understood, and Paxton and I quite naturally took the children. That left the school for Mrs. Rakich. She could handle the preliminary survey there, and also give Beran any help he might need. The transportation problem was solved as soon as the matter came up. Dr. Saltzman had his own work to attend to—his practice and the health-department routine. But his nurse, Miss Whitmore, had a car, and, with his permission, she volunteered to drive Paxton and me around. So Mrs. Rakich and Beran kept ours. We all drove out to the school together. I wanted to make sure of their reception. The school was three or four blocks from the square—a one-story brick building, built since the war and in good repair, with twelve classrooms, an office, and an auditorium. The principal was expecting us. He and his staff, he said, were eager to cooperate in every possible way. That settled that. I left him to Beran and Mrs. Rakich, and Miss Whitmore and Paxton and I pulled out.

"We didn't see Beran and Mrs. Rakich again until evening. Paxton and I spent the day with the sickest children on the list. We examined all thirteen of them. Thirteen calls is a lot of calls, and they were scattered all over town, but with Miss Whitmore there to make the introductions and generally pave the way we managed. That's about all, however. I can't say we learned very much. The clinical picture we put together and the one we had had from Saltzman were practically identical. High fever. General malaise. Lassitude. Nonproductive cough. Occasional chest pains. A few complaints of nausea and vomiting. Those were the symptoms. In three or four cases, we found some enlargement of the liver and the spleen. Aside from that, the physical findings were essentially negative. The epidemiological picture also stood unchanged. No parties. No trips. No unusual group activities of any kind. The only environment common to all the children was the school—the building and the grounds. Nevertheless, it wasn't a wasted day. It was merely an unproductive one. We had done what had to be done. We had made a start.

"So had Beran and Mrs. Rakich. Only theirs was more than just that. They were waiting for us when we finally got back to the courthouse, and it was obvious that one or the other of them had turned up something interesting. They both had, in fact. Beran's contribution was a sick parakeet. There were six birds in all. Four of them had been bought in December and given as pets to the children in four lower-grade rooms—1-A, 1-B, 1-C, and 2-B. Just after the Christmas vacation, another bird was bought, and given to Room 2-C. The sixth parakeet was bought on January 24th. It went to Room 2-A, and that was the sick one. Its name was Liberace. The children named him that because he sang so prettily, but he wasn't singing now. According to Beran, he was a miserable sight. Ruffled feathers. Dull eyes. Listless stance. Loose green droppings all over his cage. The other parakeets, he said, were as bright and lively as crickets. Not, of course, that that meant much. An autopsy might show different. Psittacotic birds are often asymptomatic. Liberace also figured in Mrs. Rakich's report. While Beran was out gathering up the birds—they had been removed from the school, on Dr. Saltzman's orders, about a week before—she had settled down in the principal's office and checked the thirty-six known cases against the relevant classroom records. Her findings made interesting reading. The first child became sick on February 1. He was followed, at intervals of a few days, by five more. Then, between February 13 and February 24, there was an explosion. Twenty-six cases. Eight of them occurred on one day—February 17. Mrs. Rakich's report went on to break down the cases by classroom. One room—5-A—had had no children on the sick list. That room, incidentally, was the senior room. The school had only five grades. Sixth-graders went to another school—a junior high. Two rooms—1-B and 1-C—each had one case. There were two cases in each of four rooms—2-B, 2-C, 4-A, and 5-B. Rooms 1-A, 3-A, 3-B, and 4-B each had three. The Liberace room—2-A—had a total of fourteen. Mrs. Rakich ended her report with a kind of postscript. In the course of her room survey, she had talked with the various teachers. One of them, it developed, had been sick for several days at about the time of the onset peak. Nothing serious—no need to call the doctor. Just malaise. A little fever. Some coughing. I've forgotten her name, but I'll call her Miss Smith. Miss Smith taught Grade 4-A. Her room was across the hall and down a couple of doors from the Liberace room. However, for some reason or other, she had the job of supervising the children as they got into the school buses at the end of each day. And the place she found most convenient for doing that job was at a window in Room 2-A.

"I went to bed that night feeling pretty good. Of course, the picture was still confused. Miss Smith and the heavy concentration of cases in Room 2-A seemed enormously significant. Until you remembered the scattered cases in the other rooms, and the five other parakeets, and the teacher who taught 2-A. Why wasn't she sick, too? But Liberace was something else. If he turned out to be as psittacotic as he looked, we almost certainly had a lead. A post-mortem answer to that was Beran's job for Saturday. Dr. Saltzman gave him a place to work in his office, and he got started right after breakfast. The rest of us spent the day making house calls. Paxton and Mrs. Rakich tackled the twenty-three convalescent or recovered cases. I worked with Miss Whitmore. Our job was to revisit the thirteen sick children we had seen on Friday, and arrange for a number of diagnostic tests. They might or might not be illuminating, but it had to be done. It was routine. In general, the nature of the suspected disease determines the kind of test. Because of the number of possibilities in this case, we had to use two kinds. One was a series of simple skin tests for antibody reaction. Each child was injected intradermally with the antigens of four of our several suspects—tuberculosis, histoplasmosis, coccidioidomycosis, and blastomycosis. The results of such tests can be read in about forty-eight hours. A positive reaction—the appearance of a characteristic induration at the site of injection—is evidence that the patient has, or has had, the disease in question. There is no skin test for psittacosis. For that, as well as the other clinical possibilities, a blood test was required. A blood test is fairly complicated. It can be done only in a specially equipped laboratory, and it takes time. The nearest Public Health Service serological laboratory was the Rocky Mountain Laboratory, in Hamilton, Montana. We took samples from each of the thirteen children, packed them in dry ice, and sent them off by airmail that afternoon, along with a note of explanation and instruction. They were to be tested for Q fever, influenza, psittacosis, and histoplasmosis. The last was for insurance. In histoplasmosis, a serological examination is apt to be more conclusive than the skin test. Well, with any luck, we would have the results in about a week. Meanwhile, Paxton and Mrs. Rakich had made a good start on their twenty-three convalescent or recovered boys and girls. Not that they had anything much to report. The histories they had obtained merely confirmed the data we already had. They would finish that phase of the job on Sunday, and then get started on a round of tests and samplings of their own. The report of the day was Beran's. I don't mean that it was any sensation—his findings were only tentative. The results of gross examination are rarely anything else. They have to be substantiated by laboratory test. But he did have something to tell us. Liberace was very definitely sick, and in a very provocative way. The other birds had no visible signs of abnormality, but Liberace more than made up for that. In psittacosis, the most striking clinical manifestation is enlargement of the liver and the spleen. Liberace's liver was considerably larger than normal, and his spleen was huge. A parakeet spleen is usually about the size of a small seed. His was as big as the end of my middle finger.

"That was all that Beran could tell us at the moment. It was also as much as he could ever tell us if he stayed in Mountain Home. His next move was to try to isolate the psittacosis organism from the parakeet material, and that couldn't be done in the field. Viruses can't be cultured, like bacteria. They can only be demonstrated through transfer to a living host, so the place for him was Kansas City, where we have a virus laboratory, with the necessary tools and animals. He left by bus on Sunday morning. The rest of us saw him off, and then went back to work. There was no point in counting the hours. His job would take about as long as the blood tests at Rocky Mountain. Paxton and Mrs. Rakich spent the day with their children. My day was a series of meetings. I spent an hour or so with Saltzman reviewing our work to date. Then he took me around to meet the other local physicians. In the afternoon, at my request, we held a kind of town meeting. The group included the school board, the superintendent of schools, the mayor, the publisher of the Baxter *Bulletin,* the local weekly—all the leading citizens. I had several reasons for calling them together. Common courtesy was one. The town was entitled to a general progress report. Another was to calm their fears. I made it clear that the cause of the outbreak was still uncertain, but I told them what we thought. Our best guess was psittacosis. In any event, I added, there was nothing to panic about. All of the patients were doing reasonably well. They would certainly all recover. Moreover, the record showed just two new cases in almost two weeks, and none since our arrival. So the worst was probably over.

"The main reason for the meeting, however, was to broaden the scope of the investigation. I wanted to organize a school-wide testing program. A comparison study would help us interpret the results of our tests on the thirty-six clinical cases. Also, since the school—the building or the playground, if not specifically the Liberace room—was obviously the focus of infection, it would be useful to have an outside group of controls. One of the grades at the junior high, for example. But none of this was something we could simply decide on and do. We're not allowed to give a child an injection or draw a sample of blood without his parents' permission. I hoped to convince the civic leaders that the program was epidemiologically essential, and persuade them to bring their influence to bear. With Saltzman's help, I did. By the time the meeting adjourned, we were practically ready to go. The necessary consent forms had been drawn up, arrangements had been made to get them mimeographed, and a company of volunteers had been recruited to handle their distribution. We even had a place to work. The school authorities would let us set up a clinic in the auditorium. I couldn't have asked for fuller cooperation. Or, as it turned out, for a more enlightened response from the parents. Only thirty withheld their consent. We got permission to test and bleed a total of three hundred and sixty-nine. That included three hundred and twenty children at the elementary school, forty-eight sixth-graders at the junior high, and Miss Smith. We started in on Monday morning, and it took us most of the week. We finished up on Friday.

"Meanwhile, of course, the results of the skin tests were coming in. By Friday, we had readings on the thirteen actual patients, on the twenty-three recoveries, and on about two hundred of the other pupils at the elementary school. And also on Miss Smith. We had them, but that's about all. It was hard to say exactly what they meant. We read the clinical-case reactions first. Every single one of them was positive for histoplasmosis. For tuberculosis, blastomycosis, and coccidioidomycosis—all negative. I won't deny that I was a little startled. Only for a moment, however. And only on account of Liberace, because histoplasmosis was fully compatible with the clinical findings, with Saltzman's chest X-rays—with all the data we had. Then we saw Miss Smith's results. She gave the same reaction—positive for histoplasmosis, and only for histoplasmosis. In the light of the other reactions, that was hardly surprising. We had assumed that she was part of the outbreak. That's why we included her in the tests. Then the results of the general student body came in, and we began to wonder. As I say, by Friday we had seen about two hundred of them—more than enough to provide a definitive contrast to the clinical-case reactions. But they didn't. They showed the same—or practically the same—reactions as the others. All negative for tuberculosis, blastomycosis, and coccidioidomycosis. And practically all—nearly ninety per cent—positive for histoplasmosis.

"Well, that brought us up pretty short. I don't mean that we were back where we started. We could forget about tuberculosis and blastomycosis and coccidioidomycosis—they were definitely out of the picture now. And probably Q fever and influenza as well, though we couldn't actually cut them off the list—not yet. Nevertheless, the way things were shaping up, there wasn't much reason to keep them on it. So at least we had narrowed the field. We had a choice between psittacosis and histoplasmosis. It had to be one or the other. But which? Liberace and that concentration of cases in his room were certainly suggestive of psittacosis. But what about the clinical-case reactions for histoplasmosis? They were equally suggestive. Or, rather, they were and they weren't. The trouble was all those other positive reactions. They had to be explained. There was a choice of several interpretations. There always is with a positive skin test for one of the fungus infections. That's largely where it differs from a blood test. A skin test will accurately reflect the presence of specific antibodies, but it has no topical value. It doesn't distinguish between a current or recent infection and one experienced sometime in the past. A blood test, on the other hand, is a diagnostic test. It is positive only during the active stage of a disease and for the following week or so. You can see what I mean by possibilities. It was possible that all of our positive reactions referred to old infections. It was possible that they all reflected current infections. It was also possible—but at that point Paxton and I let it drop. The answer—or, at any rate, the beginning of an answer—rested with the Rocky Mountain Laboratory and with Beran. It was foolish to try to guess. Instead, I put in a call to Furcolow. I told him where we stood, and said that unless he had some objection, we thought we'd come home for the weekend. He said come ahead. It was time we talked things over.

"That was four or five o'clock on Friday afternoon, and I was home by midnight. I got down to the office the next day around noon. On my way to talk with Furcolow, I dropped in on Beran. No news. He had run some preliminary tests on the parakeet material, but the results were at best inconclusive. It was still too soon to say about the isolation studies. I got the impression, though, that he wasn't very optimistic. He seemed to feel that the first run of tests told the story. They did. As it turned out, we never did determine what was wrong with Liberace. Except that it wasn't psittacosis. But, as it also turned out, we were morally sure of that long before Beran made it certain. I knew it almost the minute I walked into Furcolow's office. We had hardly said hello when his secretary came in with a telegram. He opened it up and read it, and passed it over to me. It was a report from Rocky Mountain on the first group of Mountain Home blood tests. It read: ALL PRECIPITIN HISTO POSITIVE. PSITTACOSIS, Q FEVER, INFLUENZA NEGATIVE."

Histoplasmosis is one of about a dozen systemic diseases now known to be caused by the venerable fungus family. It is thus, unlike the overwhelming majority of infections that afflict the human race, of vegetable rather than animal origin. All the deep mycoses, as these fungus invasions have come to be called, are relatively new to medicine (none was recognized much before the turn of this century), all are serious disorders—some are invariably fatal—and all are rapidly growing in stature. Histoplasmosis is the newest member of this upstart group, and although it most often appears in a tractable form, it can be as deadly as any of the others. It is also by far the most increasingly common. Until little more than a decade ago, histoplasmosis was universally considered a medical curiosity of almost incomparable rarity. In 1945, according to a survey published in the *Archives of Internal Medicine,* only seventy-one authenticated cases (almost all of them fatal) were known to medical literature. Its record is now more imposing.

The first recorded case of histoplasmosis occurred in Panama in 1906. Its discoverer was Samuel T. Darling, chief pathologist at Ancon Hospital, in the Canal Zone, and a protégé of William Crawford Gorgas, the celebrated United States Surgeon General who directed the sanitation program that made possible the construction of the Panama Canal. On December 7th of that year, as Darling subsequently recounted in the pages of the *Journal of the American Medical Association,* "while examining smears [made in the course of an autopsy] from the lungs, spleen, and bone marrow in a case that appeared to be miliary tuberculosis of the lungs, I found enormous numbers of small bodies generally oval or round. Most of them were intracellular in the alveolar epithelial cells, while others appeared to be free in the plasma of the spleen and rib marrow. Tubercle bacilli were absent." A few months later, his report continued, chance provided him with two more post-mortem cases of the same distinctive stripe. He thus ventured to conclude that he had come upon a new infectious disease. Darling, however, failed to grasp the precise nature of his find. It was his belief that the "small bodies" he had observed, and to which (exercising the prerogative of a pioneer) he gave the name *Histoplasma capsulatum,* were a species of protozoa. He also believed, with somewhat better reason, that histoplasmosis (as he chose to call an invasion of *H. capsulatum*) was "a fatal disease of tropical America" and, in all probability, a most uncommon one. The first of Darling's misconceptions, though sensed (by the Brazilian biologist Henrique da Rocha Lima) as early as 1912, was not corrected until 1934, when W. A. DeMonbreun, professor of pathology at Vanderbilt University School of Medicine, successfully isolated a colony of *H. capsulatum* spores and cultivated them to definitive fungoid maturity. Darling's other errors eluded detection even longer—until 1945. Their exposure was the joint accomplishment of three now classic studies. One of these was the *Archives of Internal Medicine* case review. The others were the work of the team of Amos Christie and J. C. Peterson, both professors of pediatrics at Vanderbilt, and a team headed by C. E. Palmer, then director of the United States Public Health Service Tuberculosis Research Office, in Washington. The *Archives* review abruptly revised the accepted view of histoplasmosis as an exclusively tropical disease. Of the seventy-one cases then on record, it showed, fifty-six had originated in the temperate United States. The rest had turned up in nearly a dozen widely separated countries. In addition to Panama, the list included Argentina, Southern Rhodesia, British Honduras, the Philippine Islands, Austria, Brazil, and England. To this cheerless revelation the Christie and Peterson study (which was published in the *American Journal of Public Health*) and that by the Palmer team (which appeared in *Public Health Reports*) added a double-barrelled postscript. Histoplasmosis, the authors independently (and almost simultaneously) announced, was a good deal less deadly than had been universally thought. The fatal form was merely its more conspicuous manifestation. It also occurred in another form, which, while diagnostically slippery (because of its close resemblance to any pneumonitis) and capable of producing much misery, was relatively benign. On the other hand, they pointed out, the disease was a good deal less rare than theretofore supposed. There was, in fact, some reason to believe that the muted (or primary pulmonary) variety might be rather common. Just how common neither Christie and Peterson nor Palmer and his associates were at that time able to say. But subsequent investigators have amply answered for them. The best contemporary opinion holds that in the United States alone some thirty million people have had primary histoplasmosis, and its annual incidence here is estimated at around five hundred thousand cases.

Although histoplasmosis is now established as a disease of nearly universal distribution, it appears to be most prevalent in the United States. Its prevalence here is not, however, uniform. There is no considerable section of the country in which histoplasmosis is wholly unknown, and a total of forty outbreaks of epidemic sweep have been reported in eighteen states—New York, Maryland, Virginia, North Carolina, Florida, Alabama, Tennessee, Arkansas, Missouri, Oklahoma, Kansas, Ohio, Indiana, Illinois, Iowa, Wisconsin, Minnesota, and North Dakota—but it occurs with menacing frequency only in the central Mississippi Valley. In most of the states that lie in that area, its incidence ranges from thirty to eighty per cent, and there are some in which it runs even higher. These are the five river states, the states that converge on the Mississippi near its meeting with the Missouri and the Ohio, and from the heart of the region—Missouri, Illinois, Kentucky, Tennessee, and Arkansas. The disease is endemic there.

Why histoplasmosis should occur so variously is a matter of some dispute. It is obvious only that geography must be the determining factor in its spread—that its range reflects a physical environment hospitable to *H. capsulatum*. There is no reasonable alternative to this supposition. For histoplasmosis, an abundance of research has established, cannot be conveyed from one human being to another. Nor, it is equally clear, can it be communicated to man by any insect or higher animal. It is a disease that human beings (and other animals) contract directly from nature—by inhaling a quantity of *H. capsulatum* spores. The specific environmental factor that controls the organism's geographical range is less apparent. Three possible explanations have been seriously advanced. Its development may depend upon a certain type of soil (a red-yellow podsolic soil is peculiar to much of the high-incidence area), it may require a certain vegetable environment (some fifty common plants are more or less indigenous to the area), or it may be governed by a certain pitch of temperature and humidity. At the moment, the evidence, such as it is, most strongly supports the last of these hypotheses. Laboratory experience has conclusively shown that an exquisitely balanced microclimate—one hundred per cent humidity and a temperature ranging no higher than eighty-six degrees and no lower than sixty-eight degrees—is essential to the optimum growth and proliferation of *H. capsulatum*. So have numerous observations in the field. In nature, *H. capsulatum* flourishes only in the darkest, dampest, warmest, most snugly sheltered places. Its favorite habitats are belfries, silos, and abandoned chicken houses (probably because a constant humid heat is generated there by decaying bird droppings), but it also finds a salubrious home in caves, storm cellars, punky stumps, and densely wooded ravines and riverbanks. (Most epidemiologists believe that certain of the many mysterious ailments whose names have long enlivened regional medical literature—cave sickness, bat fever, speleologists' disease, Tingo Maria fever—are actually histoplasmosis.) Such places, of course, abound throughout the United States. Not everywhere in this country, however, is the microclimate they provide consistently warm and damp. In the central Mississippi Valley, where the mean temperature in summer is around seventy-five degrees and the average annual rainfall totals nearly fifty inches, violent extremes of rain and drought and heat and cold are notably infrequent. They are most infrequent in the five states that constitute the endemic area.

The second phase of the 1955 Mountain Home investigation differed from the first in two important respects. It was, for one thing, planned and actively headed by Dr. Furcolow. For another, it was undertaken with a much augmented team. In addition to Dr. Chin, Dr. Paxton, and Mrs. Rakich, this included not only Dr. Furcolow but three other members of the Kansas City station staff —Mrs. Dorothy Calafiore, a nurse; Peter Ney, a statistician; and Howard W. Larsh, a professor of mycology and bacteriology at the University of Oklahoma who serves as a consultant in mycology to the Public Health Service—and also Miss Whitmore (the Baxter County nurse) and two recruits from Arkansas. These were Mrs. Mildred Ware, the local district nursing supervisor, and V. E. Medlock, an X-ray technician on the staff of the Arkansas State Department of Health. The Kansas City contingent and the others met at Mountain Home on the evening of Monday, March 14. They got down to work the next morning.

"My decision to join the team had nothing to do with Dr. Chin," Dr. Furcolow says. "That is to say, it was not in any sense a criticism of his handling of the preliminary inquiry. He had done an excellent job. And if the trouble had turned out to be psittacosis —or any viral or bacterial disease—he would, of course, have continued to run the show. But histoplasmosis is a little out of his field. It is, however, very much in mine. It's been my special interest practically all my professional life. So I could hardly be expected to sit it out in Kansas City. Especially this particular outbreak, which had all the earmarks of being something quite out of the ordinary. In size, in kind, in almost every respect. The biggest histo outbreak on record at that time was one at Camp Gruber, Oklahoma, in 1944. It totaled twenty-seven cases. Another, involving twenty-five cases, occurred at Foreman, Arkansas, in 1947. One at Camp Crowder, Missouri, in 1943, involved twenty-two cases. And there was one in 1948 at Madison, Wisconsin, that involved a dozen. But the victims in all those outbreaks were almost invariably adults. This was children. It was, in fact, the first major outbreak of histoplasmosis ever observed among children. And not only that. With thirteen certain cases and twenty-three probable ones—to say nothing of that wave of skin-test positives in the general student body—it looked as if we might be up against the biggest histo outbreak of any kind in history.

"We were. By Wednesday evening, there was no longer any doubt about that. We had a record outbreak on our hands. I'd heard from our Rocky Mountain Laboratory on the second set of blood tests, and the results were the same as the first—all twenty-three of the recovered children were serologically positive for histo. That brought the total of certain cases up to thirty-six, or nine more than the Gruber outbreak. But it didn't add much to our store of useful knowledge. We still knew practically nothing.

To come right down to it, we weren't far past the point where most investigations start. The only thing we had clearly established was the nature of the outbreak. In the circumstances—I mean, in view of the fact that histo is an extremely difficult disease to diagnose —that was no small accomplishment. Nevertheless, we had only just begun. The full extent of the outbreak, and just how and where the children had come in contact with *H. capsulatum,* were still to be determined. The major question was, of course, the source. But before we could hope to answer that, it was necessary to explain all those other positive skin-test reactions. The scope of the problem came first.

"When Dr. Chin gave me his progress report in Kansas City, he and Dr. Paxton had completed the skin-testing survey at the school, and they had read most of the reactions—about two hundred, as I remember, out of a total of three hundred and twenty. Or three hundred and twenty-one, counting the teacher, Miss Smith. They had also tested a control group of forty-eight sixth-graders at the Mountain Home junior high. And they had made a start on the job of sending the blood samples taken in the two surveys off to Rocky Mountain for laboratory test. The first thing on the general program was to finish up those jobs. Chin and I did the rest of the skin-test readings. The bloods were handled by Paxton, Ney, and the nurses. Larsh and Medlock had special programs of their own to set up. We started our readings where Chin had left off the week before—at the elementary school. Reading and recording the results took us a couple of days. We then moved over to the junior high and had a look at the controls. By the end of the week, or thereabouts, we had the total skin-test picture. I'm bound to say I've seldom seen a stranger one. Not that the final results at the school were especially startling. Chin's earlier findings had naturally prepared us for a high percentage of positive reactions there, and we got them. Of the three hundred and twenty children tested, two hundred and seventy-nine—or about eighty-seven per cent—were histo positive. But we certainly weren't prepared for the reactions we found at the junior high. The controls scored almost as high as the others. Thirty-three of the forty-eight—sixty-nine per cent—were also positive for histo. The picture was already cloudy enough. Now it was practically invisible.

"So, at the end of another week, we hadn't really made much progress. With the scope of the outbreak still unknown, we still had insufficient data to attempt to determine the source. In a sense, we had even less than before. We had more or less assumed that the focus of infection was in or near the elementary school. It had been our expectation that the control study would result in a high percentage of negative reactions, and thereby make it certain. That was largely the point of the study. I don't mean to say that those thirty-three positive contrtols at the junior high made us actually change our thinking. The study had failed to confirm our assumption. It didn't deny it, though. It didn't necessarily prove we were wrong. The meaning of a positive histo skin test is too equivocal for that—particularly in an area where the disease is known to be endemic. But it made us wonder a little. It also made us wait with some impatience for Rocky Mountain to come through with the blood-test results. Those would be definitive. They would tell us exactly where we stood. Then we could get down to fundamentals.

"The first report on the serological tests arrived on Saturday morning. It referred to the elementary-school survey, but that's about all I remember of it. No matter. It was far too small a sampling to signify—a couple of dozen cases at most; and merely the first of many. Another arrived on Monday, and after that, as I recall, reports turned up at the rate of one or more a day throughout the rest of the week. I passed them along to Ney to work into proper shape. There was nothing to be gained from puzzling them out piecemeal. It was better to wait until all the results were in and he had them correlated by grade and classroom and school. We didn't just wait, of course. There was plenty else to do. While Chin and I were reading skin tests and the others were packing and mailing the rest of the bloods, Medlock had made the necessary arrangements for a comprehensive X-ray study. He had fetched a mobile seventy-millimeter X-ray unit in from Little Rock, and had arranged with Dr. Saltzman and the other local physicians to use their standard fourteen-by-seventeen apparatus in case his smaller plates turned up something tricky. And, working through the State Tuberculosis Association, he had raised an appropriation to cover the cost of the film and other equipment. Chin and I talked to the school authorities, and they once more very kindly gave us permission to use the auditorium for a clinic. All we needed now were the customary releases from the parents. That took a little more time. But by the first of the week we had enough to provide an adequate study—all thirty-six of the established clinical cases, two hundred and fifty-seven children from the general student body, thirty-two of the sixth-grade controls, and Miss Smith.

"The X-ray study itself went off with reasonable dispatch. We finished up around noon on Thursday—films, prints, readings. Ney took it from there. His analytical breakdown of our data was ready the following morning. One look, and the clouds began to lift. The clinical cases, to nobody's surprise, all showed definite lung lesions. So did Miss Smith. And so did some of the general student body—thirty out of the two hundred and fifty-seven, or roughly ten per cent. The controls, however, were the thing. This time, they acted like controls. All but one of the thirty-two were as clean as a whistle. The next day—Saturday, March 26—we got Ney's blood-test correlation. Rocky Mountain confirmed the X-ray study at every significant point. Two controls—one of them the X-ray-positive child—gave positive reactions. The other controls were all negative. Miss Smith, as expected, was positive. So were approximately thirty per cent of the general group. Considering the X-ray findings—an X-ray study is never quite as comprehensive as a serological survey in histo—that was also about what we'd expected. Well, those were the high spots of the studies. Merely the high spots, I should say. But, taken with the skin-test data, they gave us the footing we needed to move into the final phase of the investigation. We knew the nature of the problem. We knew the general focus of infection. In spite of the two positive controls, it was somewhere around the elementary school. A couple of outside cases had no epidemiological significance. They could be rationalized in various ways—this was, after all, an endemic area. And we knew the size of the outbreak. The thirty-six clinical cases had been only a hint—like the minaret of an iceberg. On the basis of the blood and X-ray studies, the actual number of recent and current histo cases was four times that. When all the data had been analyzed and all overlaps accounted for, it added up to a hundred and forty-six children. And Miss Smith.

"So we had a really champion outbreak to explain. As a matter of fact, it is still by far the biggest of its kind on record, and I doubt very much if it will even be equaled for a long time to come. It was also, in my experience, epidemiologically unique. The deeper into it we went, the more painfully apparent that became. Most outbreaks of histoplasmosis are puzzling, but only up to a certain point—only until the diagnosis is established. Once that is determined, it seldom requires much exertion to find and confirm the source of infection. A couple of questions will generally reveal that a couple of weeks before the victims got sick—the incubation period in histo is around ten to fourteen days—they had spent a day cleaning out an old chicken house. Or exploring a cave. Or cutting brush in a creek bottom. Or, often enough, just visiting somebody's farm. Samples of soil are then collected at the site for laboratory examination. The usual result is isolation of *H. capsulatum*. And that closes the case. Mountain Home was a maverick. The approximate date of exposure was easy enough to determine. The clinical case histories and the school attendance records placed it in the neighborhood of February 1. But there the pattern broke. As Chin and Paxton had already found, the usual questions didn't yield the usual answers. There wasn't any cave or creek or chicken house. Not even a trip to a farm. There was only a school —a well-kept school—with a big, sunny lawn and a playground. Which meant that the children hadn't, in the usual way, exposed themselves to *H. capsulatum*. They hadn't gone to it. It had come to them.

"But how? One possibility occurred to us almost at once. Northern Arkansas has a good-sized broiler industry, and Larsh picked up the fact that chicken manure is widely used in the area as a lawn fertilizer. While the rest of us were still occupied with our diagnositc studies, he had made some inquiries—and sure enough! Two years before, in the spring of 1953, the school had bought six or seven truckloads of the stuff. It was dumped on the edge of the grounds and left there to rot. Some time later, when the manure was well rotted, a gang of workmen had moved in and spread it on the lawn. Larsh was jubilant, and so were we. Old chicken manure and *H. capsulatum* are as akin as fire and smoke. But then the promise began to fade. The manure hadn't simply been spread. It had been thoroughly diked into the ground. Moreover, the job had been done way back at the end of the previous summer. The last load had been dug in just before the opening of the present school year, in September. That piece of information practically finished it off. It was most unlikely that a source that disappeared in September could have caused a February outbreak. Nevertheless, as a matter of routine, Larsh collected some samples of soil and sent them home for laboratory analysis. In a couple of weeks, the results came back. No soap. No sign of *H. capsulatum*.

"I can't say that we were greatly disappointed. To tell the truth, the laboratory report was something of a relief. A positive soil analysis would have raised more questions than it answered. We'd come to realize by then that the fertilizer couldn't possibly be in the picture. And not only because it failed to fit the established onset date. It also failed to explain the most peculiar aspect of the outbreak. I mean, of course, the extraordinary concentration of cases in Room 2-A—the Liberace room. The school records listed a total of twenty-eight children enrolled in the class. Of that number, as Chin's study had noted, fourteen—or fifty per cent— were clinically ill. The average for the school as a whole was nine per cent. That was odd enough. But the diagnostic studies told an even stranger story. In the skin-test study, ninety-six per cent of the tests on the children in 2-A gave positive histo reactions. The school average was eighty-eight. In the blood-test study, sixty-eight per cent of the children in 2-A were positive. The school average was thirty-four. In the X-ray study, thirty-eight per cent of the children in 2-A showed definitive lung involvement. The school average was twenty-three. That couldn't be accounted for by the fertilizer episode. And it had to be accounted for. It couldn't be ignored. It couldn't be merely coincidence. It had to signify. It had to mean—it could only mean—that the source of infection centered on the Liberace room. More than that. It obviously meant that something out of the ordinary had happened in or near that room. The clinical data told us when. It had probably happened within a day or two of February 1.

"Something *had* happened around that time. It took a bit of probing—most people have rotten memories—but the school janitor finally dredged it up. On January 29, a truckload of coal—about six tons—was delivered to the school and dumped alongside a chute that led down to a cellar bin. And where was the chute? He walked us around to the rear of the building and pointed it out—a little iron door right under the windows of the Liberace room. A few more questions and a bit of checking produced a few more facts. The chute was only a dozen steps from where the school buses loaded. January 29th had been a Saturday. For some reason to do with labor, the coal remained where it was dumped through Monday. On Tuesday and Wednesday—February 1st and February 2—it was shoveled into the bin. The weather that week was dry and clear and windy. Everybody remembered the wind. It combined with the shoveling and the buses to raise a storm of dust that covered the rear of the building and seeped into most of the classrooms. The dust was particularly heavy, the janitor recalled, in Room 2-A. It was fascinating; everywhere we looked, we found another piece of circumstantial evidence. And they all seemed to fit together like the parts of one of those puzzles. The time was right. The place was right. The conditions were right. They even explained Miss Smith. The habit she had of standing at an open window in Room 2-A to direct the children onto the buses would have exposed her to the dust at its thickest. Everything was right except the only thing that mattered. If the truck had dumped a load of chicken manure—or topsoil or leaf mold or brush—I wouldn't have hesitated. I would have felt we had the answer. But coal! *H. capsulatum* doesn't grow in the bowels of the earth.

"On the other hand, of course, we couldn't just drop a lead as persuasive as that. It went against the grain. We sat down in Saltzman's office at the courthouse and discussed the possibilities. It was at least conceivable that the coal had accidentally served as a vehicle for *H. capsulatum.* It might have become contaminated at some point en route from the mine. In some coal-yard, perhaps. Or in the truck that brought it to the school. Maybe the truck had been used for carting chicken manure. We had the name of the trucking company, and Chin and Paxton went off to ask the people there some questions. Larsh and I drove over to the school and went down cellar and had a look at the coal. It wasn't very promising. The bin was practically empty, and what little coal there was looked as clean as coal ever does. We couldn't find any signs of anything but coal. It really looked hopeless. However, Larsh scrambled around and gathered some samples of dust for analysis. The laboratory might tell us something different. Then we went back to the courthouse and waited for Chin and Paxton. They came in looking rather thoughtful. Our truck-contamination theory was out. The company did general trucking, but none of its trucks had ever hauled manure. In the past few months, the records showed, its business had been largely confined to coal, with a few odd loads of sand and gravel. But Chin and Paxton had learned something interesting. Or, at any rate, unexpected. The coal delivered to the school hadn't come from one of the big established mines. It was more or less wildcat coal. It had come from an old strip mine in Washington County, not far from Fayetteville. I remember glancing at Larsh. His expression was as thoughtful as Chin's. A strip mine is an open pit. There's all the difference in the world between an open pit and a modern deep-shaft mine of the sort we'd had in mind.

"Larsh and I spent the next day in Washington County. That was Tuesday, March 29. It was also, as it turned out, our last chance. The laboratory report on Larsh's coalbin scrapings, when it finally came back, was negative. We got to Fayetteville, which is only about a hundred and fifty miles from Mountain Home, well before noon. Then we had to find the mine. Even with the most explicit directions, that took another couple of hours. Old was no word for it. The place was overgrown and tumble-down and practically abandoned. But we found it—not only the mine but also the strip that had last been worked. And then our luck ran out. We were a month or two too late. The pit from which the school's coal had come was gone. Rain and snow and winter weather had weakened the overhang and dropped it into the pit. It was like a cave with the roof fallen in. I've seldom seen a likelier place for *H. capsulatum,* but likely was all it could be called. We couldn't prove it. We couldn't even reach it. It was buried under tons of earth and rock."

[*1960*]

CHAPTER 7

Impression:
Essentially Normal

NOW, AFTER AN INTERVAL of many months, Rosemary Morton, as I'll call her, is willing to remember the half year she spent immersed in gravitational anarchy. Until only a few weeks ago, she preferred not to think about it. The risk—the chance that memory might have power to draw her back again into that capricious world of tilted buildings and rubbery streets—was too great. Her fears were not, perhaps, quite rational. But then, neither was the episode itself.

Mrs. Morton, a trim little woman with horn-rimmed spectacles and a bun of graying hair, and her husband, Frank, are in their early fifties, and childless. Both have responsible jobs. Mr. Morton is the editor of an industrial trade journal. His wife has been for many years the librarian of a large Wall Street law firm. She is also an accomplished, though amateur, pianist. They live on the top floor of an old brownstone house in the Murray Hill section of Manhattan, and it was there, around six-thirty one February evening in 1957, that her equivocal experience equivocally began.

"I'd been home about an hour," Mrs. Morton recalls. "Dinner

was ready and waiting in the oven, and I was sitting at the piano. Not really playing—just amusing myself. That's something I often do at the end of the day. It helps me relax. My husband was in the kitchen making us a cocktail. Which is another Morton custom. We usually have a drink or two before dinner. So everything was quite ordinary and normal. Until Frank came in with the drinks. I got up to join him on the sofa, and as I did—as I started across the room—I felt the floor sort of shake. Maybe sink would be a better word. It only lasted a moment—less than that, I suppose. Just an eyeblink. But the floor very definitely moved. I remember looking at Frank. 'Good heavens!' I said. 'What was that?' Frank just looked at me. His face was a perfect blank. It was obvious that he didn't know what I was talking about. So I told him, as calmly as I could. 'The floor,' I said. 'It shook.' I suppose Frank believed me. I know he did. But he wasn't much impressed. He made some remark about old buildings' stretching and settling, and handed me my drink. In a way, that was reassuring. If Frank hadn't felt anything, it must have been very slight. And what he said was true. Old houses—and ours goes back to the Civil War, almost—do all kinds of strange things. Besides, we have a tremendous amount of weight in our living room. Not only my piano but about a thousand books and any number of phonograph records. So I took my drink and sat down, and Frank began to talk about some repairs that had to be made on a cottage we have up in the Catskills, and by the time we were ready for dinner, I'd forgotten all about it. No, that isn't true. It was too peculiar to forget. But I had put it out of my mind.

"That was early in the week. I remember because it was two or three days before Washington's Birthday, which fell on a Friday last year. And on Washington's Birthday the same thing happened again. The only difference was where it happened. The place was my office. It was a holiday, of course, and the office was closed, but I had some work I wanted to finish up. Some cataloguing. Also, in a way I couldn't explain—it was so unlike me—I wanted to be alone. Frank had turned on WQXR at lunch, and then, while I was doing the dishes, he started playing records—just the sort of lazy afternoon we both love. But somehow I wasn't in the mood. It made me nervous. Anyway, around two o'clock I went downtown. I worked at my desk for about an hour, and it was heaven. So quiet. So peaceful. Then I got up—I've forgotten for what. To get a book from the stacks, or a drink of water, or something. And it happened. The floor gave a shake, and sank. It went down and up, sort of sagging away to the left, and maybe a little more pronounced than the first time. Just one lurch, however, and then everything was back to normal. Except for my state of mind. I didn't know what to think. The best I could do was tell myself that this was an old building, too. It was built around 1900. It never occurred to me that there might be any other explanation. I suppose I didn't want it to. And that probably explains why I didn't mention the incident to Frank. It was on the tip of my tongue a dozen times that night, and over the weekend, but I didn't. Or, rather, that explains part of it. There was another reason why I held back. I don't know how to describe it, but I had the feeling that my sense of touch was getting more and more acute. Especially in the soles of my feet. I could feel little tremors that other people couldn't. But Frank is so matter-of-fact. I was afraid that if I told him—well, he wouldn't understand.

"I didn't tell Frank until the middle of the following week. On Wednesday night, to be exact. By then, I had to. I couldn't keep it to myself any longer. I'd had an errand uptown that morning. On the way back, I dropped in to see a friend in the building at 575 Madison Avenue. She works for a publishing house, and we had a nice visit—not that it matters. What matters came after I left her. I was standing just outside her office, waiting for the elevator, when the corridor suddenly sank. The same sensation as before, only a little worse—almost a jolt. Oh, this one was ghastly. I hardly remember leaving the building, I was so frightened. I was petrified. Because there was really a big difference—575 Madison is practically brand new. It's less than ten years old. I couldn't rationalize this time. I had to face the truth. There was something wrong with *me*.

"Frank agreed. But I don't think he was much upset. He seldom

is, about anything—that isn't his way. I've never known a more
thorough stoic than Frank. He agreed, though, that my experiences could hardly be considered normal. More than that, he
suggested I see the doctor. That was Dr. Dodge [as he will here
be named]—a very good man in the neighborhood, whom both of
us knew and liked—and I called him the following morning. He
gave me an appointment for Monday, March 4, at six o'clock. All
Dr. Dodge knew when I arrived at his office was the general nature
of my trouble. I'd told him over the phone that I'd been having
dizzy spells. That seemed the simplest way to describe it. But that
was on Thursday. By Monday, I hardly knew what to say. There
just wasn't any word for the awful sensations I'd been having. The
floor-shaking feeling was only one of them. I don't know how
many times that happened over the weekend—seven or eight at
least. But even that began to have a different feeling. At first, the
floor had moved or sagged as a whole, as a plane. It still did, only
now I could feel another movement, too. A kind of counterpoint.
Sometimes it was as though I were sinking into the floor. The
room would tilt and I'd take a step, and the floor was like snow.
It would give under my foot and I'd sink what felt like an inch,
and other times it was just the reverse—the floor would rise to
meet me. That was one new feeling I told Dr. Dodge about.
Another was something I'd noticed for the first time on Sunday.
By then, it wasn't simply the floor that moved. When the floor
tilted, the walls of the room tilted with it. And the ceiling. I mean,
the shape of the room never changed—only its position in space.
I began to realize that the direction of the movement was always
the same. The movement was always from right to left. I don't
think Dr. Dodge was very much interested in my case at first. Not
until I began to describe those latest symptoms. Then his manner
quite noticeably changed. He began to listen very carefully. They
seemed to make a serious impression on him. But just how serious,
and what he thought they meant, I couldn't tell. He didn't say and
I didn't ask. I wanted to, but somehow the visit was over before
I ever had an opportunity. The only comment he made was after
he'd taken my blood pressure and listened to my chest and

checked my reflexes—all the routine tests. He said a series of consultations with specialists seemed advisable. To pinpoint the source of the trouble, I gathered. He would arrange the appointments, he said, and let me know when and where. That was all, except that he wanted to see me again in two weeks. Same time. Same day.

"My first special appointment was with an ear, nose, and throat man. I saw him three days later—on Thursday, March 7. The next day, I saw an oculist. Then, on Thursday of the following week, I saw a gynecologist. They were all top men. I've never had such thorough examinations. And they couldn't have been nicer. But they told me exactly nothing. Especially the gynecologist. He was completely noncommittal. The ear, nose, and throat man gave me a prescription for Dramamine. I tried it—one fifty-milligram tablet twice a day—for a couple of days, but it didn't help. In fact, it made me slightly sick, and my dizzy spells got worse and worse. My vertigo, I should say. I've learned there's a difference. Dizziness is just a giddy feeling in the head. Vertigo is far more violent. It's a feeling outside of yourself. You usually feel that the world is moving—up and down, or around, or sliding off to one side. But you can also feel as if you were whirling with it, somewhere out in space. I did, at least. However, as I say, the ear, nose, and throat man suggested Dramamine, and that was all he said. The only one who made any sort of comment was the oculist. He said he thought my trouble might be partly visual, and recommended new glasses. When he told me that, my heart absolutely leaped. I wanted to believe him. It was such a plausible explanation. And I *had* been having some trouble with my eyes. I'd noticed lately that I couldn't seem to make them focus properly. Any quick movement, like raising my eyes from a book to look at something across the room, made me almost bleary for a minute. But he was mistaken. I had the new lenses by the time I went back to Dr. Dodge, and they didn't really help a bit. Oh, maybe there was some improvement, but, as I told Dr. Dodge, if there was, it was very, very slight. I remember he nodded, as if he already knew. Then he brought out some papers. They were reports from the specialists I'd been to. He read me their conclusions, and they were all the same. They even used the same phrase—'Impression: Essentially normal.'

"I'll never forget that phrase. I must have heard it a dozen times. It got to be a kind of litany. Even Dr. Dodge used it. He gave me a complete physical examination during my second visit. The next time I saw him—the following Monday—he told me the results. Essentially normal. Meanwhile, he had arranged for two more consultations with specialists, and I had been to see them, too. One was a neuropsychiatrist and the other was an internist, and Dr. Dodge had their reports in my folder. Impression: Essentially normal. It sounds so reassuring. Normal—essentially normal. So comforting. But it isn't. At least, it wasn't to me. It was terrifying. While I hadn't the vaguest notion of its medical meaning, I was sure of one thing. It couldn't mean that I was normal at all, because I wasn't. I was just the reverse. I became convinced it meant I was miserably sick and nobody had the faintest idea why."

The terror that mounted in Mrs. Morton during those weeks of inconclusive consultations was excited by more than uncertainty. Suspense merely sharpened its edge. Its substance was the growing vigor and variety of her hallucinations. The chronology of their proliferation is no longer very clear in Mrs. Morton's mind. Her memory of their nature, however, is cruelly complete. "I can still feel them," she says. "I can close my eyes and feel myself standing in the apartment or the office and the waves of movement rolling under my feet. At first, it was only in the afternoon or evening. Then—within a couple of weeks, I think—it began to happen earlier and earlier in the day. And more frequently. I have a mental picture of myself lying in bed on a cold morning—which would probably make it early March—and knowing what would happen the minute I stood up. It was never as pronounced in the morning, though, as it was later in the day. My vertigo seemed to build up hour by hour. To accumulate. Or, possibly, sleep and rest gave me a certain amount of resistance that gradually wore off.

But the bathroom was always an ordeal, at any time. Any small room, I should say—my closet, the kitchen, a narrow hallway. I particularly dreaded taking a bath. It was bad enough to close the door and feel the room begin to tip. But when I got into the tub —it was like being in a tiny boat in a heavy swell. Even in the morning.

"I never bathed at night—after the first time. No power on earth could have made me go through that again. In fact, every-thing was a trial at night. By the end of the day, I seemed to fall apart. It wasn't only vertigo. There was the problem of focusing. I got so I couldn't judge the spatial relationship of things with any kind of precision. I was especially baffled when it came to anything close at hand. Sewing, for example. I had to give up darning and mending. And cooking. Frank finally took over in the kitchen. I simply couldn't cope. By the time I'd finished peeling a potato, there'd be practically no potato left. I'd break an egg—I mean, I'd try to. But instead of just cracking the shell, I'd smash the egg all over the stove. I couldn't even wash the dishes. I'd finish with a plate and stick it in the drying rack. Crash! A complete miss. I'd dropped it on the floor. I broke more dishes in the month of March than I had in the whole of my life before. I also did more walking. At first, I spent every evening just walking—outdoors, where I didn't feel closed in.

"Outdoors and bed—those were the safe places. For a while, at least. The outdoor phase didn't last long, but I was always com-fortable lying down. At the office—I don't know how I managed to keep on working, but I did—there was a couch in the rest room I could use when things got too much for me. Although the act of lying down was sometimes distinctly unpleasant. I would stretch out on the rest-room couch or the sofa, or get into bed, and instantly feel myself falling. The distance I fell always varied. I might fall three or four inches. Or it might be a drop of ten feet. Then I'd slowly drift back up. But sometimes there was more to it than that. I'd bounce, for what seemed like several minutes. One of those times, it went on and on, and all of a sudden I recognized the rhythm. The fall and rise was in exact synchronization with

my breathing. Maybe that explains the whole lying-down phenomenon—I mean the getting-into-bed part. The relief I got from
lying down was something else, of course. When you're flat on
your back, I suppose, there's no strain on the equilibrium.

"The relief I got from walking the streets—I don't know how
to explain that. But, as I say, it was only temporary anyway. It
didn't last much beyond the middle of March. Then the buildings
started to lean and sway, and the sidewalk began to tilt, and it was
almost as bad as being indoors. In some respects, it was worse. The
sense of motion was somewhat less violent. I think it was less wild
and erratic, too. But the scale was so much bigger. Every block
was a downhill slide or an uphill climb, and the intersections
looked two miles wide. Crossing a street took all the courage I
had. I didn't dare glance to the left or right. Any sudden movement of my head and the world turned upside down. My system
was to fix my eyes on some distant object. That usually kept me
on balance. Then I'd make a dash for it. It was always better to
keep focused on something in the distance. At close range, everything blurred. Perhaps I should never have gone out alone. I
imagine some people would have refused to. But I'm not like that.
I was determined not to be an invalid. I'm glad to say that Dr.
Dodge agreed with me. And so did Frank. All either of them ever
did was caution me to be careful—to cross with the light, to use
the handrail on the subway stairs, and so on. Of course, they didn't
have much choice. They knew very well that my mind was made
up. But, even so, I had some really horrible experiences. At least,
they seemed awful then. I remember one evening coming home
from work. I was walking north on Lexington Avenue when I
noticed a man at the end of the block. He was one of those
derelicts, reeling drunk, and he was headed my way. There were
just the two of us—nobody else from where I was to the corner.
The minute I saw him, my system collapsed. I couldn't seem to
focus my eyes on anything, and I began to reel right with him. If
he lurched toward the curb, the sidewalk tilted toward the curb.
When he caught himself and staggered back across the sidewalk,
it tilted in that direction. I was sure we were going to collide. I

could almost feel the impact. But we didn't. We came face to face —so close I could smell his breath. And then I lost him. Everything blurred. The next thing I knew, he was gone—dancing on down the street. I had only one real collision. And it wasn't with a drunk. It was with a perfectly innocent young man, and I was entirely to blame. I was on my way to lunch. The sidewalk began to drop away from under me, and down it I went, practically at a gallop—straight for the front of a store. Fortunately, there was this young man standing there. If I hadn't toppled into him, I might have gone right through the window. He didn't say anything. Neither of us did. I was too numb. I just slunk away. But it wasn't hard to imagine what he thought.

"The most horrible experience I had of that kind wasn't actually on the street. It happened in that underground passage that runs from Grand Central Station to the Roosevelt Hotel. The time was about six o'clock one evening in early April. A woman who had been a classmate of mine at Radcliffe was in town, and we had arranged to meet at the Roosevelt for dinner and go on to a concert at Town Hall. I suppose it sounds odd for someone in my groggy condition to plan a long, unnecessary evening like that, no matter how determined I was to keep going in my day-to-day routine. Well, I wanted to see my friend. I hadn't seen her for several years. That was only part of it, though. The basic reason was me—my state of mind. It didn't seem possible that anyone could be as sick as I was and live. There were times in March and early April when I was absolutely certain I was going to die. But my reaction to death was peculiar. I don't remember feeling afraid. All I remember is an overwhelming sense of urgency. So little time. So little done. So much I wanted to do. I've almost never read imaginative literature, but one Saturday morning I rushed out of the house and bought two recent novels—Rebecca West's *The Fountain Overflows* and *The Lost Steps*, by Alejo Carpentier. It didn't matter that I could scarcely see. I devoured them both. It was like being possessed. I craved every kind of diversion. I dragged Frank to the theater more than once, and I never thought of refusing when he suggested the Philharmonic or the Metropolitan. My response to music had never been so complete. I spent hours listening to records. I'd play some old favorite —like Beecham conducting Haydn's 'London' Symphony—and it was amazing. It seemed to me that I could hear the inner structure more clearly than ever before. It was the same with food. I'd always been a rather casual eater, but I began to understand how gourmets must feel. Good food was pure delight.

"So the idea of dinner and a concert wasn't at all unusual. My only mistake was deciding to take that dreadful underground passage. It was raining and I was in a hurry, but even so I should have realized. When I did, it was too late. The passage was jammed with commuters, shoving and pushing and surging toward me. But I didn't dare turn back. The floor was beginning to wobble, and I knew if I tried to swing around, it would tip me head over heels. All I could do was go on. The traffic was still all against me. People kept looming up—towering up. They came charging at me like giants. I don't know whether it was my eyes or the way the passage was rocking or the murky glare of light down there, but they really looked like giants. Their heads were brushing the ceiling. And then I felt something right out of a nightmare. It began with a sensation I'd often had—the feeling that the floor was lifting to meet my foot and, at the next step, sinking under my weight. I was almost at the end of the passage when I felt the movement change. It was as if someone had pulled a lever. There was a little jolt, and the floor was moving very slowly backward down the passage. I was walking on a treadmill. Only for a minute, though. Then I reached the stairs. I drove myself up to the lobby and collapsed in a chair. I was jelly.

"I thought the treadmill must surely be the end. I didn't see how I could possibly stand anything more. It had to be the end. I didn't see how there could be anything more unearthly than that. But it wasn't the end—far from it. The treadmill was only another beginning. From that point on—from early April—I began to move in a different world. Two worlds, rather. In a way, it was like a personality split. I felt more and more removed from everything and everybody. I knew they existed. I lived and worked among them, but always at one remove. I stood apart. Actually, it was more of a triple split. There was me. There was the rest of the world. And there was the me of the past, when reality had been real. It wasn't a schizoid split—not as I understand that term. I never lost touch with reality. I always knew the difference between what actually was and what was wild illusion. Also, my two selves were never coexistent. The me of the past existed only in memory. The feeling I had was metamorphic. I'd moved into a new self. In a sense, of course, I had. My personality did change. I wasn't the same person then that I'd been before and that I am now. The craving I had for diversion was wholly out of character, and when that wore off, as it did in a very short while, I was even less like my normal self. I never had a pleasant thought. I lived to sigh and grumble. I couldn't read. I was too restless, too self-absorbed, too empty. My ability to focus on print dropped to practically nil. My limit was the absolute minimum my job required. And that was painful—physically painful. It was the same with music. Music, especially recorded or broadcast music, began to hurt my ears. The upper partials literally felt like needles. I gave up the piano very early. It required a coordination of eye, ear, nerves, tendons, and muscles that I no longer had, and besides, it just plain bored me. The only thing that interested me was me, and most of the time I felt too dull to even concentrate on that. I was nothing. I just sat in my shell and sulked.

"But that was only part of the new sense of self I had. The change was more than a matter of mind or emotion. It was a total metamorphosis. I was conscious of a new dimension—a new plane. I had a new relationship to space. My legs, my arms, my face, my whole body felt different. I suppose everyone remembers the Hall of Mirrors at Coney Island. That's the way my body felt. It had no permanent shape. It changed by the minute. I seemed to be completely at the mercy of some outside force—some atmospheric pressure. I was amorphous. Whatever form I had was determined by the contractions or expansions of the force that confined me. And its pressure was never evenly distributed. I mean, I never simply grew or shrunk. My left leg would seem to lengthen. Or my right arm. Or my neck. Or one whole side of me would double or treble in size. And yet that doesn't fully describe it. There were times when the force seemed to be the rotation of the earth. I would have the feeling that I was vertically aligned with the earth's axis. I could feel a sort of winding movement start up inside me. Then one of my legs would begin to shorten, as if it were an anchor being drawn slowly up by a winch. The other leg would dangle. Not for long, however. After a minute, the winch would shift. It would engage the dangling leg, and just as slowly bring it up to match the other. Sometimes the movement alternated back and forth. One leg would be raised perhaps a foot and then gently released, and as it sank back, the other would rise. The only analogy I can think of is treading water. That was exactly the sensation. But my relation to the rotation of the earth took other forms, too. I often moved in opposition. The problem then was to somehow resist—to hold on. It took every ounce of strength I had to keep from being spun off into outer space. At such times, I was always teetering on the edge of space. One sensation I had on innumerable occasions was of floating in space on a tiny platform. I was somewhere between the earth and no-where. If I was standing still, or had just stood up, the platform bounced. If I was actually moving, it tipped or tilted. Or slid away behind me—like the treadmill. The only sensation more terrifying than the platform was what happened when I had to go down a flight of stairs. It meant stepping off into space. But I don't want to talk about that. I can't even bear to think about it.

"The only person I ever attempted to tell what it felt like going downstairs was Dr. Dodge. That was on April 10. I hadn't seen him for over two weeks, and he wanted to know exactly how I felt. So I told him. I told him all my symptoms, all my experiences, all my fears and worries—everything that had happened since our last meeting. How glad I was that I did. I still am. Because, for once, his reaction was more than merely sympathetic. He actually did something. He picked up his pen and wrote out a prescription. It was for Equanil, he said, one of the tranquilizing drugs, and the dose was three four-hundred-milligram tablets per day. Then he added that there was one more specialist he wanted me to see. Another physiologist, whose particular field was electroencephalography. Dr. Dodge would make the arrangements, and let me know. He was very offhand about it. Deliberately so, I realize now. He made it seem like just another routine test. I'd heard of electroencephalography—vaguely. I knew it had something to do with recording electrical currents in the brain. But that was all I knew —or cared. My mind was fixed on Equanil. I'd heard a great deal about tranquilizers. Who hasn't? I knew they had helped and saved so many people—people much worse off than I was. I was really prepared for something not far from a miracle. What happened was not quite that. Equanil had no effect on anything fundamental. But it did something for my morale. It made me feel less like screaming. It kept my shirt on.

"Dr. Dodge phoned me the following day. My EEG test—as he called it—was set for the next afternoon, April 12, at four o'clock. Our next appointment would be on April 15. He would have my EEG report by then. I don't remember much about that test. I arrived in such a state of confusion. The doctor's office was in a huge block of buildings, and everything went wrong. I misread his name on the directory. I took the wrong elevator. I got lost in a maze of corridors. And when I finally got straightened out, I had to wait in a dim little box of a foyer that reduced me to a pulp. But apparently that didn't matter. The doctor gave me the test, and he must have been satisfied that the findings were accurate, because his report reached Dr. Dodge on schedule. He was reading it when I walked into his office. Or so it turned out. I knew he was reading something that had to do with me. Something important, too. His expression told me that. Then, as soon as I sat down, he told me so himself. He was really in a most communicative mood. He said he was more relieved than it was possible to say. Our talk at my last visit—some of the more outrageous things I'd described—had made him uneasy. It had led him to suspect that the root of my trouble might be a brain tumor. The EEG test, as perhaps I knew, was diagnostic for that. Some doctors, he said, might have ordered one earlier, but he had held off, because he wanted to see the results of other tests first. A brain tumor was a rather outside possibility. Well, anyway, here was the electroencephalographist's summary of his findings. He read it off—'Impression: Essentially normal.'

"You can imagine the chill that ran up my spine. Even the thought of brain tumor is frightening. For a moment, I was cold to the bone. I just sat there. So did Dr. Dodge. Then he tucked the report away in my folder and leaned back in his chair. He said he was sorry to have kept me in suspense so long. I realized, he hoped, that his noncommittal stand had been unavoidable. My case had presented certain problems. It still did. But, in his considered opinion, the basic outline was now sufficiently clear to justify a diagnosis. And one that held no cause whatever for alarm. My trouble was a disturbance of the internal ear called labyrinthitis. He explained it to me in some detail. Enormous detail, in fact. He couldn't have been more considerate. But I'm afraid that at the moment I didn't grasp much of what he said. Except at the end —and this was very encouraging—that it wouldn't be necessary for me to come in again for several weeks. May 8, he said, would be soon enough. Meanwhile, of course, I was to continue taking Equanil."

Mrs. Morton's recollection of that diagnostic discussion, though limited, is entirely correct. The explanation that Dr. Dodge considerately gave her was a thorough review of the subject. It was also, he told me during a recent talk I had with him in his office, a thoroughly candid one. He began with a few words on the nomenclature of middle-ear disorders. Labyrinthitis, he informed her, was only one of several more or less synonymous names by which such disturbances are known. Another was Ménière's syndrome. It takes its name from Prosper Ménière, a nineteenth-century French otologist, who first described the condition. Until 1861, when Ménière published a revolutionary paper entitled "Maladies de l'Oreille Interne Offrant les Symptomes de la Congestion Cerebrale Apoplectiforme," vertigo had been considered an invariable symptom of brain damage or disease. His paper, which was based on definitive clinical findings, demonstrated that the true seat of the phenomenon was the internal ear. Other names for Mrs. Morton's condition included aural vertigo, vestibulitis, and, of course, labyrinthitis. Dr. Dodge preferred—at least in this instance—the last. The suffix "itis," as he supposed she was aware, meant "inflammation." So the meaning of labyrinthitis, as a word, was simply an inflammation of the aural labyrinth. The specific nature of the disturbance in her case was something that neither he nor his colleagues had been able to determine. As she knew, the results of her various examinations had all been essentially normal. In other words, they were negative. That didn't mean that no disturbance existed, though. Quite the contrary. It simply meant that none of the techniques available to medicine had been able to detect it. Labyrinthitis is a slippery thing. Its possible causes are extremely numerous. The triggering mechanism can be a tumor or a cyst. It can be a bacterial or a viral infection. It can conceivably be a metabolic abnormality. Some investigators believe that it can be traced—in part, at any rate—to psychogenic factors. Or even to an allergy. And there may be other factors, still unknown to science. It was at this point, Dr. Dodge told me, that he informed Mrs. Morton that she could put her mind at rest. She could be quite sure that he had eliminated all the more serious possibilities. But he then went on to make it clear to her that the condition was considered chronic. It was also accepted, he felt obliged to add, that no known treatment was effective. Nevertheless, he felt justified in suggesting that in all probability she would soon begin to feel more like herself again. Long remissions were frequent, and there was some reason to believe that the condition was often self-limiting. In some instances, the symptoms simply vanished and never recurred. That, he told Mrs. Morton, in conclusion, had been his experience with several cases closely resembling hers. It was not, however, as it turned out, a prognosis whose accuracy was immediately confirmed.

"I often think about that visit," Mrs. Morton says. "My reaction was so peculiar. I mean, of all the things that Dr. Dodge told me, the most important made the least impression. I'd been living for weeks with the thought of death. It had haunted me day and night. I'd felt literally doomed. I'd been so sure of dying that I'd accepted it as fact. Yet when I learned that my condition wasn't even serious, I hardly reacted at all. It was as if I'd never had a moment's apprehension. What left me simply weak with relief was knowing the truth at last. My trouble was no longer a mystery. At least, it had a name. It was astonishing what a difference that made. And I wasn't alone. There were many other people in the same boat—thousands of them, apparently. Labyrinthitis is really quite common, I gathered. Or, rather, I made it my business to find out. I got hold of a medical text and read about its incidence with the most morbid avidity. The old saying is only too true, I'm afraid—misery does love company. As a matter of fact, several people I knew, or people friends of mine knew, turned out to be companions in misery. They'd had the same trouble off and on for years. One of them, poor thing, was a dancer. Another was a concert pianist. He had a recurrent experience, that was truly appalling. It would happen during recitals. He'd be playing along when he'd suddenly have the feeling that the piano was sliding away from him. The only thing he could do when that happened was to concentrate on trying to will the movement to stop. At times it would work and at times it wouldn't, but somehow he always managed to keep going.

"It was also comforting to think that I would most likely be feeling better soon. I didn't expect it to happen overnight, naturally, or even in a matter of days. Dr. Dodge had merely said it would probably be soon, and the people I talked to—my fellow-victims—said the same. Improvement took time. So at first I wasn't disappointed. I dragged myself along from day to day, and tried to be patient. And I was—through the rest of April and the first part of May. How I did it I'll never know. Those weeks were the ghastliest of all. The stairs, the treadmill, the platform in space, the awful amorphous feeling—everything seemed to be closing in at once. There was hardly an hour that wasn't a total nightmare. By that time, the mornings had become almost the worst—because each one was such a let-down. I'd wake up and lie there and wonder if this would be the day when I'd begin to improve. And pretty soon I'd get up and start for the bathroom —and then I'd know it wasn't. To be teased like that was torture.

"I went back to Dr. Dodge on May 8, as arranged. When I told him that nothing had changed, he seemed quite surprised. But all he did was renew my prescription for Equanil and repeat what he'd said during my last visit. The only new thing he did was not give me another appointment. It wasn't necessary, he said—although, of course, I could call him whenever I liked. But I never did call him, even though I went home from that appointment about as low as I've ever been in my life. About a week later, I broke down. I'd reached the end of my rope. I couldn't go on any longer. So I decided to go up to our place in the Catskills for a couple of weeks, and rest. The office was very understanding. They told me to go right ahead. But my husband had his doubts about leaving me all by myself in an isolated country cottage for a couple of weeks. After all, I still wasn't able to cook or wash dishes—a problem that hadn't even occurred to me. Frank didn't mention his misgivings to me, though. He called Dr. Dodge and asked him what he thought of my plan. Dr. Dodge said that a change of routine and scenery might indeed be helpful, and he wanted to know if it would be possible to get someone to look in on me once a day—fix my principal meal, clean up, and so on. As it happened, Frank knew just the person—the teen-age daughter of a farmer in the neighborhood. For a couple of seasons, she'd been going over to open and air out the cottage for us just before we went up for our vacation—a nice, quiet, unobtrusive girl. So Frank got hold of her on the telephone, and she agreed to come over for an hour or so every afternoon—while I could be taking a nap. We drove up on Saturday, May 18, and Frank stayed over Sunday. When he left, I was nicely settled.

"It was a glorious time to be in the country. Everything was beautiful. The weather was perfect and the foliage was bursting out. And the birds—I hadn't seen so many birds in years. I arrived at the peak of the spring migration, and they were there by the thousand. It took me back to the springs when I was a girl in Iowa.

That's where I grew up. My father was an amateur ornithologist, and we used to go on long bird-watching walks together. It was wonderful how that almost forgotten interest revived. Not that I did any great amount of walking. The least little movement still sent me reeling. In that respect, the country was just like the city. It was all I could manage to crawl from bed to a chair out on the lawn. But that was far enough. The birds were everywhere. I just sat on the lawn and watched them. They almost seemed to pose for me. I saw at least a dozen species that I'd never seen before, including a blue grosbeak and a pileated woodpecker, both of which are fairly rare. They probably mistook me for part of the scenery. I guess I was, to all intents and purposes. I spent hours at a time just sitting there. It was all I did, except eat and sleep, but I wasn't bored for a minute. It was as if all my ambition and energy had drained away. I was perfectly content to simply vegetate."

Mrs. Morton returned to New York on Sunday, June 2. Her husband went up for the weekend and accompanied her home. She felt relaxed and rested, but otherwise much the same. "The thought of going back to work was agony," she says. "I really doubted if I could make it. I was afraid that the traffic and the subway and the tension would be too much for me. And it almost was, those first few days—the first week of work—but I kept going. I felt I had to. It was a matter of pride. I was determined not to collapse again. And, little by little, the strain began to ease. The change I noticed first was in my mental attitude. I don't mean morale. It wasn't a sense of lift or an Equanil tranquillity. My mind began to function. It was something like getting over a severe attack of the flu. I began to feel less dull and self-contained—less empty in the head. Things outside myself began to penetrate. They had meaning. I think I noticed it first at home, in my relationship with Frank. Even when I'd felt my worst—in late April and early May—I'd tried not to be an absolute lump. I'd made an effort to talk and listen and seem to take an interest. Not very successfully, I'm afraid, but I'd tried. I wasn't aware of any turning point when

I got back from the Catskills. But I kept trying to be something more than a lump, and then, after a while, it gradually dawned on me that I wasn't consciously trying any more. I wasn't pretending. The interest I showed was genuine.

"That must have been around the middle of June. Then, even more slowly, I began to notice other differences. My sense of self was one—the split I'd felt between my new self and my old grew less and less distinct. I also began to have permanent form and shape. Then, at some point, some week—sometime in late June or early July—I suddenly felt intact again. I don't know what came next. I mean, there wasn't any sequence of change. I simply began to have a firmer control over motion. I could stop the treadmill. I could keep from teetering out in space. I could make the stairs be there and stand up under my weight. And yet there was a sort of pattern. The wildest sensations were the first to go. I had them less often, and then less intensely, and finally not at all—a slow simmering down to normal. By the middle of August, I was back where I'd been in February. The floor gave a lurch every now and then, and I still had a little trouble focusing, but that was all. And then, as well as I could judge, those remnants vanished, too. It's impossible to say exactly when it all ended. But I think it was Frank who really sensed it first. It was after dinner one night in late August, and he suddenly smiled and remarked that I must be feeling much better. I asked him what he meant. 'You never look scared anymore,' he said."

[*1958*]

CHAPTER 8

A Swim in the Nile

THE FIRST TIME VERNON BERRY (as I'll call him) took sick on
his trip around the world was in a Salvation Army hostel in the
Kowloon section of Hong Kong. Berry woke up that morning—
it was a Monday morning toward the end of February, 1958—
with a rocking headache, a sore throat, and the shivers. He wasn't
much alarmed. He was young (just twenty-five), his health had
always been excellent, and only a week or ten days earlier a doctor
back home in Jersey City had immunized him against tetanus,
smallpox, typhoid fever, yellow fever, and cholera. What he
mostly felt was annoyance. He was traveling on an extremely tight
budget (a few hundred dollars in savings, and a recent small
inheritance), and he couldn't afford a serious illness. Accordingly,
he decided to take no chances. He dragged himself out of bed and
down to the Salvation Army clinic, which adjoins the Kowloon
hostel. The doctor on duty there heard his unexceptional complaint and made the usual diagnostic soundings. They indicated,
Berry was told, a touch of acute bronchitis. The doctor gave him
a shot of penicillin and instructions to come back the next day for

another. Meanwhile, he was to stay in bed and rest. Berry stayed in bed, except for visits to the clinic, until Thursday. On Friday morning, feeling fully recovered, he left by plane for Singapore.

Berry spent the weekend in Singapore. He slept at a YMCA hotel on Orchard Road, had his meals at an inexpensive restaurant in the neighborhood, and saw the sights of the port. The next stop on his itinerary was Bangkok, and he chose to go by train—a journey of some fourteen hundred miles through the mountain jungles of Malaya and southern Thailand. He left on Monday morning. That night, somewhere between Singapore and the Thailand border, he took sick for the second time. His feet began to itch and swell. Then he began to itch all over. Then his joints began to hurt. He couldn't imagine what had happened, and he didn't know what to do. There was no one he could even talk to. That alone was unsettling. He began to feel a bit frightened. By morning, he was in a kind of panic, and when the train pulled into Songkhla, in southern Thailand, he grabbed his bag and got off. He found an English-speaking doctor and asked for an explanation. The doctor was noncommittal. It was possible, he said, that Berry had developed a calcium deficiency, and he gave him a restorative injection. It didn't seem to help. The next morning, after another wretched night, Berry resumed his journey, as miserable as ever. He itched and ached, and his feet were so swollen that it was all he could do to walk. Also, he had diarrhea.

Berry arrived in Bangkok around noon on Saturday, March 1. His first concern was his health. A clerk at a YMCA hostel where he arranged for a bed and left his bag directed him to a hospital operated by the Seventh Day Adventist Church, on Pitsannloke Road. He made his way there, and into the outpatient clinic. The examining physician looked him over, noted his recent history, and offered an opinion. It was his belief that Berry was suffering from two unrelated ailments. One appeared to be a mild attack of bacillary dysentery. That was a common complaint among Western travelers in the East, and it usually responded satisfactorily to sulfadiazine. The other was almost certainly an allergic reaction. There were, of course, any number of possible allergens. The list included foods, drugs, dusts, and pollens. In this instance, however, the circumstances seemed clearly to implicate a drug—the penicillin injections that Berry had been given in Honk Kong. Allergic symptoms, too, could be satisfactorily controlled, with cortisone and antihistamines: But Berry must also do his part. He was to rest in bed, drink plenty of water, and report to the clinic daily for treatment and observation until further notice.

Berry did as he was told. He reported to the Adventist clinic every day for nine days. At the end of that time, he was discharged to convalesce on his own. The following day, he took a train to Rangoon. Except for a little swelling around the eyes, his allergic symptoms had largely subsided, but he still was somewhat diarrheic. He spent the next week traveling by plane and train and bus through Burma and East Pakistan to Calcutta. He reached there feeling tired and apprehensive. He had lost twenty-five pounds in the past two weeks and had had to force himself to eat. He rested for a day in a Salvation Army hostel and then sought out a doctor. The doctor found him nervous and undernourished, and recommended vitamins. Berry took his advice. A few days later, having exhausted the sights of Calcutta, he moved on, by train, to New Delhi. From there, he flew to Karachi. It had been his intention to spend several days in Karachi, but he stayed only overnight. He awoke in a knot of restlessness, threw on his clothes, and caught a plane for Damascus. On the way, he became aware of a leaden ache low in his back and flanks. It smoldered there throughout the rest of the flight. Then, as the plane came bumping down at Damascus, it shifted around to his abdomen and sharpened into a twitching pain. Berry rode uncomfortably into the city with the glum conviction that he was going to be sick again. A French doctor confirmed his fears. Berry had a temperature of just over a hundred, and there was a trace of blood in his urine. Those findings, together with the location of the pain, suggested a kidney infection. Either that or renal colic.

Whatever the nature of Berry's illness, it kept him in a nursing home in Damascus for a little over a week. He emerged, wobbly but well, on April 14, and had a final consultation with the doctor.

He was advised to take it easy. The next day, he caught a plane to Egypt. After a couple of post-convalescent days in Cairo, he traveled up the Nile as far as Luxor. He spent three days in Luxor, wandering among its many antiquities and swimming in the Nile, and then returned to Cairo. From Cairo, he flew to Amman, from Amman to Beirut, from Beirut to Ankara, and from Ankara to Athens. He halted there, at a students' hostel, for two nights and a day. On April 28, he left by plane for Zurich and a tour of Western Europe. In the course of the next six or eight weeks, he visited Geneva, Marseille, Barcelona, Madrid, Paris, Brussels, Amsterdam, Frankfort, Munich, and Vienna. An allergic puffiness around the eyes plagued him off and on, but otherwsie he felt encouragingly well. On his second morning in Vienna, he was awakened by a twinge of abdominal pain. It came again while he was dressing, and again on the way to breakfast. Then it went away. Nevertheless, it made him uneasy, and that afternoon he had himself directed to the office of a urologist. The urologist questioned him closely about his Damascus experience, and did a urinalysis. The results, he reported, were entirely normal. Thus reassured, Berry resumed his travels. Between the end of June and the end of July, he visited Berlin, Copenhagen, Stockholm, Oslo, and London. Early in August, he sailed from Southampton on the *Queen Elizabeth* for New York.

Berry settled down in Jersey City again, and after a week or two of looking around he found a job as a draftsman with an architectural firm. He lived with his parents and saw his old friends and did as he had always done. It was not, however, as if he had never been away. The allergic puffiness around his eyes persisted, and the itches and swellings elsewhere on his body continued to come and go. In addition, early in September he began to feel a vague and vagrant nausea. The feeling was particularly pronounced just before meals. He told himself that it would soon go away, and when it didn't, he tried to ignore it. At the end of a week, he gave up and called the family doctor for an appointment. The doctor examined him with care, and admitted that he was stumped. In every respect but one, he said, Berry seemed to be in his usual excellent health. The exception was revealed by a blood count. It showed a morbidly significant increase in white blood cells of a kind known as eosinophiles. That tended to suggest the presence of infection. The possibilities were too numerous to mention, but a consultation with a specialist might help to sort them out. He would make an appointment for Berry with an internist of his acquaintance. A few days later, Berry emerged from another thorough examination with another equivocal report. Except for a notably elevated eosinophile count, all tests and soundings performed by the internist had had negative results. The internist, however, was not completely deterred. He had formed, he said, a certain impression—a hunch. Eosinophilia, he reminded the family doctor, is characteristic of trichinosis. So are allergic-like swellings around the eyes, and urticarial itches. He wondered if a mild trichinal infection might not be the cause of the patient's complaint. The family doctor passed this tenuous diagnosis on to Berry, adding that he was generally inclined to concur.

There is no specific treatment for trichinosis, but in the great majority of cases time will effect a cure. The curative powers of time did next to nothing for Berry. Although the swelling around his eyes subsided, and the occasional hives and itches gradually ceased to recur, the dragging mealtime nausea and the ambiguous eosinophilia continued unabated—for weeks, for months, for a year. Then, one Saturday afternoon in October, 1959, Berry passed in his stool a ten-inch worm. He submitted the specimen, in consternation, to the family doctor. The latter was able to reassure him. The worm was a species of round worm *(Ascaris lumbricoides),* an intestinal parasite only slightly less common than a tapeworm. It was also, the doctor went manfully on to say, the probable cause of what had seemed to point to trichinosis. Nor was that all. Ascariasis, unlike trichinosis, was readily responsive to treatment. The drug of choice was piperazine citrate, and a single dose was usually sufficient. It was in Berry's case. Within a week, his stomach had steadied and settled, and he felt almost like himself again. The only trouble was his eosinophile count. It had dropped, but it still stood abnormally high.

Around the middle of November, Berry was seized with a new affliction. It hurt him to urinate. He endured this puzzling complaint for several uneasy days, and then went down to see the doctor. The doctor heard his symptoms with relief, and an examination, which disclosed an inflammation of the prostate gland, confirmed his first impression. Berry had prostatitis. The doctor then turned his attention to treatment. It consisted of a prostatic massage, a prescription for sulfisoxazole, and a warmly encouraging prognosis. He was hopeful that by the end of the week Berry would be greatly improved. It didn't turn out that way. The pain continued as before. A cystoscopic, or direct, examination of the urinary tract was made, and these findings, too, were inconclusive. The doctor saw no reason to alter his diagnosis. Or his treatment. He was convinced that periodic prostatic massage would eventually have a salutary effect.

And so, before long, it apparently did. Berry's occasional abdominal cramps became increasingly occasional, and presently— early in 1960—they vanished altogether. This was followed by an encouraging change in his original urinary complaint. For a while, it seemed to Berry that his troubles might be almost over. But only for a while. The remissions, it soon became clear, were nothing more than remissions. In time, the pain always returned. Toward the end of 1960, Berry gave up any real hope of a cure and resigned himself to a life of chronic prostatitis.

Berry's life as a chronic invalid lasted almost a year. It ended, with a jolt, one December evening in 1961, when he noticed a stain of blood in his urine. The next morning, it had noticeably deepened and darkened. Sick with dismay, he telephoned the family doctor. The doctor told him to come to the office at once. From there, on the doctor's referral, he went to see a urologist. The urologist asked some careful questions, and then led him into his surgery for another cystoscopic examination. At nine o'clock the following morning, Sunday, December 17, Berry was admitted to Memorial Hospital, in Manhattan, with a diagnosis of suspected cancer of the bladder. Berry spent six days in Memorial Hospital. On December 22, he was discharged in the care of Dr. Harry Most, chairman of the Department of Preventive Medicine at New York University-Bellevue Medical Center, and an authority on tropical diseases. The report of his stay read:

> Physical examination on admission revealed a WDWN [well-developed, well-nourished] white male with . . . prostate minimally enlarged and non-tender.
> Urine microscopically showed a few RBC [red blood cells] and few WBC [white blood cells]. Hgb. [hemoglobin] was 15 gms. Hct. [hematocrit] 42%. WBC [white blood count] 5.9 with 4% eosinophiles on one differential and 20% eosinophiles on a second. Eosinophile count showed 98 eosinophiles per cc./mm. Blood sugar [was] within normal limits. Urinary cytologies were negative for malignant cells. . . .
> Cystoscopy and biopsy were carried out. Bladder was of normal capacity and contour. Scattered throughout the bladder were numerous small, virtually punctate elevations of the bladder mucosa without any visible change in the normal lemon-yellow color of the bladder lining. These lesions appeared submucosal in location. High on the posterior wall of the bladder were irregular, partially confluent, reddened, raised areas with a surface midway in appearance between that of papillary neoplasm and bullous edema, but not characteristic of either. The latter areas totalled several sq. cm. but were resected completely transurethrally.
> Pathologic examination revealed unequivocal evidence of schistosomiasis.

The cause of schistosomiasis is a small but readily visible endoparasitic worm of the genus *Schistosoma* that has an essential environmental predilection for human blood. Its name derives from the Greek *schistos,* meaning "cleft," and *soma,* meaning "body," and refers to a deep longitudinal crevice in the body of the male worm in which the female more or less permanently resides. The schistosomes, though humbly placed in the evolutionary scale, lead highly complicated lives. They reach maturity only after three total metamorphoses in three distinctly different settings. The organism first manifests itself as a newly hatched larva in a pond or lake or river. If all goes well, the larva soon finds and enters the body of a suitable species of fresh-water snail. It is there transformed into an aggregation of reproductive spores, and these, after several weeks of ceaseless proliferation, emerge from the snail as a multitude of tadpole-like wrigglers. The wrigglers must promptly find a suitable animal host, and one host to which they have irrevocably adapted themselves is man. Wrigglers fortunate enough to come upon the necessary host attach themselves to the skin, discard their tails, and (with the help of a tissue-dissolving enzyme) burrow through the epidermal barrier. Penetration usually takes about twenty-four hours and is unperceived by the host. Once through the skin, the wrigglers launch themselves in the current of the peripheral blood vessels and are carried along to the veins and through the heart and into the systemic circulation. The survivors of this intricate voyage (which may take several days) then gather in the richly nutritious blood that flows from the gastro-intestinal tract to the liver and beyond. There they grow to maturity (about an inch in length for the females, and less than half that for the males), and mate. They then retire to the comfort of a tiny backwater vein, where the female, still enclosed in the male's embrace, deposits—for incubation and eventual excretion with the body's wastes—the first of an almost infinite number of eggs. It is not unheard of for a female schistosome to live and breed for thirty years.

The site that a schistosome chooses for its eggs is not a haphazard choice. It is predetermined by the species of worm. There are three species to which man is warmly hospitable—*Schistosoma haematobium, Schistosoma mansoni,* and *Schistosoma japonicum.* Gravid *S. haematobium* worms are drawn to the urinary bladder. Those of the other species prefer the bowels, *S. japonicum* lodging in the small intestine, and *S. mansoni* in the large intestine. The species differ in other respects as well. They have different intermediate hosts, different geographical ranges, and different rates of growth. Each schistosome species has so evolved that it can pass from the larval to the wriggler stage in

the body of only certain species of snail, and the snails acceptable to one are unacceptable to either of the others. It is thus the
snail that determines the distribution of the worm. The distribution of *S. japonicum* is largely confined to coastal China and the
Yangtze Valley, parts of Japan, and the Philippines. *S. haematobium*, though probably indigenous to the Nile Valley (its
calcified eggs have been found in Egyptian mummies of the
twelfth pre-Christian century), occurs throughout most of
humid Africa, in much of the Middle East, and at the southern
tip of Portugal. The range of *S. mansoni* extends from the Middle East (Arabia and Yemen), through much of Africa (the Nile
Delta, the southeast coast, and the rain forests of the Congo), to
Brazil, Venezuela, and the Caribbean islands. Its extension to
the Western Hemisphere is generally attributed to the slave
trade, but there is more to the matter than that, for the African
slaves were infested with *S. haematobium* as well. The decisive
factor was the presence here of a species of snail *(Australorbis
glabratus)* to which *S. mansoni* could adapt. Just why the different species have different rates of growth is yet to be satisfactorily explained. The only certainty is that *S. japonicum* matures
in four or five weeks, *S. mansoni* in six or seven, and *S. haematobium* in ten or twelve.

For all their several differences, the schistosomes are essentially much alike. So are the varieties of schistosomiasis they
produce. A schistosome is not a toxic organism in the usual
sense. Except for certain allergic reactions (of the sort that any
foreign protein may excite), its impact is largely mechanical. The
eggs of a schistosome are equipped with clawlike spines to hold
them in their venous incubator against the constant pull of the
circulating blood, and it is from this constant tugging and tear-
ing that the chief discomfort of schistosomiasis stems. For reasons not entirely clear, schistosomiasis japonica (as an infestation of *S. japonicum* is called) tends to be the most destructive of
the three forms, but all are serious diseases—always unpleasant,
often debilitating, and not infrequently fatal—and the symptoms
of their presence are fundamentally the same. They typically in-

clude, as in the case of Vernon Berry, an outbreak of hives, an elevated eosinophile count, and internal (gastro-intestinal or genito-urinary) bleeding.

Berry's referral to still another physician was unusual but in no sense exceptional. It was arranged, with the enthusiastic approval of the Jersey City urologist, by the head of the urological service at Memorial Hospital, and was made for the best of reasons. Dr. Most is not just another physician. He is one of the few physicians in the New York metropolitan area with a particular knowledge of schistosomiasis, and one of even fewer experienced in its sometimes tricky treatment.

"I was very glad to have Berry commended to my care," Dr. Most says. "Nothing could have interested me more. I thought I could help him, and I was eager to try. That, of course, was the first consideration, but it wasn't the only one. I was eager to talk to him—to learn everything I could about the case. Because I knew enough already to know that it signified something important. It seemed to confirm the development of a new and potentially rather ominous trend. I'll tell you what I mean. Schistosomiasis has always been among the rarest of rare diseases in this country. It doesn't exist here, and in the absence of a suitable species of snail to serve as intermediate host it can't. The only cases that are ever seen in the United States are importations. A lot of American soldiers came home from the Philippines after the Second World War with schistosomiasis japonica, and a certain number of Puerto Rican immigrants with schistosomiasis mansoni turn up in the Eastern-seaboard cities every year, and that's been about the story. Until recently. It seems to be taking a new twist now. A new kind of victim has appeared. It isn't a soldier stationed in an endemic area and it isn't a native exposed from earliest childhood. It's a tourist—someone just passing through.

"Berry was the fourth of these to come to my attention in less than two years. The others were a physician and his wife and a spinster friend of theirs who took a three-week Caribbean cruise in the winter of 1960. Their exposure occurred—it could only have occurred—in a fresh-water pool on St. Lucia, one of the Windward Islands. The physician was the first to become sick. He thought for several days that he had the flu. Then he was hospitalized, and a colleague took over. A high eosinophile count indicated a parasitic infection, and the attending physician sent an account of the clinical findings to the health officers of the various islands visited and asked for diagnostic suggestions. The replies he got came out most strongly for typhoid fever, hookworm, ascariasis, amoebic dysentery, and strongyloidiasis. The possibility of schistosomiasis was never even considered until I came into the picture. I was consulted because of my interest in tropical medicine. Schistosomiasis was suggested to me by a careful review of the itinerary, and a series of tests confirmed my hunch. A rectal biopsy identified the variety as schistosomiasis mansoni. That was exactly two weeks after onset. In the circumstances—classic clinical manifestations, a known recent visit to an endemic area, and first-rate hospital facilities—I think one might say that it could have been diagnosed a bit sooner. The two other cases were fairly mild at onset, and the symptoms were not particularly pronounced. If there hadn't been that obvious link with the sick physician, they might have gone undetected indefinitely. As Berry practically did. And yet the essential clues—a visit to an endemic area and eosinophilia—were there in his case, too. All that was needed was an awareness of the possibility.

"The trouble is that the possibility of schistosomiasis just doesn't enter an American physician's mind. That's understandable enough—or, rather, it used to be. In my opinion, it isn't anymore. Berry and the physician and his wife and their friend may be merely isolated cases, but I don't think so. I think more likely there are others that I haven't heard about. That nobody has. And I think there are many more to come. I'll tell you why. The American tourist has changed in the past few years. There's a whole new breed. Our tourists are no longer mainly the rich or the well-to-do from a few big cities. They're everybody, and they come from everywhere. The reason, of course, is cheaper and faster air travel. There are more tourists now, and they do more traveling, and they visit more distant places. Including, increasingly, the tropics. And every year the chances are greater that some of them will be exposed to a serious tropical disease. Schistosomiasis is only one such disease, and it is far from being the worst. I'm not thinking of plague and cholera and that sort of thing. I mean bizarre diseases, like kala-azar and African trypanosomiasis, or sleeping sickness, and malignant tertian malaria. Malignant tertian malaria is an extremely serious disease with a high mortality rate, and the others can also be fatal, but they have their redeeming features. They are all amenable to treatment. They can be successfully treated and cured. They can, that is, if they are diagnosed correctly and in time. But first they have to be suspected. They have to be brought to mind. Brian Maegraith, of the Liverpool School of Tropical Medicine, has proposed an up-to-date addition to the routine interrogation of a patient. He suggests that at some point the doctor ask, 'Where have you been?' I second the motion.

"Maegraith assumes that the doctor will hear and heed the answer. That, of course, is quite as important as the question. It could have made all the difference in Berry's case. I realize there were complicating factors there. The early allergic manifestations of schistosomiasis were confused with the even earlier penicillin reaction, and certainly that intestinal parasite was a most unfortunate coincidence. However, since the doctors concerned were fully aware from the very beginning that Berry had just returned from a rugged trip through some of the world's least hygienic countries, one can hardly say that his case was handled with much sophistication. Or so it seems to me after talking and working with Berry.

"I saw Berry for the first time about a month after his discharge from Memorial Hospital. Sooner would have been better, but that couldn't be helped. I wanted to be sure just what I was treating before I started treatment. The treatment of schistosomiasis depends to some extent on the type. Or types. I knew that Berry had made a trip around the world, so it was possible that he had more than one type. He might even have all three. First, I wanted to see the slides from which the hospital diagnosis had been made, and there was some delay in getting them down to my office. Then I wanted to examine some stool and urine samples for the microscopic eggs that would identify the type, or types, and that meant another delay. We finally got started on January 29. I had the results of the sample tests by then, and they held no ugly surprises. They simply confirmed the original bladder-biopsy findings. Berry had schistosomiasis haematobium, and only haematobium. Where he got it is a little hard to say. He visited half a dozen different places in the endemic areas in Egypt and the Middle East, and most of the washing and bathing he did there was in tubs or basins filled by hand with water from some possibly contaminated source. It's my guess, though, that he picked it up in Egypt—at Luxor, when he went swimming in the Nile.

"The treatment of schistosomiasis involves some form of antimony. Potassium antimony tartrate, the most potent tolerable form, is generally used for schistosomiasis japonica. Its administration is intravenous, and is an exceedingly delicate process. A slip can cause real trouble. Moreover, there have been some serious, and even fatal, toxic reactions to the drug. A less powerful form containing sodium antimony and called stibophen is effective in both schistosomiasis haematobium and schistosomiasis mansoni, and it can be given intramuscularly. It isn't as bad as the other, but it can cause its share of trouble. January 29 was a Monday. I gave Berry a stibophen injection of five cc.s that afternoon, and the same amount each afternoon for the next four days. We skipped the weekend to give him a breather, and then did the same the following week. He stood it pretty well. We got through seventeen injections—or a total of eighty-five cc.s of the drug—before he had any important toxic reaction. On February 22, the day of the seventeenth injection, he had a wave of nausea and vomited. Eighty-five cc.s was enough to do the job, so I stopped. He wasn't cured, but he was well on the way. Tests during the course of treatment showed his urine increasingly clear, and by the end of the course it contained no schistosome eggs and only a few microscopic traces of blood. His eosinophilia also showed a healthy change. At the beginning of treatment, he had an eosinophile reading of fifteen per cent. When I withdrew the drug, it was down to ten. A month later, when he came in for a checkup, it was down to four, and in the next few months it continued that encouraging decline. Another cystoscopy was performed on August 22. It showed no evidence of acute inflammation, and in general the look of the bladder was compatible with the other signs of improvement. Seven months later, on March 13, 1963, a final cystoscopy was done, and this time there was no doubt about it. We could safely say he was cured."

[*1964*]

CHAPTER **9**

The Orange Man

AROUND ELEVEN-THIRTY on the morning of December 15, 1960, Dr. Richard L. Wooten, an internist and an assistant professor of internal medicine at the University of Tennessee College of Medicine, in Memphis, was informed by the receptionist in the office he shares with several associates that a patient named (I'll say) Elmo Turner was waiting to see him. Dr. Wooten remembered Turner, but not much about him. He asked the receptionist to fetch him Turner's folder, and then, when she had done so, to send Turner right on in. The folder refreshed his memory. Turner was fifty-three years old, married, and a plumber by trade, and over the past ten years Dr. Wooten had seen him through an attack of pneumonia and referred him along for treatment of a variety of troubles, including a fractured wrist and a hip-joint condition. There were footsteps in the hall. Dr. Wooten closed the folder. The door opened, and Turner—a short, thick, muscular man— came in. Dr. Wooten had risen to greet him, but for a moment he could only stand and stare. Turner's face was orange—a golden, pumpkin orange. So were both his hands.

Dr. Wooten found his voice. He gave Turner a friendly good morning, asked him to sit down, and remarked that it had been a couple of years since their last meeting. Turner agreed that it had. He had been away. He had been working up in Alaska—in Fairbanks. He and his wife were back in Memphis only on a matter of family business. But, being in town, he thought he ought to pay Dr. Wooten a visit. There was something that kind of bothered him. Dr. Wooten listened with half an ear. His mind was searching through the spectrum of pathological skin discolorations. There were many diseases with pigmentary manifestations. There was the paper pallor of pituitary disease. There was the cyanotic blue of congenital heart disease. There was the deep Florida tan of thyroid dysfunction. There was the jaundice yellow of liver damage. There was the bronze of hemochromatosis. As far as he knew, however, there was no disease that colored its victims orange. Turner's voice recalled him. In fact, he was saying, he was worried. Dr. Wooten nodded. Just what seemed to be the trouble? Turner touched his abdomen with a bright-orange hand. He had a pain down there. His abdomen had been sore off and on for over a year, but now it was more than sore. It hurt. Dr. Wooten gave an encouraging grunt, and waited. He waited for Turner to say something about his extraordinary color. But Turner had finished. He had said all he had to say. Apparently, it was only his abdomen that worried him.

Dr. Wooten stood up. He asked Turner to come along down the hall to the examining room. His color, however bizarre, could wait. A chronic abdominal pain came first. And not only that. The cause of Turner's pain was probably also the cause of his color. That seemed, at least, a reasonable assumption. They entered the examining room. Dr. Wooten switched on the light above the examination table and turned and looked at Turner. The light in his office had been an ordinary electric light, and ordinary electric light has a faintly yellow tinge. The examining room had a true-color daylight light. But Turner's color owed nothing to tricks of light. His skin was still an unearthly golden orange. Turner stripped to the waist and got up on the table and stretched out on his back. His torso was as orange as his face. Dr. Wooten began his examination. He found the painful abdominal area, and carefully pressed. There was something there. He could feel an abnormality—a deep-seated mass about the size of an apple. It was below and behind the stomach, and he thought it might be sited at the liver. He pressed again. It wasn't the liver. It was positioned too near the center of the stomach for that. It was the pancreas.

Dr. Wooten moved away from the table. He had learned all he could from manual exploration. He waited for Turner to dress, and then led the way back to his office. He told Turner what he had found. He said he couldn't identify the mass he had felt, and he wouldn't attempt to guess. Its nature could be determined only by a series of X-ray examinations. That, he was sorry to say, would require a couple of days in the hospital. The pancreas was seated too deep to be accessible to direct X-ray examination, and an indirect examination took time and special preparation. Turner listened, and shrugged. He was willing to do whatever had to be done. Dr. Wooten swiveled around in his chair and picked up the telephone. He put in a call to the admitting office of Baptist Memorial Hospital, an affiliate of the medical school, and had a few words with the reservations clerk. He swiveled back to Turner. It was all arranged. Turner would be expected at Baptist Memorial at three o'clock that afternoon. Turner nodded, and got up to go. Dr. Wooten waved him back into his chair. There was one more thing. It was about the color of his skin. How long had it been like that? Turner looked blank. Color? What color? What was wrong with the color of his skin? Dr. Wooten hesitated. He was startled. There was no mistaking Turner's reaction. He was genuinely confused. He didn't know about his color—he really didn't know. And that was an interesting thought. It was, in fact, instructive. It clearly meant that Turner's change of color was not a sudden development. It had come on slowly, insidiously, imperceptibly. He realized that Turner was waiting, that his question had to be answered. Dr. Wooten answered it. Turner looked even blanker. He gazed at his hands, and then at Dr. Wooten. He didn't see anything unusual about his color. His skin was naturally ruddy. It always looked this way.

Dr. Wooten let it go at that. There was no point in pressing the matter any further right then. It would only worry Turner, and he was worried enough already. The matter would keep until the afternoon, until the next day, until he had a little more information to work with. He leaned back and lighted a cigarette, and changed the subject. Or seemed to. Had Turner ever met the senior associate here? That was Dr. Hughes—Dr. John D. Hughes. No? Well, in that case . . . Dr. Wooten reached for the telephone. Dr. Hughes's office was just next door, and he arrived a moment later. He walked into the room and glanced at Turner, and stopped—and stared. Dr. Wooten introduced them. He described the reason for Turner's visit and the mass he had found in the region of the pancreas. Dr. Hughes subdued his stare to a look of polite attention. They talked for several minutes. When Turner got up again to go, Dr. Wooten saw him to the door. He came back to his desk and sat down. Well, what did Dr. Hughes make of that? Had he ever seen or read or even heard of a man that color before? Dr. Hughes said no. And he didn't know what to think. He was completely flabbergasted. He was rather uneasy, too. That, Dr. Wooten said, made two of them.

Turner was admitted to Baptist Memorial Hospital for observation that afternoon at a few minutes after three. He was given the usual admission examination and assigned a bed in a ward. An hour or two later, Dr. Wooten, in the course of his regular hospital rounds, stopped by Turner's bed for the ritual visit of welcome and reassurance. Turner appeared to be no more than reasonably nervous, and Dr. Wooten found that satisfactory. He then turned his attention to Turner's chart and the results of the admission examination. They were, as expected, unrevealing. Turner's temperature was normal. So were his pulse rate (seventy-eight beats a minute), his respiration rate (sixteen respirations a minute), and his blood pressure (a hundred and ten systolic, eighty diastolic). The results of the urinalysis and of an electrocardiographic examination were also normal. Before resuming his rounds, Dr. Wooten satisfied himself that the really important examinations had been scheduled. These were comprehensive X-ray studies of the chest, upper gastrointestinal tract, and colon. The first two examinations were down for the following morning.

They were made at about eight o'clock. When Dr. Wooten reached the hospital on a midmorning tour, the radiologist's report was in and waiting. It more than confirmed Dr. Wooten's impression of the location of the mass. It defined its nature as well. The report read, "Lung fields are clear. Heart is normal. Barium readily traversed the esophagus and entered the stomach. In certain positions, supine projections, an apparent defect was seen on the stomach. However, this was extrinsic to the stomach. It may well represent a pseudocyst of the pancreas. No lesions of the stomach itself were demonstrated. Duodenal bulb and loop appeared normal. Stomach was emptying in a satisfactory manner." Dr. Wooten put down the report with a shiver of relief. A pancreatic cyst—even a pseudocyst—is not a trifling affliction, but he welcomed that diagnosis. The mass on Turner's pancreas just might have been a tumor. It hadn't been a likely possibility—the mass was too large and the symptoms were too mild—but it had been a possibility.

Dr. Wooten went up to Turner's ward. He told Turner what the X-ray examination had shown and what the findings meant. A cyst was a sac retaining a liquid normally excreted by the body. A pseudocyst was an empty sac—a mere dilation of space. The only known treatment of a pancreatic cyst was surgical, and surgery involving the pancreas was difficult and dangerous. Surgery was difficult because of the remote location of the pancreas, and dangerous because of the delicacy of the organs surrounding the pancreas (the stomach, the spleen, the duodenum) and the delicacy of the functions of the pancreas (the production of enzymes essential to digestion and the secretion of insulin). Fortunately, however, treatment was seldom necessary. Most cysts—particularly pseudocysts—had a way of disappearing as mysteriously as they had come. It was his belief that this was such a cyst. In that case, there was nothing much to do but be patient. And careful. Turner was to guard his belly from sudden bumps or strains. A blow or a wrench could cause a lot of trouble.

Nevertheless, Dr. Wooten went on, he wanted Turner to remain in the hospital for at least another day. There was a final X-ray of the colon to be made, and several other tests. In view of this morning's findings, the examination was, he admitted, very largely a matter of form. The cause of Turner's abdominal pain was definitely a pseudocyst of the pancreas. But prudence required an X-ray, and it would probably be done the next day. It was usual, for technical reasons, to let a day elapse between an upper-gastrointestinal study and a colon examination. Two of the other tests were indicated by the X-ray findings. One was a test for diabetes —the glucose-tolerance test. Diabetes was a possible complication of a cyst of the pancreas. Pressure from the cyst could produce diabetes by disrupting the production of insulin in the pancreatic islets of Langerhans. Such pressure could also cause another complication—a blockage of the common bile duct. The diagnostic test for that was a chemical analysis of the blood serum for the presence of the bile pigment known as bilirubin. Dr. Wooten paused. The time had come to reopen the subject that he had tactfully dropped the day before. He reopened it. It was possible, he said, that the bilirubin test might help explain the unusual color of Turner's skin. And Turner's skin *was* a most unusual color. He held up an adamant hand. No. Turner was mistaken. His color *had* changed in the past year or two. It wasn't a natural ruddiness. It was a highly unnatural orange. It was a sign that something was wrong, and he intended to find out what. That was the reason for a third test he had ordered. It was a diagnostic blood test for a condition called hemochromatosis. Hemochromatosis was a disturbance of iron metabolism that deposited iron in the skin and stained it the color of bronze. To be frank, he didn't hope for much from either of the pigmentation tests. Turner's color wasn't the bronze of hemochromatosis, and it wasn't the yellow of jaundice. The possibility of jaundice was particularly remote. The whites of Turner's eyes were still white, and that was usually where jaundice made its first appearance. But he had to carry out the tests. He had to be sure. The process of elimination was always an instructive process. And they didn't have long to wait. The results of the tests would be ready sometime that afternoon. He would be back to see Turner then.

Dr. Wooten spent the next few hours at the hospital and his office. He had other patients to see, other problems to consider, other decisions to make. But Turner remained on his mind. His first impression, like so many first impressions, had been mistaken. It now seemed practically certain that Turner's color had no connection with Turner's pancreatic cyst. They were two quite different complaints. And that returned him to the question he had asked himself when Turner walked into his office. What did an orange skin signify? What disease had the power to turn its victims orange? The answer, as before, was none. But perhaps this wasn't in the usual sense a disease. Perhaps it was a drug-induced reaction. Many chemicals in common therapeutic and diagnostic use were capable of producing conspicuous skin discolorations. Or it might be related to diet.

The question hung in Dr. Wooten's mind all day. It was still hanging there when he headed back to Turner's ward. On the way, he picked up the results of the tests he had ordered that morning, and they did nothing to resolve it. Turner's total bilirubin level was 0.9 milligrams per hundred milliliters, or normal. The total iron-binding capacity was also normal—286 micrograms per hundred milliliters. And he didn't have diabetes. When Dr. Wooten came into the ward, he found Turner's wife at his bedside and Turner in a somewhat altered state of mind. He said he had begun to think that maybe Dr. Wooten was right about the color of his skin. There must be something peculiar about it. There had been a parade of doctors and nurses past his bed ever since early morning. Mrs. Turner looked bewildered. She hadn't noticed anything unusual about her husband's color. She hadn't thought about it —the question had never come up. But now that it had, she had to admit that he did look kind of different. He did look kind of orange. But what was the reason? What in the world could cause a thing like that? Dr. Wooten said he didn't know. The most he could say at the moment was that certain possibilities had been eliminated. He summarized the results of the three diagnostic tests. Another possible cause, he then went on to say, was drugs. Medicinal drugs. Certain medicines incorporated dyes or chemicals with pigmentary properties. Turner shook his head. Maybe so, he said, but that was out. It had been months since he had taken any kind of drug except aspirin.

Dr. Wooten was glad to believe him. Drugs had been a rather farfetched possibility. The color changes they produced were generally dramatically sudden and almost never lasting. He turned to another area—to diet. What did Turner like to eat? What, for example, did he usually have for breakfast? That was no problem, Turner said. His breakfast was almost always the same—orange juice, bacon and eggs, toast, coffee. And what about lunch? Well, that didn't change much, either. He ate a lot of vegetables— carrots, rutabagas, squash, beans, spinach, turnips, things like that. Mrs. Turner laughed. That, she said, was putting it mildly. He ate carrots the way some people eat candy. Dr. Wooten sat erect. Carrots, he was abruptly aware, were rich in carotene. So were eggs, oranges, rutabagas, squash, beans, spinach, and turnips. And carotene was a powerful yellow pigment. What, he asked Mrs. Turner, did she mean about the way her husband ate carrots? Mrs. Turner laughed again. She meant just what she said. Elmo was always eating carrots. Eating carrots and drinking tomato juice. Tomato juice was his favorite drink. And carrots were his favorite snack. He ate raw carrots all day long. He ate four or five of them a day. Why, driving down home from Alaska last week, he kept her busy just scraping and slicing and feeding him carrots. Turner gave an embarrassed grin. His wife was right. He reckoned he did eat a lot of carrots. But he had his reasons. You needed extra vitamins when you lived in Alaska. You had to make up for the long, dark winters—the lack of sunlight up there. Dr. Wooten stood up to go. What the Turners had told him was extremely interesting. He was sure, he said, that Turner's appetite for carrots was a clue to the cause of his color. It was also, as it happened, misguided. The so-called "sunshine vitamin" was Vitamin D. The vitamin with which carrots and other yellow vegetables were abundantly endowed was Vitamin A.

There was a telephone just down the hall from Turner's ward. Dr. Wooten stopped and put in a call to the hospital laboratory. He arranged with the technician who took the call for a sample of Turner's blood to be tested for an abnormal concentration of carotene. Then he left the hospital and cut across the campus to the Mooney Memorial Library. He asked the librarian to let him see what she could find in the way of clinical literature on carotenemia and any related nutritional skin discolorations. He was elated by what he had learned from the Turners, but he knew that it wasn't enough. He had seen several cases of carotenemia. An excessive intake of carotene was a not uncommon condition among health-bar habitués and other amateur nutritionists. But carotene didn't color people orange. It colored them yellow. Or such had been his experience.

The librarian reported that papers on carotenemia were scarce. She had, however, found three clinical studies that looked as though they might be useful. Here was one of them. She handed Dr. Wooten a bound volume of the *Journal of the American Medical Association* for 1919, and indicated the relevant article. It was a report by two New York City investigators—Alfred F. Hess and Victor C. Myers—entitled "Carotinemia: A New Clinical Picture." Dr. Wooten knew their report, at least by reputation. It was the original study in the field. The opening descriptive paragraphs refreshed his memory and confirmed his judgment. They read:

About a year ago one of us (A.F.H.) observed that two children in a ward containing about twenty-five infants, from a year to a year and a half in age, were developing a yellowish complexion. This coloration was not confined to the face, but involved, to a less extent, the entire body, being most evident on the palms of the hands. . . . For a time, we were at a loss to account for this peculiar phenomenon, when our attention was directed to the fact that these two children, and only these two, were receiving a daily ration of carrots in addition to their milk and cereal. For some time we had been testing the food value of dehydrated vegetables, and when the change in color was noted, had given these babies the equivalent of 2 tablespoonfuls of fresh carrots for a period of six weeks.

It seemed as if this mild jaundiced hue might well be the result of the introduction into the body of a pigment rather than the manifestation of a pathologic condition. Attention was accordingly directed to the carrots, and the same amount of this vegetable was added to the dietary of two other children of about the same age. In the one instance, after an interval of about five weeks, a yellowish tinge of the skin was noted, and about two weeks later the other baby had become somewhat yellow. There was a decided difference in the intensity of color of the four infants, indicating probably that the alteration was in part governed by individual idiosyncrasy. On omission of the carrots from the dietary, the skin gradually lost its yellow color, and in the course of some weeks regained its normal tint.

The librarian returned to Dr. Wooten's table with the other references. Both were contributions to the *New England Journal of Medicine.* One was entitled "Skin Changes of Nutritional Origin," and had been written by Harold Jeghers, an associate professor of medicine at the Boston University School of Medicine, in 1943. The other was the work of three faculty members of the Harvard Medical School—Peter Reich, Harry Shwachman, and John M. Craig—and was entitled "Lycopenemia: A Variant of Carotenemia." It had appeared in 1960. Dr. Wooten looked first at "Skin Changes of Nutritional Origin." It was a comprehensive survey, and it read, in part:

The carotenoid group of pigments color the serum and fix themselves to the fat of the dermis and subcutaneous tissues, to which they impart the yellow tint. . . . Edwards and Duntley showed by means of spectrophotometric analysis of skin color in human beings that carotene is present in every normal skin and is one of the five basic pigments that determine the skin color of every living person. Clinically, therefore, carotenemia refers to the presence of an excess over normal of carotene in the skin and serum. . . . In most cases carotenemia results simply from excess use of foods rich in the carotenoid pigments. Individuals probably vary in the ease with which carotenemia develops, which is evidenced by the fact that many vegetarians do not develop it. It is said to develop more readily in those who sweat profusely. Except for the yellow color produced, it appears to be harmless, even though present for months. It eventually disappears over several weeks to months when the carotene consumption is reduced.

Dr. Wooten moved on to the third report. ("This investigation concerns a middle-aged woman whose prolonged and excessive consumption of tomato juice led to the discoloration of her skin.") He read it slowly through from beginning to end, and then turned back and reread certain passages:

> Although carotenemia due to the ingestion of foods containing a high concentration of beta carotene is a commonly described disorder, a similar condition secondary to the ingestion of tomatoes and associated with high serum levels of lycopene has not previously been reported. . . . Lycopene is a common carotenoid pigment widely distributed through nature. It is most familiar as the red pigment of tomatoes, but has been detected in many animals and vegetables. . . . It is also frequently found in human serum and liver, especially when tomatoes are eaten. But lycopene is not well known medically because, unlike beta carotene, it is physiologically inert and has not been involved in any form of illness.

Dr. Wooten closed the volume. Turner was not only a heavy eater of carrots. He was also a heavy drinker of tomato juice. Carrots are rich in carotene and tomatoes are rich in lycopene. Carotene is a yellow pigment and lycopene is red. And yellow and red make orange.

Dr. Wooten completed his record of the case with a double diagnosis: pseudocyst of the pancreas and carotenemia-lycopenemia. The results of the X-ray examination of Turner's colon were normal ("Terminal ileum was visualized. No pathology was demonstrated in the colon"), and the carotene test showed a high concentration of serum carotenoids (495 micrometers per hundred milliliters, compared to a normal concentration of 50 to 350 micrometers per hundred milliliters). The diagnosis of lycopenemia was made from the clinical evidence. Turner was discharged from the hospital on December 17. His instructions were to avoid abdominal blows, carrots (and other yellow vegetables), and tomatoes in any form. Four months later, on April 16, 1961, he reported to Dr. Wooten that his skin had recovered its normal ruddiness. Two years later, in 1963, he returned again to Memphis and dropped in on Dr. Wooten for a visit. His abdominal symptoms had long since disappeared, and a comprehensive examination showed no sign of the pseudocyst.

Elmo Turner was the first recorded victim of the condition known as carotenemia-lycopenemia. He is not, however, the only one now on record. Another victim turned up in 1964. She was a woman of thirty-five, a resident of Memphis, and a patient of Dr. Wooten's. He had been treating her for a mild diabetes since 1962, and had put her at that time on the eighteen-hundred-calorie diet recommended by the American Diabetes Association. She had faithfully followed the diet, but in order to do so had eaten heavily of low-calorie vegetables, and (as she confirmed, with some surprise, when questioned) the vegetables she ate most heavily were carrots and tomatoes. She ate at least two cups of carrots and at least two whole tomatoes every day. Dr. Wooten was unaware of this until she walked into his office one October day in 1964 for her semi-annual consultation. He greeted her as calmly as he could, and asked her to sit down. He would be back in just a moment. He stepped along the hall to Dr. Hughes's office and looked in. Dr. Hughes was alone.

"Have you got a minute, John?" Dr. Wooten said.

"Sure," Dr. Hughes said. "What is it?"

"I'd like you to come into my office," Dr. Wooten said. "I'd like to show you something."

[*1967*]

CHAPTER 10

The Dead Mosquitoes

————————◆————————

DR. JOHN P. CONRAD, JR., a senior associate in a suburban pediatric group practice in Fresno, California, excused himself to the mother of the young patient in his consultation room and crossed the hall to take a telephone call in his office. The call was a request from a general practitioner on the other side of town named Robert Lanford to refer a patient to Dr. Conrad for immediate hospitalization and treatment. That morning—it was now around four o'clock in the afternoon (on October 4, 1961)—an eight-year-old boy whom I'll call Billy Cordoba had been brought to Dr. Lanford's office by his mother. Billy had been sent home sick from school. He was pale, his eyes had a glassy look, and his heart was a little fast. Dr. Lanford had examined him, found nothing significantly out of order, and sent him home to rest. But Billy was now back in his office, and there was no longer any doubt that he was sick. The manifestations of his illness now included a ghastlier pallor, a glassier look, a notably faster heart, rapid and irregular breathing, muscle twitches, diarrhea, nausea, vomiting, and abdominal pain. He was also confused in mind and almost comatose.

Something about this inharmonious symphony of symptoms had prompted Dr. Lanford to make a urine-sugar test, and the results were strongly positive. That suggested a frightening possibility. He was afraid that Billy was a hitherto unsuspected diabetic on the brink of diabetic coma. In any event, he said, the boy was in urgent need of sophisticated help. Dr. Conrad agreed. He told Dr. Lanford that he shared his sense of urgency, and that he would arrange at once for Billy's admittance to Valley Children's Hospital.

Mrs. Cordoba drove her son to the hospital. Billy was admitted there at five o'clock. He was put to bed, and a sample of blood was taken for immediate laboratory analysis to confirm or deny the presence of diabetes. That had been ordered by Dr. Conrad when he made the admittance arrangements. When he himself reached the hospital, at a little before six, the results of the blood studies had been noted on Billy's chart. Dr. Conrad read them with a momentary lift of spirit. The relevant values (blood glucose, blood carbon dioxide, blood sodium, blood potassium, blood pH) were close enough to normal to make it comfortably certain that despite the earlier positive urinalysis the boy was not a diabetic. But that was all. Or practically all; the studies did show a morbid elevation in the white-blood-cell count. Other than that, the studies had no positive diagnostic significance. Dr. Conrad replaced the chart and went into Billy's room to take his first look at his patient. It was anything but reassuring. The boy was clearly sicker than he had been two hours before. Dr. Conrad sat down and began with care the standard physical examination. His findings were even more discordant than those recorded by Dr. Lanford. Billy's pulse was fast, his breathing was fast, his temperature was 100 degrees, his skin was pale and clammy, the pupils of his glassy eyes had shrunk to pinpoints, his face and arms were twitching, he was drooling saliva, and he appeared to be in almost constant abdominal pain. Twice during the short examination the pain was so great that he screamed. He was still confused, still comatose, still nauseated, still diarrheic. Dr. Conrad finished the examination and sorted out his impressions. They led in two distinctly different directions. One possibility was shigellosis, or bacillary dysentery. The other was chemical poisoning.

"I didn't particularly favor the idea of shigellosis," Dr. Conrad says. "It was simply suggested by some of the clinical evidence—the high white-cell count and the gastrointestinal symptoms. And I didn't favor it at all for very long. A shigella infection produces a rather distinctive kind of damage that can be detected by microscopic examination of a stool specimen. It isn't conclusive, but it's reliable enough to be useful. Well, I asked the laboratory for a report and the answer came back in a matter of minutes. Negative. I wasn't much surprised. Chemical poisoning had always been by far the stronger possibility. The very bizarreness of the symptoms was suggestive of poison. Certain particular symptoms were even more suggestive. Stupor. Abdominal pain. Salivation. But the real tipoff was those pinpoint pupils. What I had in mind was an insecticide—specifically, one containing an organic phosphate. That isn't as inspired as it may sound. Fresno County is a big agricultural county. It produces everything from cantaloupes to cotton, and it uses tons of highly toxic chemicals. Including organic phosphates. Then Mrs. Cordoba said something that seemed to make my hunch a certainty. I was asking her the usual questions for Billy's personal history, and she remembered a remark that Billy made when he came home sick from school. The Cordobas live on the edge of town, and there are cultivated fields all around the stop where Billy waits for the bus. That morning, Billy said, there was a spray rig working in one of the fields and a spray plane flying back and forth overhead. Organic phosphates can enter the body in various ways, but the commonest route is absorption through the skin. Also, they work very fast. Symptoms can begin within a couple of hours of exposure. And it doesn't take much of the stuff to cause a lot of trouble. The fatal skin dose is only about five drops.

"I was practically certain that Billy had been poisoned by some organic-phosphate insecticide. I was sure enough to start treatment on that assumption. I followed the standard procedure. I ordered intravenous fluids to restore the loss of body fluids through sweating, salivation, and diarrhea, and a regimen of atropine—one milligram injected intramuscularly every two hours. Atropine is a lifesaving drug in organic-phosphate poisoning, because it relieves the threatening symptoms. It doesn't, however, get at the source. It doesn't eliminate the poison. The next step in the treatment involves a drug called PAM—pralidoxime chloride. But I couldn't take that step—not until I was absolutely certain. PAM is a little too specific to prescribe on mere suspicion. The definitive test for organic-phosphate poisoning is a blood test that measures the levels in the plasma and the red cells of an enzyme called cholinesterase. Cholinesterase is a kind of neural moderator. Its presence controls the accumulation of an ester that governs the transmission of impulses of the parasympathetic nervous system. Organic phosphates destroy cholinesterase, and the destruction of cholinesterase allows an excessive accumulation of the ester. The result is a powerful overstimulation of the parasympathetic nerves. The cholinesterase test is too elaborate for the average small hospital laboratory. The only laboratory equipped to do that kind of thing here is in the Poison Control Center at Fresno Community Hospital, down in the center of town. I drew a sample of blood and rounded up a messenger and got on the telephone to Dr. Bocian—Dr. J. J. Bocian, the director there. That was around seven o'clock. Dr. Bocian called me back around eight-thirty. He had the results of the test. Billy's plasma cholinesterase level was only forty percent of normal, and his red-cell level was a scant seventeen. His illness was definitely organic-phosphate poisoning.

"It was gratifying to know that I'd made a good guess. And that I'd been able to make it in time. But the really gratifying thing was Billy's response to atropine. By the time I had Dr. Bocian's definite diagnosis, Billy was just as definitely out of danger. His vital signs were all good. Moreover, he was beginning to look more alert. His pupils were coming back to normal size. And he wasn't salivating the way he had been. I was so satisfied that I decided to hold off on PAM. Atropine would continue to counteract the potentially dangerous neuromuscular symptoms, and time would do the rest. It would gradually bring the cholinesterase levels back to normal. I stayed at the hospital until about ten o'clock, and went home feeling pretty good. I had diagnosed the nature of Billy's illness, and he was responding well to treatment. And I thought I knew just how his illness had come about.

"But I was wrong about that. It wasn't the spray rig or the spray plane at the bus stop. It couldn't have been either of them. Mrs. Cordoba or her husband or somebody made some inquiries. Those rigs weren't spraying an organic phosphate. Or any kind of insecticide. The fields they were working were cotton fields, and they were spraying a defoliant to strip the plants for mechanical picking. But I wasn't mistaken about Billy. He continued to do just fine. I kept him on atropine and intravenous fluids for a total of forty-eight hours. His symptoms all subsided and his serum cholinesterase levels began to improve. At the end of the second hospital day, he showed a plasma level of forty-two per cent of normal and a red-cell level of almost thirty-two. By the sixth day, the plasma level had risen to ninety-two per cent of normal. The red-cell concentration is always slower to recover. It requires the formation of new cells. But it was up to forty per cent. There was no reason to keep him in the hospital any longer. I could follow him the rest of the way as an out-patient. So I ordered his discharge."

Billy was discharged from Valley Children's Hospital to convalesce at home on October 9. That was a Monday. He remained at home, sleeping and eating and resting, until the following Monday, October 16. That afternoon, by prearrangement, Mrs. Cordoba drove him back across town to Dr. Conrad's office for what was expected to be a final physical examination and dismissal. Their appointment was for four o'clock, and they were on time.

"Billy looked fine," Dr. Conrad says. "And he was fine. Blood count, blood pressure, chest, pupils—everything was completely normal. So that was the happy ending of that. I walked Billy and his mother out to the waiting room and said goodbye and went back to my office and closed the case and rang for my next patient. I saw that patient and then the next, and then the receptionist called. She sounded almost frightened. Mrs. Cordoba was in the waiting room and she was practically hysterical. Billy was sick again. He was out in the car—too sick to even walk.

"It was true. I found Mrs. Cordoba and we went out to the car, and there he was, and he looked terrible. He looked shocky. His skin was cold and clammy with sweat, and he was salivating and breathing very fast, and he didn't seem to be able to move his legs. I didn't even go back in the building to say I was leaving. I just slid in beside Billy and told Mrs. Cordoba to head for the hospital. The hospital was only a block up the street, but on the way she told me what had happened. There wasn't much to tell. They had started home from my office, and they were almost there when all of a sudden Billy said he was sick. That was all she knew. She had turned around and driven right back to see me. But it was perfectly plain that this was the same thing all over again. Only worse —much worse. Dr. Bocian confirmed it later on in the evening. The serum cholinesterase levels were very low. The plasma level was down to twenty-seven per cent of normal, and the red-cell level was only twenty. I got Billy started on atropine and intravenous fluids, but he didn't respond as he had before. Two hours after I got him into the hospital, he was seized with severe abdominal cramps and began to vomit. Then he developed diarrhea. It was time for PAM. I ordered an intravenous injection of five hundred milligrams. The next three hours were a little anxious, but then he began to improve. And the next morning he was very much better. He had had another five hundred milligrams of PAM, and his cholinesterase levels were up enough to show that he was improving.

"That gave me a chance to think. Organic-phosphate poisoning is not a notifiable disease in California, so there had been no reason for me to report Billy's case to the Fresno County Public Health Department, but now I thought perhaps I should. I thought I had a lead that they might want to follow up. The lead was this: For a week at home, Billy had been as good as well. Then he got up and drove over here to my office, and less than an hour later he was critically ill again with organic-phosphate poisoning. I'm not an epidemiologist, but it seemed to me that the probable source of his exposure wasn't far to look for. It almost had to be either something in the family car or something he was wearing. When I got to my office on Wednesday morning, I called the Health Department and talked to Mary Hayes. Dr. Hayes has since left the Department but she was then the acting health officer, and she was very interested in my story. She said she would have somebody look into it. She called me back on Friday afternoon. They had the answer—or part of it, anyway. The source was Billy's clothes—his blue jeans. They were brand new blue jeans that his mother had bought at a salvage store, and he had worn them only twice. He had worn them to school on the morning of October 4 and to my office on the afternoon of October 16. The Department had had the jeans tested and had found them contaminated with some form of organic phosphate.

"By that time, of course, Billy was recovering very nicely, and I could relax and begin to think about him as a case. It fascinated me. I'd never had a more dramatic experience in all my years of practice. Well, I'm on the staff at Fresno General Hospital and I make teaching rounds there on Monday, Wednesday, and Friday mornings, and I was so fascinated by Billy and his poisoned blue jeans that I told the interns and residents about them on my next rounds. That was on Monday—Monday, October 23. The next day, I got a call from one of the residents, a doctor named Merritt C. Warren. He had a new patient on his service—an eight-year-old boy. We can call him Johnny Morales. Johnny had become sick at school that morning and had been admitted to the hospital by his family physician around noon. His initial symptoms were sweating, dizziness, and vomiting. He reached the hospital in a stumbling, mindless stupor. His pulse was fast, his respiration was weak and shallow, his face was contorted by muscular twitches, and the pupils of his eyes were contracted to pinpoints. He also had abdominal cramps. The family physician had tentatively diagnosed Johnny's trouble as acute rheumatic fever. Dr. Warren thought differently. He said he thought it was another case of poisoned pants. That was the way he put it. I thought he was probably

right. And he was. Dr. Bocian confirmed it by a serum cholinesterase test a couple of hours later."

The inquiry by the Fresno County Public Health Department
into the case of Billy Cordoba was conducted by an investigator
in the Division of Environmental Health named R. E. Bergstrom. Mr. Bergstrom, who was then senior sanitarian in the
Division (he is now its director), received the assignment within
an hour of Dr. Conrad's report to Dr. Hayes on the morning of
October 18. He and a colleague named Tiyo Yamaguchi were at
the Cordoba house within an hour.

"We spent the rest of the day out there," Mr. Bergstrom says.
"There and around the neighborhood. Mrs. Cordoba told us about
the spraying operation near the bus stop. We followed that up and
confirmed what she had learned herself. It was a standard cotton-
defoliation spray—magnesium chloride and dinitrose. We went
through the Cordoba house and the garage out back looking for
anything in the way of a garden spray or insect bomb that might
include an organic phosphate. Nothing. We examined the family
car. Nothing. That left Billy's clothes, and Mrs. Cordoba showed
us his blue jeans. She told us about them. They had been bought
new about a month before at the salvage depot of the Valley Motor
Lines. They were cheap, and she bought five pairs. But Billy had
worn only one pair. And he had worn them only twice—to school
that day and then to Dr. Conrad's office. I looked at Yamaguchi
and he looked at me. We knew we had found what we were
looking for. It had only to be proved. We wrapped up the jeans
—all five pairs—for laboratory analysis. The Bureau of Vector
Control of the California State Department of Public Health has
a research station here, and we took the jeans over there the next
morning. The first thing we wanted to know was whether they
were contaminated. The Bureau had a quick and easy test for that.
They breed mosquitoes at the station for experimental purposes,
and they simply tossed the worn pair of jeans in with one of the
colonies. I tell you, it was a sight to see. Those mosquitoes just
curled up and died. It took only fifteen minutes. At the end of that

time, every mosquito in the colony was dead. Not only that. There was another breeding colony about twenty feet away, and in about five more minutes all *those* mosquitoes were dead, too. The poison was that volatile.

"The next thing we wanted to know was the identity of the poison. We thought it was an organic phosphate, but was it? There is a color-reaction test that reveals the presence of phosphate. It takes a little longer than the mosquito test, but the Bureau had the chemistry to do it. We left the jeans with them to work on, and drove back in to town and down to the office of the Valley Motor Lines. It wasn't a very satisfactory visit. About all we learned was that there had been a sale of blue jeans at their salvage depot in September, and that all the jeans had been sold. They supposed the jeans had been damaged, but they didn't know in what way. They didn't know where the jeans had come from. They didn't know the number of jeans in the batch. All company records were stored at their main office, in Montebello, down in Los Angeles County. And, of course, they had no idea who had bought the jeans at the sale. We left them with the understanding that they would recover the relevant records. When we got back to the office, I called our friends at the Bureau of Vector Control. They were a lot more helpful. They had run the color-reaction test, and they had the result. It was positive for phosphate.

"That wasn't any great surprise, of course, but it was crucial. It established that Billy's blue jeans were in fact the source of his phosphate poisoning. All we needed to establish now was the source of the poison. And not just where it came from but also what it was. There are at least twenty-five commercial phosphate pesticides in common use. Like Parathion, for example. And Malathion. And Fenthion and Phosdrin and Diazinon and Dicapthon and Trithion and TEPP. And so on. So it might be easier to find out where it came from if we knew what particular phosphate pesticide we were looking for. Well, that kind of information can be got. It takes a little time, but it's possible by certain tests to identify an unknown phosphate pesticide. The Bureau couldn't do the analysis, but they knew who could—the Division of Chemistry of the California State Department of Agriculture, up in Sacramento. They said they would make the necessary arrangements. We should have a report in a week or ten days. The following day, we looked in at the Valley Motor Lines again. They still hadn't recovered the blue-jeans records. And the day after that it was the same. Apparently, it wasn't easy to get records out of Montebello. And then we heard about Johnny Morales. Dr. Conrad must have telephoned the news to Dr. Hayes. At any rate, we had the simple facts by the morning of October 25. We went over to the hospital —it's just across the street—and talked to Dr. Warren and to Mrs. Morales, and finally to Johnny himself. Johnny was still pitifully sick, but he had been treated in time with atropine and PAM, and he was off the critical list. His story was Billy Cordoba's story all over again. There was a new pair of blue jeans. They came, like Billy's, from the Valley Motor Lines' salvage depot. They carried the J. C. Penney label. So did Billy's. And, as we very soon found out, they were also heavily contaminated with an organic phosphate. Johnny had worn the jeans for the first time on October 20. He wore them to school that day and got sick around midmorning and was sent home. His mother put him to bed, and in a few days he was well. Then he put on his jeans again and went back to school, and ended up at Fresno General Hospital.

"Johnny's new jeans brought the total accounted for up to six. Mrs. Morales had bought only one pair. We still didn't know how many jeans had been sold in the sale, but it was certain that there were more than that. Dr. Hayes got in touch with all the local media. She called in the *Bee* and radio station KMJ and KMJ-TV, and it was all in the paper and on the air that evening, with a warning about the still unaccounted-for jeans and an appeal to the buyers to bring them in to the County Health Department for examination. The response was immediate, and good. As a matter of fact—although we didn't know it for a couple of weeks or more —it was one hundred per cent. We received a total of ten pairs of J. C. Penney jeans, from six different buyers. They represented five families and an institution for children. We checked them out for recent illness and found four cases with much the same clinical picture. Four boys, in four of the five families. They were all recovered now, and they had all been differently diagnosed. Brain tumor was one diagnosis. Another was bulbar polio. One of the others was encephalitis. In retrospect, however, the signs and symptoms were unmistakably those of organic-phosphate poisoning, and when their jeans were tested, that confirmed it. But it was also a little peculiar. Not because they all recovered without specific treatment. That could be explained by light contamination or brief exposure, or both. The peculiar thing was that only those four got sick. What about the fifth family and the institution? They had each bought two pairs of jeans, and the jeans had been worn, but none of the boys who wore them had been even mildly ill. As I say, it seemed a little peculiar—until it turned out that those jeans were not contaminated. And the reason they were not contaminated was that they had been washed. And the reason nobody got sick was that they had been washed before they were worn. Billy and Johnny and the four other boys had worn their jeans the way most kids do. Just as they came from the store."

The transformation of Billy Cordoba's solitary seizure of organic-phosphate poisoning into a looming epidemic also changed the stature of the investigation. It was now imperative that the records of the Fresno blue-jeans sale be recovered from the Montebello office of the Valley Motor Lines, but doing so appeared to be beyond the strength of the Fresno County Public Health Department. Its exhortations did not carry across the state and into Los Angeles County. What was needed was the stronger voice of the California Department of Public Health. Accordingly, on October 26 Dr. Hayes invited that agency to take over the direction of the larger investigation, and her invitation was accepted. It was, however, immediately obvious to the Department of Public Health that in this instance the interrogational powers of a more specialized state agency would be even more compelling. That agency, whose assistance it sought and at once received, was the Public Utilities Commission, which at that time was charged with enforcing motor-carrier safety regulations.

The Public Utilities Commission's investigation was carried out by members of its Operations and Safety Section. They began their inquiry on October 27. Six days later—on Thursday, November 2—they were pleased to receive from the Division of Chemistry of the Department of Agriculture (by way of the Bureau of Vector Control of the Department of Public Health) the ultimate test report on Billy Cordoba's blue jeans. It read, "The stained portion of the jeans contained Phosdrin, 4.8% by weight. The contaminant was specifically identified as Phosdrin by its characteristic infra-red absorption curve. . . ." This was useful information. They now were looking for a particular pesticide. That would make a difference in their progress through the labyrinth of bills of lading, manifests, and invoices. It remained only to link the contaminated J. C. Penney jeans in time and place with a quantity of Phosdrin.

They did so in just two weeks. The chain of circumstances that led to the poisoning of Billy Cordoba and the others had had its innocent beginning some eight months before at the Bayly Manufacturing Company, in the nearby town of Sanger. On February 3, 1961, a shipment of Bayly blue jeans—two large bales and a carton—consigned to a J. C. Penney store in Los Angeles was picked up at the Bayly plant by the Triangle Transfer Company, a Sanger trucking firm, and taken to the Fresno terminal of the Valley Motor Lines for transshipment south. Within an hour or two of its arrival in Fresno, the shipment was loaded aboard a Valley Lines trailer with a conglomeration of other freight. This freight consisted of machinery, machine parts, metal pumps, and a hundred twenty gallons of emulsifiable concentrate of Phosdrin, in one-gallon and five-gallon cans. The Phosdrin was the product of De Pester Western, Inc., a Fresno manufacturer, and was consigned to the Valley Chemical Company, at El Centro, down on the Mexican border.

The Valley Lines trailer left Fresno the following morning with this miscellaneous load, and that evening it reached the company terminal at Montebello, where the Phosdrin was unloaded for transshipment. Two days later, on February 6, it was put on board a truck operated by the Imperial Truck Lines, a Los Angeles firm, for the final leg of its journey. The Imperial driver made the usual precautionary inspection of his load before signing the delivery receipt, and found that one of the Phosdrin cans had sprung a leak. He traced the leak to a little puncture about three inches below the top of a five-gallon can. After some discussion, he signed the delivery receipt, but noted a formal exception to the shipment on the grounds that around a gallon of Phosdrin concentrate had been lost from the punctured can. (How the puncture occurred was never determined, but the loss was estimated in a subsequent claim by the Valley Chemical Company at one and one-eighth gallons, valued at twenty-four dollars and fifty cents.) Meanwhile, the shipment of blue jeans was delivered that same day by the original Valley Lines trailer to a J. C. Penney store in the Los Angeles suburb of Westchester. A shipping clerk there noticed a dark stain on the paper wrapping of one of the bales of jeans. He asked the driver about it, but the driver didn't know. He had never seen it before. The clerk went in and brought out the manager, and the manager told the Valley Lines driver that a damage claim would be filed if any of the jeans turned out to be soiled. Sixteen pairs of jeans were found to be stained with some unknown oily substance, and a claim for damages was filed on February 8. The claim was acknowledged by the Valley Motor Lines, and the sixteen pairs of jeans were stored in the J. C. Penney warehouse for pickup by the Valley Lines. They remained there all spring and all summer—until September 6. Then they were finally picked up and returned to Fresno. On September 19, they were put on cut-rate sale at the company's salvage-depot store. The jeans by then apparently looked all right. They might also by then have been as safe as they looked. It is possible. Seven months of storage in a warehouse subject to swings of heat and cold and damp and dry might well have caused much of the Phosdrin to volatilize and vanish. But the J. C. Penney warehouse was a new and modern warehouse. It was air-conditioned.

The Public Utilities Commission's report of these findings to the State Department of Public Health ended on a reassuring note. It concluded, "The staff's investigation of the personnel records and waybills of the two carriers involved failed to disclose any evidence of employee illnesses on the days in question or subsequent thereto, and failed to disclose any evidence that foodstuffs or other personal effects, including clothing, had been contaminated."

The Commission's report was not, however, the end of its interest in the matter. It at once instituted an investigation into the general operations, safety practices, equipment, and facilities of the Valley Motor Lines and the Imperial Truck Lines, and on February 14, 1962, a public hearing on the results of that investigation was held at Fresno. Both companies were found guilty of carelessness, and admonished and fined. The Valley Lines was fined five thousand dollars—the maximum penalty—and the Imperial Lines was fined twenty-five hundred dollars.

[*1969*]

CHAPTER **11**

Something a Little Unusual

AROUND NOON ON OCTOBER 28, 1963, five people—two men, two women, and a child—sat down to midday dinner in the kitchen of a house they shared on a tobacco farm in the Caney Valley hills of Hawkins County, Tennessee, about fifteen miles southwest of Kingsport. They were (I'll say) Homer Mason and his wife, Louise; the latter's sister and her husband, Grace and Leroy Smart; and the Smarts' son, a boy of three called Buddy. Mason and the two women had spent the morning in the barn stripping and bundling cured tobacco leaves for the market. Smart had had other work to do, and it was he who prepared the meal. It consisted of split-pea soup, spaghetti with meat sauce, sliced tomatoes, sweet milk, and corn bread. Mason was the first to finish eating. He told his wife and Mrs. Smart that he would meet them down at the barn. He was going to stop by the cowshed for a look at an ailing calf. He left the house and crossed the yard, and suddenly began to stagger. He lurched against a tree. The barnyard tilted and the sky reeled. For a moment, Mason clung to the tree, and then it, too, began to sway. He turned and staggered back across the pitching barnyard toward the house. He stopped and blinked and shook his head. Something was happening to his eyes. The house was hardly fifty yards away, but he could only just make it out. He seemed to be going blind.

So did Mrs. Smart. When Mason stumbled into the kitchen, she was sitting there at the table with her head in her hands and moaning. Her husband stood beside her. Mrs. Mason stood on the other side of the table with Buddy in her arms. He was staring at his mother and whimpering. Mason dropped into a chair, and Mrs. Smart raised her head. "Homer, I can't hardly see," she said. "I think I'm going blind."

"Me, too," Mason said.

"And I'm dizzy," she said. "I tried to stand up a few minutes ago and everything was just spinning. I felt like a drunk man."

"She like to fell," Smart said. "I had to catch her."

"Same here," Mason said. "I feel the same way."

"Then it isn't just me," Mrs. Smart said. "I thought it was just me. Because Leroy feels all right, and so do Louise and Buddy."

"You'll be all right, honey," Smart said. "Everything's going to be all right."

"I don't feel as good as I did," Mrs. Mason said, and abruptly sat down. "There's something funny in my head. I feel kind of goofy."

Mrs. Smart gave a little wail. "Oh, my gosh!" she said. "I even feel sick to my stomach! What is it, Homer? What's the matter?"

"I don't know," Mason said. "Maybe we've been poisoned. There's three of us don't feel right—you and me and Louise. And there was only just the three of us working down in the barn this morning." He stopped and licked his lips. They felt stiff and cracked, and his mouth was dry. "Maybe we've got tobacco poisoning," he said. "We've all of us got the nicotine stain on our hands. Some of it maybe come off in our food."

"I never heard of tobacco poisoning," Smart said.

"I have," Mrs. Smart said. "Oh, my gosh, Homer. What are we going to do?"

Mason looked at Smart. He could see him, but his features were out of focus, and there was something moving just above his head. It was like a play of light and shadow—or a cloud of smoke. It became a swarm of bees. They swarmed silently around and around and around. "Leroy," Mason said, "I think you better drive us in to the doctor."

The medical needs of Caney Valley residents are served by a middle-aged general practitioner whom I'll call Francis Craig and a young associate whom I'll call Henry Rathbone. These two physicians operate a clinic at Church Hill, a roadside village about midway between Kingsport and the Mason farm, and it was there that Smart drove his wife and son and the Masons. They arrived at the clinic at two o'clock.

"I was alone that afternoon," Dr. Rathbone says. "Dr. Craig was in Kingsport at the hospital—the Holston Valley Community Hospital—making rounds. Lucy, our nurse, called me on the interoffice phone and said that five people had just come in from Caney Valley. Three of them, she said, were acting very strange. She sounded rather agitated. I was with a patient, but I finished up as quickly as I could. Lucy isn't a flighty girl. I went out to the waiting room—and she was entirely correct. They were acting very strange indeed. The women were twitching and jerking and moaning, and one of the men—Mason, it turned out to be—was waving his arms and talking a wild blue streak of gibberish. I thought for a minute that he was simply scared. But then I looked again. He was peering into space and making grabs at the air, and I realized that scared wasn't it at all. He was hallucinating. The other man, the man named Smart, came up and introduced himself and told me what he could about the matter. I gathered that his wife and Mason had been taken right after eating. Then, a little later, Mrs. Mason took sick. Smart and his little boy, however, were perfectly all right. The symptoms, as I made them out, were vertigo, blurred vision, dry mouth, generalized weakness, nausea, and—in Mason's case, at any rate—hallucinations. Smart added that his wife and the Masons had spent the morning stripping tobacco. He wondered if the tobacco could in some way have poisoned them. I told him no on that. Tobacco poisoning could be only nicotine poisoning, and that couldn't happen from that kind of superficial contact. But they had almost certainly been poisoned. My guess was food poisoning—food intoxication. And, to judge from those clearly central-nervous-system symptoms, it was something pretty serious.

"Smart helped me get his wife and the Masons into the examining room. Mason really needed help. He couldn't even stand alone. We got him stretched out on a cot and as comfortable as possible, and settled the two women in chairs. The blurred vision they all complained of was easy to understand. They all had widely dilated pupils—very glassy-looking. The immediate problem as far as Mason and Mrs. Smart were concerned was nausea. They were sick as dogs, and, to make matters worse, they both had an insatiable thirst. I gave them each an intramuscular injection of trimethobenzamide hydrochloride. That seemed to help the nausea a bit, but it was obvious that they were getting sicker by the minute. Mason was so wild it was hard to keep him from falling off the cot. He kept reaching for imaginary doorknobs, as if he wanted to get out of the room. Sometimes he would be fighting off a swarm of bugs. Then at times he seemed to calm down, and he would point across the room or up at the ceiling and say something about all the beautiful flowers. But most of the time nothing he said made any kind of sense. And Mrs. Smart was almost as bad. She was beginning to hallucinate, too, and raving and thrashing around in her chair. It was unnerving. I really didn't know what to think. Or, rather, there was only one thing I could think of, and that possibility was almost too frightening to contemplate. I mean botulism. Botulism, as you probably know, is the most dangerous of the bacterial food poisonings. It has a mortality rate of about sixty-five per cent. It is also, of course, a pretty rare bird. But it happens. As a matter of fact, it had happened right here in eastern Tennessee—in Knoxville— only a couple of weeks before. You probably read about it in the paper. There were seven cases in the outbreak, and two of them were fatal. So botulism was more or less on my mind. I didn't have far to reach. I thought of it the minute I saw those central-nervous-system symptoms.

"I don't mean to say I was certain. Not at all. There were several points that didn't quite fit a diagnosis of botulism. The onset, for one thing, seemed a little too sudden. And the symptoms were not exactly right. The central-nervous-system symptoms that Mrs. Smart and Mason had were more pronounced—more violent —than the central-nervous-system symptoms that are classically characteristic of botulism. But they were close enough. They were certainly too close to ignore. Botulism can be treated, you know. There's an antitoxin, and if it's given in time it can make all the difference. Dr. Craig agreed with me. I called him at the hospital and gave him the facts and asked him what he thought, and he wasn't for taking any chances, either. He proposed that I call the ambulance service and get them right into the hospital. I was glad to take his advice."

Mason and Mrs. Smart were carried into the emergency room of the Holston Valley Community Hospital in Kingsport at five minutes after four. Mrs. Mason arrived at four-fifteen with Smart and his little boy. They were received by Dr. Craig and (such was his aversion to taking any chances) three hurriedly recruited consultants—a neurosurgeon; the hospital pathologist, Dr. William Harrison; and an internist whom I'll call Richard Johnson. Both Mason and Mrs. Smart were now wildly delirious and almost totally helpless. They were also deeply flushed, dry of mouth, and tormented by an unquenchable thirst, and Mason was shaken by frequent muscle spasms. Mrs. Mason, however, was still only weak, dry-mouthed, and vertiginous. The emergency-room examination was diagnostically uninstructive. Mason's pulse rate was a hundred, or about thirty beats faster than normal for a man, and Mrs. Smart's was eighty-eight, or only slightly faster than normal for a woman, and both had a temperature of ninety-nine degrees. Mrs. Mason had no fever, and her pulse rate was normal. All three had widely dilated pupils that reacted sluggishly to light. The results of the other routine tests

—blood pressure, blood count, urinalysis—were normal in all three cases.

"I think we were all inclined to accept Dr. Rathbone's first impression," Dr. Johnson, the internist, says. "The trouble was obviously some kind of poisoning. What kind was hard to say. It looked like botulism, and yet it didn't. Hallucinations and disorientation very seldom occur in botulism, and when they do, they tend to be rather late-developing symptoms. Still, it wasn't a possibility that any of us were willing to rule out of the picture entirely. Even a hint of botulism is unsettling. We were standing there in the emergency room and feeling very unsettled when Smart got up from where he was sitting with his son and came over. He said his little boy was complaining about his eyes. Everything looked funny, Buddy said. That sounded like what had happened to his wife and the others, so he thought we ought to know. Also, Smart said, he wasn't feeling too good himself. His eyes were all right, but he had a cramping pain in his stomach, and he was beginning to feel a little nauseated. Well, that decided us. Botulism antitoxin isn't something you can get at any drugstore. Or at any hospital, for that matter. There isn't that much demand for it. The nearest possible source we could think of was the Poison Control Center at the University of Tennessee Memorial Hospital, in Knoxville. Robert Lash, the director of the Center, had laid in a supply of antitoxin during the botulism outbreak they had over there earlier in the month, and maybe some of it was left. I went to the phone and gave Dr. Lash a ring, and we were in luck. He still had several hundred thousand units on hand. He said he would get it off at once by a special highway-patrol messenger. It was now about a quarter to five. We should have it by seven o'clock."

It took Dr. Lash about ten minutes to arrange with his dispensary and the Tennessee highway patrol for the dispatch of some five hundred thousand units of polyvalent botulinus antitoxin to the Holston Valley Community Hospital. He then returned to his desk and put in a call to Nashville—to Cecil B. Tucker, director of the Division of Preventable Diseases of the Tennessee State

Department of Public Health, in the Cordell Hull State Office Building there. Botulism is a communicable disease, and consequently a notifiable one. In Tennessee, as in all other states, its appearance (proved or suspected) must be reported to the state health authorities for prompt investigation. When Dr. Tucker came on the line, Dr. Lash gave him the required report.

"I wasn't as startled as I might have been by that call from Dr. Lash," Dr. Tucker says. "Botulism was no great novelty in Tennessee that month, you know. My only thought was something like 'Here we go again.' I thanked him and hung up and put in a call to Kingsport—to Dr. Johnson. I wanted a few more facts before sending an investigator up there. But I had Dan Jones standing by. Dr. Jones is an Epidemic Intelligence Service officer assigned to us by the U.S. Public Health Service through its Communicable Disease Center, in Atlanta. I got Dr. Johnson, and he described the cases. He gave me the clinical picture and what he could of the epidemiology, and I began to have my doubts. It just didn't sound like botulism. But that, of course, was only an opinion. Botulism was still a possibility, so we had to go and see. I told Dr. Johnson that Dr. Jones would be up there in the morning, and started to say goodbye. And Dr. Johnson said, 'Wait a minute.' I waited. Then Dr. Johnson came back on. 'By the way,' he said. 'One of the doctors here has been talking to one of the patients, and he says he just mentioned something about eating Jimson weed.' I don't remember what I said to that. Except that I would call him back.

"Dr. Jones had heard what I heard. He was listening in on an extension. We lit out down the hall and up the stairs to the chemical lab. That's where we keep our file on poisons. I pulled out the card on Jimson-weed poisoning, and no wonder I'd had my doubts. I'll read you what it says under 'Symptoms and Findings': 'Pupils dilated, dry burning sensation of mouth, thirst, difficulty in swallowing, fever, generalized flushing, headache, nausea, excitement, confusion, delirium, rapid pulse and respiration, urinary retention, convulsions.' It was almost word for word the clinical picture that Dr. Johnson had given me on the two patients more seriously stricken. We went back downstairs to my office. I got Dr. Johnson on the phone again and told him what we had found. I said it very much looked to me as though Jimson-weed poisoning was the answer to his problem. Dr. Johnson said he thought I was right. They had done some checking themselves, he said, and they had come to that same conclusion."

Jimson weed (or stinkweed, or thorn apple, or devil's-trumpet) is a big, hardy, cosmopolitan annual of Asian origin. It is known to science as *Datura stramonium* and is a member of the large and generally noxious Solanaceae, or nightshade, family of plants. Other members of this family include tobacco, horse nettle, henbane, belladonna, the petunia, the tomato, and the Irish potato. All these plants, including the tomato and the potato, are at least in some respects pernicious. Jimson weed is entirely so. Its leaves, its seeds, its flowers, and its roots all contain a toxic alkaloid called hyoscyamine. Hyoscyamine is closely related to atropine and is, if anything, more toxic. Jimson weed is distributed throughout most of the United States. It made its first appearance here in the early seventeenth century, possibly as early as 1607. Some authorities think it may have been introduced in ballast and other rubbish discharged from the ships that landed Captain John Smith and his fellow Virginia colonists at Jamestown in that year. In any event, it seems reasonably certain that it entered this country there. Early records indicate that the Powhatan Indians of coastal Virginia called it the "white man's weed." The white colonists of Virginia and elsewhere, on the other hand, called it "Jamestown weed." Robert Beverly refers to it as Jamestown weed in his *History and Present State of Virginia,* of 1705, and gives a recognizable, if somewhat excessive, depiction of its hallucinatory powers:

This being an early Plant, was gather'd very young for a boil'd salad by some of the Soldiers . . . and some of them ate plentifully of it, the Effect of which was a very pleasant Comedy; for they turn'd natural Fools upon it for several Days. One would blow a Feather in the Air; another would dart Straws at it with much Fury; and another stark naked was sitting up in a Corner, like a Monkey grinning and making Mows at them; a Fourth would fondly kiss and paw his Companions, and snear in their Faces, with a Countenance more antik than any in a Dutch Droll. In this frantik Condition they were confined, lest they in their Folly should destroy themselves; though it was observed that all their Actions were full of Innocence and Good Nature. Indeed, they were not very cleanly; for they would have wallow'd in their own Excrements, if they had not been prevented. A Thousand such simple Tricks they play'd, and after Eleven Days, return'd themselves again, not remembering anything that had pass'd.

The other names by which Jimson weed—which is, of course, a corruption of "Jamestown weed"—is sometimes known are more conventionally descriptive. They call attention to one or another of its several notable characteristics. The plant gives off a fetid smell; its fruit, or seed pod, is barbed with thorns, like a chestnut bur; and its poisonous flowers—milky white and sometimes streaked with purple—are trumpet-shaped. Jimson weed is in every sense a weed. Like beggar's-lice and tumbleweed and the cocklebur, it flourishes almost everywhere and is everywhere detested. It sprouts early (as Beverly noted), it grows fast, and it blooms until late in the fall. Its size, for an annual, is considerable. It often reaches a height of six feet, and it averages around four. Like most other successful weeds, Jimson weed can exist on even the poorest land, but its existence there is no more than dogged survival. It does well only in fertile soil, and when it finds itself so placed it feeds voraciously—as voraciously, and as destructively, as corn or cotton. It is also, however, among the easiest of weeds to control. A couple of swipes with a scythe or a hoe before the seeds are formed will clear the most firmly established Jimson-weed jungle. Its presence on cropland or in pasture is thus traditionally taken as a sign of indifferent farming. Mark Twain was aware of its reputation, and in *Tom Sawyer* he turned it to effective atmospheric use: "She [Aunt Polly] went to the open door and stood in it and looked out among the tomato vines and 'jimson' weeds that constituted the garden." Livestock are repelled by its smell, which is so rank that only animals addled by hunger are rash enough to ignore it. Jimson-weed poisoning in livestock is almost entirely limited to the ingestion of hay or ensilage accidentally contaminated with Jimson-weed seeds or leaves, and such cases are relatively few. Man is less instinctively prudent. Most people find the smell of Jimson weed repellent, but it frequently fails to repel them. Jimson-weed poisoning in man, though hardly commonplace, is anything but rare. "During the past five years at the University of Virginia Hospital, which services a large southern rural area, *Datura* has accounted for approximately four percent of pediatric patients admitted because of ingestion of a toxic substance," Joe E. Mitchell and Fred N. Mitchell, both members of the Department of Pediatrics of the University of Virginia School of Medicine, reported to the *Journal of Pediatrics* in 1955. "Although distinctly less frequent than kerosene or salicylate intoxication, *Datura* has had about the same incidence as lead, barbiturates, alcohol, rodenticides, and insecticides as a source of poisoning." Jimson-weed poisoning in children can usually be laid to innocence. The seeds are mistaken for nuts, or are used in play as "pills." Its adult victims are more variously poisoned. Some of them are victims of homespun credulity (folk medicine recommends a tea of Jimson-weed leaves for the relief of asthma, constipation, and certain other ills), and some are victims of a credulous sophistication (the street-corner pharmacopoeia recommends Jimson-weed seeds for a liberating hallucinatory experience). A few of them—including, as it turned out, Homer Mason and his family—are victims of simple ignorance.

The doctor to whom one of the patients in the emergency room of the Holston Valley Community Hospital that October afternoon in 1963 said "something about eating Jimson weed" was William Harrison, the hospital pathologist, and the patient was Leroy Smart. That wasn't, however, exactly what he said.

"I'd been talking to Smart about the meal he had fixed that noon," Dr. Harrison says. "I wanted to know just what had been eaten and just how it had been prepared. It bothered me. I don't mean I had any ideas. It was only that Mason and Mrs. Smart had taken sick almost immediately after eating. That was a little suspicious. Either that or a rather odd coincidence. But the trouble was, of course, that ordinary bacterial food poisoning doesn't act that way. It doesn't come on that fast. It takes hours, and even days. The same is true of most other kinds of food poisoning. About the only poisons that hit in a matter of minutes are chemical poisons, like antimony and sodium fluoride, and the symptoms they produce are nothing like those we had here. So I was simply floundering. But something I said must have struck a chord in Smart. His whole expression changed. Come to think of it, he said, there was something he hadn't thought to mention about that meal—something a little unusual. The tomatoes they'd had weren't ordinary tomatoes. They were grafts. They were grown on a tomato stalk that Mason had grafted onto a Jimson-weed plant. He wondered if that might have had anything to do with the trouble.

"That was when I spoke to Dr. Johnson. I knew he was talking to Nashville, and I thought they ought to know. The way he heard it was a little confused. Or maybe I was a little confusing. I probably was. I mean, I knew without any question that Jimson weed was the answer. It answered all our questions—the central-nervous-system symptoms, the sudden onset after eating, everything. But the whole thing was so fantastic. It was also such a relief. Jimson-weed poisoning can be extremely serious. It can be fatal. Still, almost anything is preferable to botulism. And not only that. It relieved our minds about Smart and his little boy. The late onset of their symptoms suggested a mild exposure. The same was largely true in the case of Mrs. Mason. Her symptoms were somewhat delayed, and they were also relatively mild. In fact, in the end we didn't even admit those three to the hospital. It was different, of course, with Mason and Mrs. Smart. They were really sick. They didn't seem to be in critical condition, but they certainly needed hospitalization. We put them to bed and started them on a course of oral pilocarpine, a nerve stimulant that would serve to counteract the action of the toxin on their vision by stimulating the parasympathetic nerves. But that was about all we could do for them. There is no specific treatment for Jimson-weed poisoning.

"I had another talk with Smart just before he left the hospital. I had no real connection with the case, of course, but I was interested—intensely interested. Smart was with his son and Mrs. Mason. Dr. Craig had given her ten milligrams of pilocarpine for symptomatic relief, and she seemed to be in pretty good shape. I was particularly interested in the nature of the tomato graft. Smart couldn't help me on that. He said I'd have to talk to Mason, since Mason did all the gardening. He was able to tell me the why of it, though. It was really a bright idea. Mistaken, to be sure, but most ingenious. Mason wanted a hardy, frost-resistant tomato— one that would ripen late in the fall. And he knew that Jimson weed was a hardy, frost-resistant plant that flourished until well into November. So he put the two together. He was right, too. The graft was completely successful. As a matter of fact, Smart said, that tomato they ate at dinner was the very first fruit of the experiment. It had only that morning turned ripe enough to pick. It was a good-sized, good-looking tomato, he said, and it had a good flavor. It tasted like any good home-grown, vine-ripened tomato. He had eaten one slice, and so had his wife and Mrs. Mason. It was possible that his wife's slice had been one of the big center slices. His own had been an end slice, and Buddy's had been only a sliver. Mason, on the other hand, had eaten three or four slices.

"I looked in on Mason the following morning. He was still sick, still flat on his back, but he seemed to be perfectly rational. Except, I should say, on the subject of his tomatoes. He knew what had happened. Somebody on the staff had already told him the cause of the family outbreak. But he didn't quite believe it. He had never heard of Jimson-weed poisoning. Jimson weed was a weed like any other weed to him. I didn't argue with him. We just talked, and I finally got him around to telling me about his grafting technique. The Jimson-weed plant he had used was growing in a fence row not far from his house. He made the graft at the first fork of a secondary branch, snipping off the branch and inserting the sharp-ened stem of a tomato plant into the pithy center of the stump. Then he fastened the parts together with a clothespin until the union healed. And that was all there was to it. It was simplicity itself. I was really quite impressed, and I told him so. I asked him how he happened to get the idea. He looked a little surprised. What idea? The idea of grafting tomatoes? Why, that wasn't his idea. He got it from a friend—a fellow over in the next valley. I'll give him the name of Clayton. Clayton had been growing tomatoes on Jimson-weed grafts for years. He was always fooling around with plants.

"Well, that was an interesting piece of news. It was flabbergasting. And it raised some flabbergasting questions. Why hadn't we heard of this before? Or had Clayton never been poisoned? And if he hadn't been poisoned, why not? And so on. It was some little time before we got any answers. By 'we' I mean the State Health Department, the Hawkins County Health Department, and the interested doctors here at the hospital. The first step was one that would have been taken in any case. Somebody from the county went out to Mason's farm and got a sample Jimson-weed tomato and sent it in to Nashville for analysis. Then Clayton was interviewed. He confirmed what Mason had said. He and his family had been growing and eating Jimson-weed tomatoes for years— since 1958, to be exact. No, he never sold any. There were only enough for home use. They ate them raw, they ate them stewed, and they ate them canned. And without any ill effects. The very idea that they might be poisonous astonished him. He, too, had never heard of Jimson-weed poisoning. As it happened, he hadn't yet sampled this year's crop of Jimson-weed tomatoes, but he was glad to give the investigator a couple for analysis.

"We got the report from the laboratory sometime the following week. It made rather curious reading. It raised as many questions as it answered. It fully confirmed the cause of the Mason family's outbreak. They were victims of Jimson-weed poisoning. There are no exact data on the toxicity of hyoscyamine, the principal stramonium alkaloid. It is known, however, that hyoscyamine is somewhat more toxic than atropine, and that as little as two milligrams of atropine will produce such symptoms as rapid pulse, dryness of the mouth, pupil dilation, and blurred vision. Well, the

Mason tomato yielded 4.2 milligrams of stramonium alkaloid per hundred grams of tomato. That worked out to 6.36 milligrams of alkaloid for the whole tomato. In other words, it was very definitely toxic. But the Clayton tomatoes were different. They averaged just 1.9 milligrams per hundred grams of tomato. Or a scant three milligrams for the whole. I couldn't understand it. None of us could. Why should Mason's Jimson-weed tomatoes be twice as toxic as Clayton's? Why should there be any difference at all? That was one question. And how was it that the Claytons could eat their tomatoes with impunity? That was another question. The toxic content of the Clayton tomatoes wasn't very high, but it was far from negligible. There must surely have been times when Clayton and his wife each ate a whole tomato, and three milligrams of hyoscyamine is quite enough to cause trouble.

"I think I can say that we found the answer to one of those questions. I can also say, I think, that I helped to find it although I certainly didn't realize it at the time. I picked up a piece of information, but I didn't know what it meant. It had to do with grafting. A few days after the laboratory report came in, I had another talk with Mason. I drove down to his farm. Both he and Mrs. Smart were home by then and fully recovered, and he took me out and showed me his Jimson-weed tomato plant. It was quite a sight—a tomato plant growing out of a big, bushy Jimson weed. I even took some pictures of it. On the way home, I dropped in on Clayton. I introduced myself and told him of my interest in the case, and we discussed it for a while. Then he showed me *his* Jimson-weed tomato patch. It didn't look much like Mason's. It seemed to be all tomato plants. They were growing out of Jimson-weed stock, but the Jimson-weed branches were practically bare. Only the tomato plants were lush and leafy. I spoke to Clayton about that, and told him how Mason's plants looked. Clayton shrugged. This was the way he did it, he said. He liked to keep the Jimson weed pretty well pruned of leaves in order to concentrate the growth in the tomato plant. He didn't know why Mason didn't do the same.

"I thought that was an interesting point. As I say, I didn't know what it meant. I didn't know if it had any significance at all. But I passed the information on to Dan Jones at the State Health Department in Nashville. He was handling the case at that end, and he was as fascinated, and as puzzled, by it as I was. And that was as far as my contribution went. Dr. Jones took it from there. I understand he read everything he could find on Jimson weed in the hope of making some sense out of the case. Finally, he wrote to an expert on *Datura stramonium*—a professor of botany at Columbia University named Ray F. Dawson. Dr. Jones and I had both been under the impression that the alkaloid in Jimson weed was produced only in the roots of the plant and then distributed to the stem, the leaves, and the fruit. Dr. Dawson straightened us out on that. That was his contribution. The way he explained it to Dr. Jones, Clayton's pruning trick made a considerable difference. Hyoscyamine is synthesized also in the *leaves* of a Jimson-weed plant. So Mason could hardly have grown a more toxic tomato if that had been his aim.

"The other question is still unanswered. We don't know how Clayton could have eaten his tomatoes with impunity, and I doubt if we ever will. There are certain facts that may have some bearing on the matter. Different Jimson-weed plants produce somewhat different amounts of stramonium alkaloid. It depends on where and how they grow. And people differ somewhat in their sensitivity to the alkaloid. But that doesn't take us very far. I'm afraid it doesn't take us anywhere at all."

[*1965*]

CHAPTER **12**

A Man Named Hoffman

AROUND TEN O'CLOCK on the morning of Wednesday, March 4, 1964, a man named Donald Hoffman presented himself for treatment at the Student Health Clinic of Miami University in Oxford, Ohio, some thirty miles northwest of Cincinnati. Hoffman was thirty-six years old, married, and a resident of Cincinnati, but, as he explained to the receptionist, he was currently employed, as an insulation installer, in Oxford—on a remodeling job at McCullough-Hyde Memorial Hospital—and his company had an arrangement with the clinic. He was here, he added, because his foreman had sent him. That was the only reason. His trouble was nothing—an itchy sore on the side of his neck. He had probably picked up a sliver of glass-wool fiber. It had happened several times before. It was a common complaint in his trade.

The doctor who saw him was inclined to agree. There was no good reason not to. Hoffman worked with Fiberglas and his lesion had the look of a Fiberglas lesion. The history of the lesion, the doctor found, was equally suggestive. It had first appeared on Monday evening as a tiny red swelling that might have been

179

caused by a chafing shirt collar. It was larger on Tuesday and somewhat sensitive. This morning it was larger still, and it alternately itched and burned. The doctor slipped a thermometer under Hoffman's tongue, and picked up a scalpel and nicked the edge of the lesion. There was no discharge. He removed and read the thermometer. Hoffman had a temperature of 99.2 degrees. He noted the reading on his record of the case, and added: "Has erythematous swollen area at base of neck anteriorly on left, extending over chest. A firm furuncle is present in the center of this area. Impression: Fiberglass dermatitis with secondary infection." The doctor then turned his attention to treatment. He covered the lesion with a bacitracin dressing, and got out a hypodermic needle. In view of the threat of infection, he said, a course of penicillin was indicated. He proposed to begin with an intramuscular injection of 300,000 units. Hoffman stood up. That wouldn't be necessary, he said. He had had all the treatment he wanted. He didn't believe in taking penicillin every time he had a little scratch. He put on his jacket and left.

Hoffman drove back to the job and resumed his work. He worked until noon and then knocked off and sat down to lunch with one of his friends. He thought he was hungry, but after a couple of bites he changed his mind. His appetite had vanished. He only wanted to sit and rest. Nevertheless, when the lunch hour was over, he went back to work and finished out the day. When he got home, a little before six, he was exhausted. He stretched out on the living room sofa for a rest before dinner and instantly fell asleep. His wife let him sleep, and he slept two hours. He awoke feeling worse than ever. His head ached, his bones ached, and it hurt him to move his neck. He looked so sick that his wife was frightened. She insisted that he see a doctor at once. The Hoffmans had no regular doctor, but Mrs. Hoffman knew the name of a general practitioner in the neighborhood who kept evening office hours, and she looked up his address. Hoffman reluctantly agreed to go. He felt too sick to argue. He dragged himself out of the house and into his car and down to the doctor's office.

A glance was enough to tell the doctor's nurse that Hoffman

was seriously ill. She spoke to the doctor, and the doctor saw him at once. He heard from Hoffman an account of his trouble, and removed the bandage and examined the lesion. The lesion was about the size of an aspirin tablet. It was brightly inflamed and firm to the touch. Surrounding it were several blistery swellings. The doctor could see that an attempt had been made to lance it. He also noted a swelling below the lesion that extended along the neck and halfway down the left side of the chest. He then took Hoffman's temperature. The reading was 102.2 degrees. That decided him. His diagnosis, he told Hoffman, was an abscess with cellulitis. Cellulitis was a potentially dangerous inflammation of the cellular tissue and could best be treated in a hospital. He suggested that Hoffman go home and pack a bag. He, meanwhile, would call Christ Hospital and arrange for his admission. It was now a little past nine. He would meet him there at ten.

Mrs. Hoffman drove her husband to the hospital. On the way, he told her that he felt even worse than he had an hour before. His voice, she thought, had a strained and scratchy sound. He was admitted to the hospital under a tentative diagnosis of abscess of neck with cellulitus and (as the apparent victim of a communicable disease) placed in isolation. The admission examination showed "marked swelling of the left side of the neck with induration; redness and swelling of the neck, with redness and swelling over the left anterior chest down to the sixth intercostal space." The routine diagnostic tests—chest X-ray, electrocardiography, urinalysis—were made, and the results of all were normal. So, with one exception, were the results of the usual blood studies. Hoffman's white cell count, a generally reliable barometer of infection, was very slightly elevated—10,800, or about a thousand beyond the normal range. His pulse rate was 92, or a little faster than normal for a man, and his temperature had risen to 103 degrees. The doctor prescribed an immediate intramuscular injection of 500 mg. of chloramphenicol, to be followed at six-hour intervals by oral doses of 250 mg. each. Hoffman received the injection at ten-thirty. Five minutes later, he was sound asleep.

Hoffman spent a comfortable night. He awoke on Thursday

morning feeling much improved. When the doctor came by on his morning round, Hoffman asked to be discharged. He was ready, he said, to go home. The doctor smiled and said he would think it over, and introduced a consultant. The consultant examined the lesion and generally confirmed his colleague's diagnosis. He added, however, that the situation of the lesion, and to some extent its appearance, suggested the possibility of erysipelas. In any event, he concluded, the patient was clearly mistaken. Hoffman was still a sick man. The attending physician agreed. He also decided to revise and increase his antibiotic attack. At one o'clock that afternoon, Hoffman was given an intravenous injection consisting of 1,000,000 units of penicillin and one gram of chloramphenicol. To this was added, as a routine supportive measure, 1000 cc. of a five per cent glucose solution. At four o'clock, his blood pressure suddenly dropped. Its fall was checked by the prompt administration of blood, plasma, and vasopressor drugs, and he remained entirely conscious. By six o'clock, when the doctor looked in with another consultant, Hoffman seemed encouragingly convalescent. His lesion was conspicuously larger— it was now about an inch in length and almost half that wide— but his only complaint was a headache. The consultant confirmed the diagnosis of abscess with cellulitis. There was nothing about the look of the lesion to make him question it.

Hoffman continued comfortably convalescent until about eight o'clock. He then began to find it difficult to breathe. The swelling around the lesion had increased and was now constricting his throat. Supplementary oxygen was administered nasally, and he soon relaxed and fell asleep. He slept for two or three hours. Just before midnight, his blood pressure dropped and he went into shock again. He was rallied as before, but this time his response was only briefly satisfactory. In spite of every attention, his blood pressure swung mercurially up and down throughout the rest of the night. Around six o'clock on Friday morning, he had a sudden attack of nausea, and vomited. A moment later, he was seized by a racking convulsion. The convulsions continued for several minutes. A little before seven, he died.

Four days later, at nine o'clock on Tuesday morning, March 10, T. Aidan Cockburn, assistant commissioner of the Cincinnati Board of Health and the executive in charge of its communicable disease section, received a telephone call from Christ Hospital. His caller identified himself as Evans A. Schmidt, the hospital bacteriologist. He was calling, he said, in response to a recent request by Dr. Cockburn that he be promptly informed of any unusual occurrence in the field of communicable disease. Well, something unusual had just turned up at Christ Hospital. Or so, at any rate, it seemed. A patient named Hoffman had died at the hospital on Friday of what the attending physician had tentatively described on the death certificate as "septic shock due to abscess with cellulitis of neck and anterior chest." An autopsy was performed within three or four hours, but its findings were equally ambiguous. However, Dr. Schmidt went on to say, blood samples taken from the patient at admission had been cultured, and this morning, when he returned to the laboratory after a short holiday, an assistant had shown him the plates. At this point, of course, he couldn't be more than suspicious, but the cultures had produced an organism that looked unpleasantly like the organism of anthrax. He thought Dr. Cockburn would want to know.

Dr. Cockburn took a deep breath. He didn't disbelieve Dr. Schmidt's report, but he couldn't help but wonder. A couple of months before, around the end of December, five laboratory goats at the University of Cincinnati College of Medicine had sickened and died of what a preliminary investigation indicated was anthrax. For several days, until a more thorough investigation unmasked an animal pathogen called PPLO (pleuro-pneumonia-like organism) as the responsible agent, the threat of an epidemic had loomed in the local press. So anthrax was a touchy subject at the Board of Health. But it was also one that couldn't be ignored, and he thanked Dr. Schmidt for his ready cooperation. He supposed, he then went on to say, that Dr. Schmidt would be sending a subculture sample to the State Department of Health, at Columbus, for further examination. Would he be good enough to send him a sample, too? And by messenger. As luck would have it, he was flying down to Atlanta that afternoon for a meeting at the Communicable Disease Center of the United States Public Health Service there, and he would like to take it along. Dr. Schmidt said he would, and did. By the time the sample—a pale lump of nutrient jelly at the bottom of a stoppered test tube—was on his desk, Dr. Cockburn had made the arrangements regarding it that protocol required. He had telephoned Columbus, and then made a call to Atlanta. Columbus had given him permission to take the culture to Atlanta. And Atlanta had agreed to receive it.

Dr. Cockburn carried the culture down to Atlanta in the pocket of his shirt. That was to keep the organism warm enough to grow and proliferate. It was received at the Communicable Disease Center by Philip S. Brachman, chief of the Investigations Section of the Center's Epidemiology Branch. Dr. Brachman examined the culture with professional interest, and passed it along to the Laboratory Branch for continued cultivation and the standard identification tests. Dr. Cockburn would be informed as soon as its identity was determined. The first progress report was telephoned to Dr. Brachman on Wednesday afternoon. It was squarely inconclusive. In spite of Dr. Cockburn's incubatory efforts, the organism had suffered an inhibiting chill and was only now resuming its growth. The second report, which came on Thursday morning, was tentatively negative. The reaction of the organism to a staining test for type had not been typically that of the anthrax bacillus. At eleven o'clock on Friday morning, the laboratory called again. The staining test of yesterday had been a little premature. That and other tests now told a more straightforward story. The culture was unequivocally positive for *Bacillus anthracis.* Dr. Brachman politely acknowledged the news. He hung up the phone and rang for his secretary. He asked her to put in a call to Columbus, Ohio—to the Ohio State Department of Health. He wanted to speak to Harold Decker, chief of its Division of Communicable Disease.

Anthrax is a disease of animals to which man occasionally succumbs. It is also one that occupies a unique position in the history of medical science. It was anthrax whose elucidation by the then obscure German country doctor Robert Koch at the University of Breslau in 1876 (in a paper entitled "Die Ätiologie der Milzbrandkrankheit begründet auf die Entwicklungsgeschichte des *Bacillus Anthracis*") provided the first proof that a specific microorganism could cause a specific disease, a demonstration that largely completed the establishment of the germ theory of disease causation. And it was from the procedure he so successfully followed in his anthrax study that Koch conceived the celebrated postulates of experimental evidence that bear his name. This formulation, which establishes four conditions that must be met before a given microorganism can be accepted as the cause of a given disease, is the keystone of modern bacteriology.

The hosts preferred by the anthrax organism are horses, goats, cattle, sheep, and swine. Its human visitations are merely accidental. Man acquires the disease by contact with an ailing animal or from a contaminated animal product—skin, hide, hair, wool, bristle, bones. *B. anthracis* enters the body in the form of a spore excreted by an animal host. The nose or an open cut or abrasion are its usual portals of entry in man. Sometimes, though rarely, it enters by way of the mouth. Among grazing animals, on the other hand, the mouth is the customary portal. Anthrax spores, like those of most bacilli, are brutally resistant to extremes of heat and cold, and can exist for many years in the harshest natural environment. Once settled in an acceptable habitat, however, the spores discard their protective casing and rapidly proliferate. They generally make their presence felt within a couple of days. The signs and symptoms of anthrax reflect the initial focus of infection. Inhalation anthrax comes on like bronchopneumonia, and the manifestations of intestinal anthrax are those of any severe intestinal disorder. Cutaneous anthrax, as the other—and most common —form is called, is classically proclaimed by a lesion that, if left to mature, develops a black, scablike crust. (It is from the look of this distinctive blemish that the disease derives its name. Anthrax is the Latin transliteration of the Greek word for coal.) But these are superficial differences. Anthrax is essentially an acute intoxication. Its victims are overwhelmed, not by a concentration of *B. anthracis* at a certain site, but by a concentration in the blood and the lymph of a toxic substance that the organism elaborates as it grows and multiplies. *B. anthracis* is destroyed by massive doses of penicillin and other antibiotics, but drugs have no effect on its toxin. The prompt destruction of the organism is thus of vital importance in the treatment of anthrax. For once the accumulation of toxin has reached a certain level in the blood, the victim is beyond salvation. The immediate cause of death in anthrax is shock.

Anthrax has been a dreaded disease for at least five thousand years. It is probable that man became aware of its existence soon after he turned from hunting to a life of farming and animal husbandry. In the judgment of some medical historians, a reference to anthrax appears in the oldest pages of the Old Testament. They profess to recognize the disease in one of the curious plagues described in the Book of Exodus with which God punished Pharaoh for holding the Jews in bondage. "Behold," the verse they offer into evidence reads, "the hand of the Lord is upon thy cattle which is in the field, upon the horses, upon the asses, upon the camels, upon the oxen, and upon the sheep: there shall be a very grievous murrain." A murrain, however, is not necessarily anthrax. The word is a general term for any grievous disease of livestock. A more striking likeness of anthrax is found in the *Georgics* of Virgil. The pertinent passage reads:

> To death at once whole herds of cattle go;
> Sheep, oxen, horses, fall; and heaped on high,
> The differing species in confusion lie,
> Till, warned by frequent ills, the way they found
> To lodge their loathsome carrion under ground:
> For useless to the currier were their hides;
> Nor could their tainted flesh with ocean tides
> Be free from filth . . .
> Nor safely could they shear their fleecy store
> (Made drunk with poisonous juice, and stiff with gore),
> Or touch the web: but, if the vest they wear,

Red blisters rising on their paps appear,
And flaming carbuncles, and noisome sweat,
And clammy dews, that loathsome lice beget;
Till the slow-creeping evil eats his way,
Consumes the parching limbs, and makes the life his prey.

But Virgil's representation of anthrax is more than merely graphic. It is exceedingly precocious, as well. In his apparent understanding of the contagiousness of the disease and the manner in which it is spread, he seems to have been almost two thousand years in advance of his time. Some thirteen hundred years elapsed between the publication of the *Georgics,* in the first pre-Christian century, and the inclusion of anthrax (along with epilepsy, leprosy, and the itch) among the diseases thought to be contagious, and the nature of its contagiousness eluded rediscovery even longer than that. As late as the seventeenth century (when a pandemic in southern Europe killed sixty thousand people and uncounted thousands of livestock), it was considered only prudent that an animal dying of anthrax be slaughtered for food and its hide (or fleecy store) be salvaged and used. It was not until the early nineteenth century, little more than a generation before Koch's epochal depiction, that medical science could much improve upon the Virgilian view of anthrax.

The scientific comprehension of anthrax, though late in taking recognizable shape, was accomplished with dispatch. Few diseases have been so thoroughly riddled so fast. In addition to being the first disease irrefutably laid to a germ, it was the first of its kind to yield to total penetration and control. The control of anthrax was initiated by Louis Pasteur in 1881. In the spring of that year, before a gathering of scientists assembled near Paris by the Agricultural Society of Melun, he demonstrated (on a flock of sheep) that an animal inoculated with a culture of heat-attenuated *B. anthracis* was rendered immune to anthrax. He also showed, in another study, that the burial of infected animal carcasses (as originally recommended by Virgil) was not enough to check the natural spread of the disease. His recommendation was that the carcasses first be burned. Buried anthrax spores, he pointed out, are eventually disinterred by the peregrinations of earthworms and other subterranean creatures. A few years later, in 1895, an Italian investigator named Achille Sclavo, inspired by Pasteur's contributions to the control of anthrax in animals, developed an antianthrax serum for the treatment of the disease in man. Since then, the treatment of human anthrax has been further refined by the introduction of a succession of drugs to which *B. anthracis* is more or less susceptible —neoarsphenamine (in 1926), the sulfonamides (1935), penicillin (1944), chloramphenicol (1947), and (around 1950) the several tetracyclines. The impact of these prophylactic, hygienic, and therapeutic innovations has been decisive. Anthrax is no longer much of a problem anywhere in the Western world. It has almost entirely disappeared from Europe, and the biggest epizootic in the United States in recent years—an outbreak in Kansas and Oklahoma in 1957—felled fifteen hundred head of cattle, sixty-eight pigs, thirty-nine sheep, and fifteen horses. The incidence of human anthrax in this country—on the farm and in the animal products trades—has dropped from two hundred and two cases in 1917 (the second year for which accurate records are available) to five in 1964.

Nevertheless, the disease persists—not all susceptible animals are regularly vaccinated, not all infected carcasses are properly destroyed, not all processors of hides and hair and bristle and bone take the trouble to sterilize their products. And, in spite of the effectiveness of penicillin and other antibiotics, it is still, on occasion, fatal. It will probably always be so. For anthrax has now become such a rarity that the great majority of doctors have never seen a case, and when one turns up—as it did in Cincinnati in March of 1964—they only too often fail to recognize it in time. "The chances of making a correct early diagnosis of anthrax," Herman Gold, a staff physician at the Chester (Pennsylvania) Hospital and an authority on anthrax, has noted in a recent monograph, "are directly proportional to the physician's index of suspicion."

Dr. Brachman's telephone call to Dr. Decker on Friday the thirteenth of March was prompted by two considerations. One, of course, was to give the Ohio authorities the laboratory report from the Communicable Disease Center. The other was to express a personal interest in the case.

"Anthrax is something of a specialty of mine," Dr. Brachman says. "That's a little these days like specializing in botulism or rabies or smallpox. I mean, it's an interest one doesn't have much occasion to indulge. For which, of course, I'm duly grateful. But when a case does come along, I like to follow it up if I can. That was no problem here. Decker and I are old friends. So after I had given him our report and asked him to pass it on to Cockburn down in Cincinnati, I said I'd like to keep in touch with the investigation. Decker couldn't have been more obliging. In that case, he said, he'd give the assignment to Peter Greenwald. I knew Peter. He was one of our own people. He was a young doctor who had chosen to do his military service in the U. S. Public Health Service instead of in the Army, and he was then on loan from CDC to Ohio as what we call an Epidemic Intelligence Service officer. I thanked Decker for his courtesy, and said I'd be glad to give Peter any help I could. On anthrax, that is. Not on the case. I knew nothing about that at this point. Except that it was a case of cutaneous anthrax—Cockburn had mentioned a skin lesion. And that the victim was a man named Hoffman.

"I had a call from Peter that afternoon. He was leaving for Cincinnati in the morning and wanted to talk things over. All he knew about anthrax was what the textbooks say. The best I could do was suggest some questions to ask. First on the list was Hoffman's occupation. Did he work with animals or animal products? Did he ride? Did he often visit the zoo? Had he done any digging or gardening lately? Grubbing around in the soil can bring up buried spores. Did he use a shaving brush? How about new rugs at home, any new upholstered furniture, new clothes? Some of those items may sound a little fanciful. I think they did to Peter. But they're not in the least far-fetched. You can find them all in the literature. Take shaving brushes, for example. There was a serious epidemic in this country around the time of the First

World War that was traced to shaving brushes made from contaminated horse hair. As for clothing as a source of infection, I can vouch for that myself. I worked on such a case back in 1956. It happened in Philadelphia. The victim turned up with a classic anthrax lesion on his left forearm—so classic that the doctor couldn't help but recognize it. But there wasn't a clue where he might have got it. He was a grocery clerk and he lived in a downtown apartment—a completely sterile environment. Then we learned two things. He had recently bought a rough woolen mackinaw, and he usually wore it with a short-sleeved shirt. And not long before the appearance of the lesion, his wife had accidentally scratched him on the left forearm. We got the coat and took some samples of the wool and cultured them, and two of them were positive for anthrax. But the strangest I ever heard of was a fatal case somewhere up in Connecticut in 1946. The victim worked for a piano manufacturer and his job was making keys out of ivory—raw ivory. It turned out that the ivory he used was contaminated.

"I didn't hear from Peter again until Tuesday. He called me Tuesday morning from Columbus, and he seemed to have the matter pretty well in hand. He had followed the trail from Cincinnati out to Oxford and back to Cincinnati again, and he had talked to a number of people—Cockburn, Mrs. Hoffman, a doctor in Oxford, several Cincinnati doctors, the company that Hoffman worked for, and even a friend of Hoffman's on the job. The answers to all but one of my questions were negative. That one affirmative answer, however, was extremely interesting. Hoffman was an insulation man, and the materials he worked with included hair felt. The hair in hair felt is animal hair. Not only that. He had been working with hair felt insulation at the time he took sick. The company—I'll just call it that—showed Peter the invoice for the felt that Hoffman had been working with, and he copied down the specification number. It was number 303. It came, they said, like all their felt, from a manufacturer I'll call the Ajax Corporation. Ajax was an old concern, with headquarters in Chicago and plants at Milwaukee and Newark. Well, that was even more interesting.

It changed the whole complexion of the case. It was no longer just a state or local problem—it was now a national affair. I told Peter I'd probably be seeing him soon, and hung up.

"But first I had to have a word with Decker. That was only common courtesy. Decker was expecting my call. He had naturally seen Peter's report and could guess what I had in mind. He supposed that I would want to stop off in Ohio and go over the evidence before calling on the Ajax people. If so, he said, come ahead. He would clear it with the Department. I told him I planned to leave that afternoon, and we arranged to meet in Columbus. The next call I made was to the headquarters of the Ajax Corporation in Chicago. I identified myself and described the case and asked permission to visit their plants. They were stunned. They simply couldn't believe it. Ajax had been in business since before the Civil War and this was the first time it had suffered even a suspicion of anthrax. However, they said, they were glad to cooperate, and would instruct the managers of the Milwaukee and Newark plants to expect me. After that, I made a couple of routine for-your-information calls—to the state epidemiologists in Wisconsin and New Jersey. Then I got my working kit together—a set of plastic bags for the samples I proposed to collect, a couple of marking pens, and a camera. Then I went home and packed my suitcase, and left.

"Peter met me at the Columbus airport. We spent the evening with Decker and Charles Croft, the chief of the Division of Laboratories up there, reviewing the case. They didn't like it much better than the Ajax people. Hoffman was the first case of anthrax in Ohio in a good many years. Since the 1952 epizootic, I think. That was an outbreak that began with swine and then moved on to cattle and sheep, and Ohio had two human cases. The trouble was traced to contaminated bone meal that had originally come from Pakistan and India. It was also arranged at the meeting that Peter would drive me down to Cincinnati in the morning. I spent the night at a motel somebody recommended, and we made an early start. Our first stop was the hospital where Hoffman died. I wasn't checking up on Peter. It was just that I wanted to get my own feel of the case. I talked with Schmidt, the bacteriologist, and the attending physician and his consultants, but I didn't learn anything that I didn't already know. Then they had a question for me. What about the people who had been in contact with Hoffman —the nurses and the orderlies and his wife and children? Were they in any danger? That was an interesting question. It gives you a pretty good idea of what a mystery anthrax has become. But I was able to reassure them. Human-to-human transmission of anthrax is unheard of. It has never happened. Besides, if anyone had been infected, he would certainly be showing some signs of it by now.

"The next stop we made was the office of the company. I wanted to see the hair felt that Hoffman had worked with. The management were as nice as could be—Peter had broken the news very smoothly—and they took me into the warehouse and showed me several big rolls of the stuff. I cut and bagged a couple of samples, and then I noticed something. All these rolls carried the specification number 318. 'Wait a minute,' I said. 'The invoice that Dr. Greenwald saw here yesterday had a different number. It was number 303.' Oh, no, they said. That was impossible. There must be some mistake. The only hair felt they ever used was number 318, and they would show me the invoice to prove it. Which they did—and it was just as Peter had said. It was marked number 303. They were flabbergasted. The invoice covered a shipment of twenty-five rolls of 303 felt—each of them fifty feet long by three feet wide by one inch thick—and it was dated December 2, 1963. That would have been in plenty of time for the hospital remodeling job at Oxford. Work began there on January 8. When the management had recovered from the shock of the invoice, we went back to the warehouse and checked out every roll of hair felt in stock. All were number 318. Then somebody suggested the shipping room. We went down to the shipping room, and there, back in a corner, was part of a roll—maybe a dozen feet—marked number 303. It was all that remained of the shipment. I took a sample for analysis. It looked exactly like the other felt, but, of course, that didn't mean a thing. I asked them to cover it up and leave it alone until further notice. Then, like the people at the hospital, they had a question to ask. Only this was a natural question. They wanted to know if their other employees were in any danger. A lot of them had handled or worked with felt. I had to admit that the danger was there. I doubted, though, if it amounted to much by now. Too much time had passed. In any event, that was something for the city health authorities to look into. They would, I was sure, do a properly thorough job.

"The people at the hospital in Oxford were also worried about infection. They had every reason to be. Peter introduced me to the administrator there, and he showed us around. The remodeling included insulating some water pipes in the kitchen ceiling and enlarging the newborn nursery and installing air conditioning there. The pipes in the nursery were hidden behind a false ceiling of acoustical tile, but those in the kitchen were exposed. All of the insulated pipes were wrapped in a protective canvas sheathing, and except for the finishing touches, the job was done. The kitchen was already back in use, and the nursery was scheduled to reopen in about a week. I wanted a sample of the installed felt, and the administrator called over Hoffman's working mate, a man named Jensen, and he climbed up and cut me a little piece from one of the pipes in the nursery. I was glad to meet Jensen, and we talked for a moment. He was terribly depressed about Hoffman. They had been good friends. But otherwise he felt fine. No lesions, no symptoms of any kind. I asked him about the roll of felt—was there any of it left? He thought there was, and took me down to a basement hall where they kept their supplies and showed me the remains of the roll, and I got myself a sample. I suggested that the rest of the roll be boxed and returned to the company in Cincinnati. And it might be a good idea to disinfect this part of the hall. That was when the administrator asked about the danger of infection. Did I think they ought to close the hospital and rip out all the insulation? I was prepared for the question and I told him what I thought. There was certainly no reason to close the hospital. Nor was there any reason to take out the insulation—not at this point. And I thought it was safe enough to reopen the nursery as planned. The insulation was well enough covered. But I advised them not to use the air-conditioning. Not until they heard from me it was safe.

"That completed that phase of the investigation. I was ready to begin on Ajax. Peter drove me back to the city and across the river and down through the Kentucky bluegrass country to the Cincinnati airport. I had a reservation on a late afternoon flight to Milwaukee, and while I was waiting, I telephoned Mrs. Hoffman and satisfied myself about some clinical odds and ends. Then I flew to Milwaukee and found a hotel and went to bed. The next morning—that would be Thursday—I took a cab out to the Ajax plant and introduced myself to the manager. He was a little stiff at first, but as soon as I mentioned their hair felt number 303, he completely relaxed. I've seldom seen a friendlier man. It seemed that his plant didn't make number 303 hair felt. Number 303 was made exclusively at the Newark plant. It was composed of several different kinds of animal hair. Only cattle hair was used here in Milwaukee. It was all domestic fiber, he said, and it came from a number of Midwestern tanneries. He led me out to the warehouse and showed me several hundred bales of cattle hair and cattle hair felt. I knew it was a waste of time, but I took a couple of pictures of the layout and some samples of hair and felt for the record, and got out as fast as I could.

"I spent that night in New York, and took the Hudson Tube over to Newark the first thing in the morning. I had a hunch it was going to be a long day. The Ajax manager there was expecting me. He was ready and waiting. Milwaukee must have called him. He allowed that his plant made number 303 hair felt, and also number 318. The composition of the latter was seventy-five per cent jute and the rest animal hair—usually cattle hair, but sometimes calf or horse. Number 303 felt was half cattle hair and half goat hair. Practically all of the animal hair they used was domestic hair. The chief exception was goat hair. That was usually imported. I found that very provocative. Animal anthrax is pretty well under control in this country, you know, but it isn't everywhere. There are plenty of places where it is still enzootic—North

Africa, southern Europe, the Middle East. And those are all big goat-raising areas. At my suggestion, the manager brought out his records and we looked up their recent goat hair purchases. The record showed that in 1963 they had bought approximately one hundred and twenty thousand pounds of goat hair—all of it imported. Moreover, only foreign hair had been used in the manufacture of 303 felt. A total of thirteen hundred rolls of 303 had been made and distributed in 1963. About sixty different customers were involved, including the Cincinnati company. Their inclusion was also explained in the records. It was quite true that Cincinnati always ordered number 318 hair felt. But last December, when a Cincinnati order came in, it so happened that Ajax was temporarily out of stock on number 318, and filled the order with 303 instead. The price was the same, the manager said, and the quality of 303 was, if anything, higher. Goat hair was the best of the animal fibers, and the most expensive. That's why it was used so little. Actually, they had done the Cincinnati company a favor. We then moved on to the warehouse. I spent the afternoon there, taking pictures and collecting samples. I took four samples of cattle hair, four of horse hair, two of buffalo hair, two of domestic goat hair, and eight of imported goat hair. All were raw, unprocessed fiber.

"I got back to Atlanta late Friday night, and on Monday morning, I carried my samples down to the laboratory to be cultured and analyzed. The first step was to wash the anthrax spores, if any, from the hair. The procedure we used was this. We immersed each sample in an Erlenmeyer flask containing a liter of distilled water and a drop or two of detergent. The flasks were stoppered and placed in a mechanical agitator and shaken for about an hour. They were allowed to stand for five minutes. The next step was to streak a sample of liquid, or wash-water, from each flask onto nutrient blood agar plates, and the plates were then incubated. But the anthrax organism isn't the only organism that will grow on such a medium, so the next step was a further screening. Two hundred and fifty cc.s of the liquid were poured into a centrifuge bottle which we had heated in a water bath at a high temperature for ten minutes to kill organisms other than *B. anthracis.* The bottle was then centrifuged at two thousand rpm.s for ten minutes. The supernatant liquid was discarded and the sediment resuspended in distilled water. Then a sample of that was streaked on nutrient blood agar plates. We then tucked the plates away in an incubator and let nature take its course. The results, if any, would be visible in the morning.

"We got some results. There were forty-nine samples in the lot we cultured, and thirteen of them showed up positive for *B. anthracis.* Four of the positive cultures were from the Newark warehouse—they were four of the eight samples I had taken from bales of imported goat hair. The other positives were samples of the number 303 hair felt. One came from the roll at the Cincinnati warehouse, five were from the roll in the hospital basement, and three were from the air-conditioning pipe in the nursery. Everything else—the number 318 felt and all the other animal fibers— was negative. I wasn't particularly surprised. I wasn't really surprised at all. What I mostly felt was relief. Number 303 was the only possible answer—the only bearable answer. Any other result would have meant starting all over again from scratch. Either that or something a great deal worse.

"It was bad enough as it was. The fact that nine of about twenty samples of 303 and half of my goat hair samples were contaminated with anthrax spores raised a number of more or less urgent problems. Some of them, of course, were pretty much routine—the insulation at the hospital, the leftover yardage there and at the Cincinnati warehouse, and the imported goat hair in stock at Newark—and I handled them routinely. I began by calling Decker and Cockburn and putting them in the picture. The purely local aspects of the matter were their responsibility. Then I called the administrator at the Oxford hospital and told him what we had found. I advised him to remove at once the insulation in the kitchen and, as soon as convenient, the nursery insulation. And I told him how to do it in the safest possible way. The workmen were to thoroughly moisten the outside of the pipes with a five per cent Clorox solution in water, remove the hair felt in disposable bags, wash the area again with Clorox, and then burn the bags and carefully wash their hands. I also instructed him to return the remains of the roll of 303 to the Cincinnati company. It would be picked up there, along with the company's yardage, by the city health authorities and burned. I then got in touch with the Ajax people. I advised them to hold the imported goat hair at Newark in stock until further notice. They were only too willing to agree.

"The other problems raised by the case were actually one big problem. It had to do with the dozen or more other insulation jobs in and around Cincinnati where the company had installed number 303 hair felt, and with those sixty other Ajax customers—their remaining supplies of 303 and the jobs on which *they* had used the stuff. My investigation indicated that roughly half of the 1963 production of 303 was contaminated. Did that require its removal wherever installed? What were the risks involved? How risky would it be to do nothing? Anthrax spores can lie dormant for many years. On the other hand, what risks would we run by removing it? The danger to the workmen was minimal—they could be fully safe-guarded. But a job of that size and scope could hardly be kept a secret. And when the news came out, just how would the public react? Would it panic? Well, questions like that, thank heavens, are not for me to decide. A committee was formed and we had a meeting in Washington. The members included an assistant surgeon general and representatives of all interested divisions of the U. S. Public Health Service, and the report we made was subsequently approved by the Surgeon General. I'll read you some of our recommendations: 'The consensus was that leaving the installed material in place involved the least risk. It was also decided not to publicize the information that it may be contaminated with *B. anthracis*. It was recommended that attempts be made to record the locations of all buildings in which this particular lot of hair felt may have been used. This data should be requested discreetly, and this activity should be under the direction of the Communicable Disease Center, working with the Division of Occupational Health. The recommendation is made that the [Ajax Corporation] should prohibit further use of contaminated hair-felt material of type 303, and locate any from the involved batches that may be in warehouses of their customers throughout the country. All finished material in their own warehouses should either be disinfected or destroyed—in either case with the assistance and supervision of the CDC. Depending upon the amount of unused material located throughout the country and its location, it will be either returned to the [manufacturer] or destroyed locally, with assistance from the CDC. The recommendation is made to either disinfect or destroy all imported contaminated goat hair in the [Ajax Corporation] warehouse, with the assistance and supervision of the CDC. Concerning the total problem of the importation of animal products contaminated with *B. anthracis,* the consensus was that the infrequency of cases and the gradual decline in the use of these materials did not justify any specific action by the Public Health Service at this time.' Those precautions, of course, have all been carried out. In addition, I understand, the various buildings where 303 was installed have all been identified, and they are under surveillance by the local authorities. If and when a building is remodeled or demolished, the workmen will be warned and instructed.

"The policy meeting officially closed the case. There were still some details, however, that bothered me a bit. They had to do with Hoffman. Why was he and not Jensen the victim—or rather, how did Jensen escape infection? The answer, once I started thinking about it, wasn't too hard to find. I found it in Peter's report. Jensen didn't have much to do with the felt. He only helped Hoffman install it. Moreover, it was Hoffman who lugged the stuff around to wherever it was needed on the job. Very well. Cutaneous anthrax thrives on contact. But *how* did Hoffman get infected? Mere exposure isn't enough. Well, suppose he carried the felt on his shoulder. That seemed a safe enough assumption. Suppose it kept rubbing and chafing against his neck until it made an abrasion? Would that explain what happened? No, I had to admit, it wouldn't. At least not very satisfactorily. The site of the lesion was much too low—it was almost down on his collarbone. His shirt would be in the way. And then I remembered something that Mrs. Hoffman had told me when I called her from the Cincinnati airport down in Kentucky. We were talking about the initial appearance of the lesion, and she said she first noticed it when Hoffman came home on Tuesday evening. That was the night before he was treated at the Miami University clinic. She said she noticed it the minute he took off his jacket. Because her husband didn't wear a shirt to work. He always worked in a sweatshirt."

[*1965*]

CHAPTER 13

Three Sick Babies

————————◆————————

DR. PAUL M. TAYLOR, an assistant professor of pediatrics at the University of Pittsburgh School of Medicine, left his office on the first floor of Magee-Womens Hospital, an affiliate of the medical school, and climbed the stairs to the premature-baby nursery, on the second floor. It was twenty minutes to eleven on the morning of July 12, 1965, a Monday, and this was his regular weekday round. He was, in a way, attending physician to all the babies in the nursery. Two pediatric residents were waiting for him in the gown room. That, too, was as usual, and while Dr. Taylor scrubbed and disinfected his hands and got into a freshly laundered gown, they gave him the customary nursery news report. The weekend had been generally uneventful. All but one of the twenty-six babies in the nursery were progressing satisfactorily. The exception was one of the smallest—a twenty-five-day-old, two-and-a-half-pound boy. He was in Room 227. Dr. Taylor nodded. He knew the one they meant. Well, early that morning, just after midnight, a nurse had noted that his breathing was unusually slow and his behavior somewhat apathetic. The resident on duty,

200

seeing these signs as suggestive of septicemia, had treated the baby with penicillin and kanamycin, but he still looked and acted sick. It was probable that he would soon need artificial respiration. Meanwhile, the usual samples (blood, stool, mucus, spinal fluid) had been taken and sent along to the laboratory for culture and analysis.

Dr. Taylor heard his residents' review with no more than natural interest. Serious illness in a premature nursery is not an unusual occurrence, and a blood infection is only one of the many diseases that may afflict a premature baby during its first days of life. Trouble is inherent in the phenomenon of prematurity. For the truth of the matter is that premature birth is itself a serious affliction. A premature baby is a baby born in the seventh or eighth month of pregnancy. Its birth weight is largely determined by its relative prematurity, ranging from around two to five and a half pounds. The average term (or nine-month) baby weighs around seven pounds at birth, and it generally comes into the world alive and kicking. Premature babies begin life almost incapable of living. There is nothing more frail and fragile. Many of them are too meagerly developed to maintain normal body temperature. One in three requires immediate, and often prolonged, resuscitation, and about the same number are unable to nurse, or even to swallow. Some of them are even unable to cry. All of them are exquisitely susceptible to infection. Mental and neurological defects are also common in premature babies. One of the commonest of these is cerebral palsy. About half of all victims of the affliction are of premature birth. As recently as twenty-five years ago, most premature babies died. The technological innovations of the postwar era have greatly improved that record, but the mortality rate among newborn premature babies is still high. In even the best hospitals, some ninety per cent of all two-pound babies die. So do about fifty per cent of those weighing two to three pounds, around ten per cent of those weighing three to four pounds, and between five and eight per cent of those weighing four to five pounds. Such babies are simply too fetal to survive outside the womb. There are many causes of premature birth (including falls and blows), but most current investigators think that malnutrition is perhaps the most important cause. Prematurity would thus seem to be a socio-economic problem, and this supposition is confirmed by statistics. The great majority of premature babies are born to women too poor to buy the nutritious food they need.

Dr. Taylor tied up the back of his gown and led the way through an inner door to the central nursing station of the nursery. Room 227 was the second room on the right. The sick baby was one of five babies being cared for there. He lay on his side in his incubator bassinet with a stomach feeding tube in his mouth, and he looked even sicker than the resident had said. There was nothing, however, that Dr. Taylor could do that hadn't already been done. He confirmed the resident's course of treatment and agreed that a respirator would probably soon be required, and then moved on to the other babies in that and the other rooms. They were all, as far as he could read the almost imperceptible manifestations of the premature, in their usual precarious but normal condition. At the end of his rounds, Dr. Taylor had lunch, and after lunch he occupied himself with his other professorial duties. Before leaving for home, he put in a call to the nursery for a report on the sick baby. The report was not comforting. The baby's condition had continued to worsen, and he had been moved to Room 229, which is reserved for babies needing constant scrupulous care. Later that evening, on his way to bed, he called the nursery again. The nurse who picked up the telephone was able to answer his question. The sick baby from Room 227 was dead.

Dr. Taylor's Tuesday-morning round was much like that of Monday. The resident's gown-room report included another sick baby. It also contained a post-mortem note on the baby from Room 227. The hospital laboratory had confirmed the general diagnosis of septicemia. A microbial culture had been grown from a sample of the baby's blood and identified as bacteria of the gram-negative type. It was one of some twenty-five species of bacteria (among them the causative organisms of typhoid fever, brucellosis, whooping cough, plague, and gonorrhea) that react negatively to the standard staining test devised in 1884 by the

Danish bacteriologist Hans Christian Joachim Gram. A more specific identification was promised for the next day. Meanwhile, of course, the laboratory had been supplied with diagnostic samples from the second sick baby. This baby, also a boy, was one of the babies in Room 229. He was a term baby, three days old, but had been assigned to the premature nursery directly from the delivery room because he required artificial respiration and other intensive care. Dr. Taylor remembered the baby when he saw him. He had begun life with a severe aspiration pneumonia, stemming from an original inability to breathe, but that had been quickly controlled with penicillin, kanamycin, and dexamethasone. His trouble now was diarrhea. Diarrhea in a newborn baby is not a common complaint, and Dr. Taylor wondered if this attack might be a septic aftermath of the earlier pneumonia. But that was a question that only the laboratory could readily answer. There was no doubt about the treatment the baby was receiving. That seemed to be entirely satisfactory.

The Wednesday-morning gown-room news review contained three major items. One was that still another baby boy had become seriously ill overnight. The next item was a preliminary laboratory report on the second sick baby, which identified the cause of his diarrhea as a gram-negative bacillus. The third item was a more or less definitive laboratory report on the dead baby. After forty-eight hours of growth, the gram-negative bacteria cultured from his blood presented the colonial configuration, the fluorescent yellow-green pigmentation, and the spearmint odor generally characteristic of the type known as Pseudomonas aeruginosa.

Dr. Taylor went to Room 229 for a look at the new sick baby. He was nine days old and weighed about three pounds. His illness appeared to be a pneumonia. This illness had come on abruptly, but, like so many other premature babies, he had never been really well. He had been unable at birth to breathe spontaneously and had spent the first five or six days of his life in a respirator. It was obvious to Dr. Taylor that the baby's condition was grave. It seemed equally clear, however, that he was receiving the best of care, and a conventional course of penicillin and kanamycin had been started. Dr. Taylor left the nursery in an uneasy state of mind. His uneasiness had to do with Room 229 and the sudden string of serious illnesses there. He was afraid that they might be more than merely coincidental.

Dr. Taylor was not kept in suspense for long. One of his fears was confirmed that afternoon by a call from the resident then on duty in the nursery. He called to report that the new sick baby—the three-pound nine-day-old—was dead. His illness had lasted a scant eight hours. The next morning brought another confirmation. There was a final laboratory report on the first sick baby: the cause of his death was definitely a Pseudomonas infection. There was a forty-eight-hour report on the second sick baby: the gram-negative bacteria grown from his blood had all the important characteristics of Pseudomonas aeruginosa. There was a twenty-four-hour report on the third sick baby: cultures grown from his blood had been identified as gram-negative bacteria. And that was too much for coincidence. It meant—it could almost certainly only mean—that a Pseudomonas epidemic had struck the premature nursery.

Pseudomonas aeruginosa is one of a group of gram-negative pathogens that have only recently come to be seriously pathogenic. Other members of this group include Escherichia coli and the several species of the Proteus and the Enterobacter-Klebsiella genera. Their rise to eminent virulence is a curious phenomenon. These micro-organisms regularly reside in soil and water, and in the gastro-intestinal tracts of most (if not all) human beings. Their presence in that part of the body is normally innocuous. Healthy adults are impervious to the thrust of such bacteria. The victims of Pseudomonas (and E. coli and Proteus and Enterobacter) infections are the very old, the very young, and the very debilitated (the badly burned, the postoperative, the cancerous), and in almost every case they have been receiving vigorous sulfonamide or antibiotic or adrenal-steroid therapy. Most antimicrobial drugs have little destructive effect on bacteria of the Pseudomonas group. Just the reverse, in fact. Their action, in essence, is tonic. In people rendered susceptible to the gram-negative pathogens by age or illness, the result of chemotherapy is the elimination not only of the immediately threatening pathogens but also of the natural resident bacteria that normally hold further incursions of Pseudomonas (or E. coli or Proteus or Enterobacter) in check. The virulence of Pseudomonas and its kind is thus a wry expression of perhaps the most beneficent accomplishment of twentieth-century medicine.

It is also cause for some alarm. "One of the great changes wrought by the widespread use of antibacterial agents has been the radical shift in the ecologic relations among the pathogenic bacteria that are responsible for the most serious and fatal infections," the *New England Journal of Medicine* noted editorially in July of 1967. "Whereas John Bunyan could properly refer to consumption as 'Captain of the Men of Death,' this title, according to Osler, was taken over by pneumonia in the first quarter of this century. During the last two decades, it has again shifted, at least in hospital populations, first to the staphylococcal diseases and more recently to infections caused by gram-negative bacilli. Most of the gram-negative organisms that have given rise to these serious and highly fatal infections are among the normal flora of the bowel and have sometimes been referred to as 'opportunistic pathogens.' . . . Before the present antibiotic era, some of them, like Escherichia coli, although they frequently caused simple urinary-tract infections, only occasionally gave rise to serious sepsis. . . . Strains of Proteus or Pseudomonas did so very rarely, and those of Enterobacter were not even known to produce infections in human beings before the introduction of sulfonamide drugs. Of great importance are the facts that most of these opportunistic pathogens are resistant to the antibiotics that have been most widely used, and that the infections they produce are associated with a high mortality." This mortality is anachronistically high. It is roughly that resulting from the common run of pathogenic bacteria some thirty years ago—in the days when the best defense against the many Men of Death was a strong constitution.

The most conspicuously troublesome of these ordinarily unag-

gressive pathogens is E. coli. It has been implicated in some ninety
per cent of the urinary-tract infections caused by members of this
group, and it is responsible for many of the more serious cases of
bacteremia, gastroenteritis, and pneumonia. It is not, however, the
most opportunistic. The organism best equipped to take advantage
of almost any chemotherapeutic opportunity is the Pseudomonas
bacillus. Pseudomonas aeruginosa is all but invulnerable to the
present pharmacopoeia. Only two antibiotics—polymyxin B and
colistin—are generally effective against most Pseudomonas
strains. Moreover, both must be used with great discretion to
prevent severe kidney side reactions. Ps. aeruginosa is also distinc-
tively lethal. Its mortality rate, as numerous recent outbreaks
(including that in Pittsburgh in 1965) have shown, may run as
high as seventy-five per cent.

An investigation into the source of the Pseudomonas infections in
the premature nursery at Magee-Womens Hospital was started by
Dr. Taylor on Friday morning, July 16. A forty-eight-hour report
from the laboratory had by then established as Ps. aeruginosa the
gram-negative bacteria that had been cultured from the blood of
the third sick baby, and the presence of an epidemic was now
beyond dispute. The investigation began with a survey to determine the scope of the trouble. There were at that time, in addition
to the surviving (or diarrheal) sick baby, twenty-eight babies in the
nursery. Samples of nose, throat, and stool material were taken
from each, to be dispatched to the laboratory for culture and
analysis. Dr. Taylor saw this work well under way, and then
walked down to his office and put in a call to a colleague named
Horace M. Gezon, at the Graduate School of Public Health. Dr.
Gezon (at the time professor of epidemiology and microbiology at
the University of Pittsburgh Graduate School of Public Health
and now chairman of the Department of Pediatrics at the Boston
University School of Medicine) is an authority on hospital infections, and Dr. Taylor wanted his help. Dr. Gezon had two immediate suggestions. One was that the investigators meet at the hospital the following morning for an exchange of information and

ideas. The other was that Joshua Fierer, of the Allegheny County Health Department, be invited to join the investigation. Dr. Fierer (now a postdoctoral fellow in infectious diseases at the University of Pittsburgh Department of Medicine) was an Epidemic Intelligence Service officer assigned to Allegheny County by the National Communicable Disease Center, in Atlanta.

"Dr. Taylor called me on Friday afternoon," Dr. Fierer says. "I know it was Friday, because that's when all investigations seem to begin—at the start of the weekend. I knew Dr. Taylor. I had met him with Dr. Gezon back in March—on a Friday in March. There had been an outbreak of diarrhea in the premature nursery that they thought might be a viral disease, and they called the county because we had the only virus-diagnostic laboratory in the area. That case turned out to be nothing to worry about. It got the three of us together, though, and I guess that was what brought me to mind when this new problem came up. I was delighted to be asked to participate. Pseudomonas is a very interesting organism these days. But I had to tell Dr. Taylor that I couldn't make the Saturday-morning meeting. Or, if I could, I'd be late. I had a firm commitment at the Pittsburgh Children's Zoo on Saturday morning. They had a chimpanzee out there with hepatitis.

"I got to the hospital, but I was more than late. It was after lunch, and the meeting was over and everybody had gone. I looked around the nursery, feeling kind of foolish, and said hello to the nurses, and they told me who had been at the meeting. There were six in the group, including Dr. Taylor and Dr. Gezon. The others were the two residents, a study nurse of Dr. Gezon's, and an assistant professor of epidemiology at the School of Public Health named Russell Rycheck. Dr. Rycheck was a particular friend of mine. I went over to the school and looked him up, and he gave me a good report. The meeting had naturally concentrated on the nursery. The big question, of course, was: Where had the infection come from? How had Pseudomonas been introduced into the nursery? Well, Pseudomonas is a water-dwelling organism. It can live on practically nothing in the merest drop of water. That suggested water as the probable source of the trouble, and the nursery had plenty of such sources. There were thirty incubator bassinets equipped with humidifiers drawing on water reservoirs, and there were fourteen sinks—one in each of the ten baby rooms, three in the central nursing station, and one in the gown room. And then there were the usual jugs of sterile water for washing the babies' eyes and for other medicative purposes. Dr. Gezon arranged for water samples from every possible source. That included two samples from each sink—one from the drain and one from the aerator on the faucet. The screens that diffuse the water in an aerator can provide a water bug like Pseudomonas with an excellent breeding place. For good measure, he took a swab of the respirator used in the ward. Also, Dr. Rycheck said, Dr. Gezon arranged for throat and stool samples from the two residents and from all the nurses working in the premature nursery. And he had called another meeting for Monday morning. The laboratory findings would be ready for evaluation by then.

"I made the Monday meeting. The laboratory reports were presented, and then we tried to decide what they meant. The human studies made pretty plain reading. There were two sets— the nurses and residents, and the twenty-eight seemingly well babies in the nursery. The laboratory eliminated the nurses and residents as possible carriers. Their cultured specimens were all negative for Pseudomonas. The reports on the babies confirmed what I think most of us had already suspected. This was a real epidemic. Twenty-two of the babies were negative for Pseudomonas, but six were positive. They weren't clinically sick. They didn't show any symptoms. They were, however, infected with Ps. aeruginosa. Why they weren't sick is hard to say. There were several possible explanations. The best one was that their exposure was relatively slight and their natural defenses were strong—they hadn't been weakened by antibiotics. The results of the environmental studies were very interesting. But they were also rather confusing. They showed five sink drains and three of the bassinet reservoirs to be contaminated. Everything else was negative for Pseudomonas—the water jugs, the respirator, the faucet aerators, and the other drains and bassinets. The contaminated drains were in Room 207, Room 209, Room 227, an unoccupied room, and the gown room. The contaminated bassinets were in 224, 227, and 229. All the infected babies were associated with just two rooms. They were, or had been, in either Room 227—the room where the first baby took sick—or Room 229, where he died and where the two other babies became sick. There was a contaminated sink in Room 227, but the sink in 229 was clean. There was a contaminated bassinet reservoir in each room, but only one of the bassinets was, or had been, occupied by an infected baby. There were no infected babies in two rooms—Room 209 and Room 207 —that had contaminated sinks, and none in Room 224, which had a contaminated bassinet. It was all very peculiar. We had a lot of contamination and we had a lot of infected babies, but there didn't seem to be any connection between the two. The only link we could think of was the nurses. The babies had no contact with each other. The bassinets were self-contained, and none of the babies shared any equipment or medication. The nurses might have carried the infection on their hands. They could do that without becoming infected themselves. Healthy adults don't succumb to Pseudomonas. But why did they carry it only to the babies in 227 and 229?

"The meeting ended on that unsatisfactory note. Dr. Gezon was as puzzled as the rest of us. But, of course, this puzzling point was only part of the investigation. We still had an epidemic to contain. We didn't understand the mechanics of its spread, but we did know what it was, and we thought we had enough information to bring it under control. We knew who was sick and we knew that the nursery was contaminated at eight specific sites. By Monday night, the nursery was as clean as Dr. Gezon and the nursery staff knew how to make it. All the sinks were scrubbed with sodium-hypochlorite disinfectant. The bassinet reservoirs were emptied and disinfected with an iodophor, and only those in use were refilled. As a further precaution, that water was to be changed every day. Certain nurses were assigned to take exclusive care of the infected babies. Also, in the hope of dislodging their infection, all the infected babies were placed on a five-day course of colistimethate given intramuscularly, and colistin sulphate by mouth. It wasn't necessary to isolate the infected babies. They were already isolated. And it was arranged that specimens be taken every day from all the babies in the nursery, and from all the sinks and bassinets and so on. That would give us a constant focus on the course of the epidemic.

"I didn't participate in the sanitation program. They didn't need me. I would have been an extra thumb. I went back to my office and back to the regular Health Department routine, but part of my mind was still out there at the nursery, and I got to thinking about something I'd read a few months before. It was a report in the *Lancet* about an outbreak of Pseudomonas in an English nursery that was traced to a catheter used to relieve throat congestion in the babies. The source of the trouble eluded detection for almost a year. What I particularly remembered about the report was a description of a new system of microbial identification. There are several different types of Ps. aeruginosa, and this report told how they could be differentiated by a laboratory procedure called pyocine typing. Pyocine typing makes use of the fact that certain strains of Pseudomonas will kill or inhibit the growth of other Pseudomonas strains, and it's a complicated procedure. Well, it occurred to me that pyocine typing might help to clarify our problem. It could at least tell us if all the sinks and all the bassinets and all the infected babies were infected with the same strain of Pseudomonas. I thought about it, and finally I called up Dr. Gezon. He saw the point at once. But, as I say, pyocine typing was then very new, and we didn't know where to turn for help. It could be that the system was being used only in England. We talked it over and decided that the best place to begin was at the Communicable Disease Center, in Atlanta. If anybody was doing pyocine typing in this country, the people there would certainly know. That was Monday evening. On Tuesday morning, I got on the phone to Atlanta and talked to one of my friends at C.D.C. and asked him what he knew about something called pyocine typing. 'Pyocine typing?' he said. 'Why, Shulman is working on that right now. I'll switch you over to him.' Shulman was Dr.

Jonas A. Shulman, and a fellow Epidemic Intelligence Service officer. He's now assistant professor of preventive medicine at Emory University. I described the case to him and asked him if he could help us out, and he was more than willing. He was eager. He wanted all the work he could get. So as soon as we finished talking I arranged for specimens of all our isolates to be airmailed down to Atlanta. Pyocine typing takes about two days. Shulman might have something for us by Thursday.

"Before I left the hospital, I went around to the premature nursery. That *Lancet* paper was still on my mind. Not pyocine typing. What interested me now was the source of the outbreak it described—that contaminated catheter. I looked up one of the pediatric residents and asked him what went on in the delivery rooms. I was thinking about contaminated equipment that the babies might have shared. For example, did they use a regular aspirator? The resident said no. The aspirators they used were all disposable and were discarded after each use. What about the resuscitators? Did they have humidifying attachments? A humidifier would mean water, and a possible breeding place for Pseudomonas. Another no. The resuscitators used in the delivery rooms were simply bags and masks attached to an oxygen line from the wall. I asked a few more questions along those lines and got the same kind of answers, and gave up. This wasn't a case like the *Lancet* case. So I was back in the nursery again. But the more I thought about those contaminated sinks and bassinets the less convinced I was that they were the source of our trouble. I just couldn't see any plausible link between those particular sites and those nine particular babies. But if the answer wasn't a piece of contaminated equipment, what else could it be? A contaminated person? And then I got a thought—maybe a contaminated mother. It sounded only too possible. A contaminated mother could very easily transmit an infection to her baby in the course of its birth. Childbirth is not a very tidy process.

"The next question was: Which mother? I thought I could answer that. It had to be the mother of the second sick baby. Not the baby in whom the infection was first diagnosed. The significant case was the second Pseudomonas baby—the term baby who came into the nursery with pneumonia and then developed diarrhea. The first sick baby had been healthy until the day before his death. He had been healthy for over three weeks. So he was actually No. 2. It wasn't hard to reconstruct the possible course of events. The infection was introduced into the nursery by the term baby and then spread to the other babies by the nurses. Pseudomonas is a difficult organism. You can't wash it off your hands with a little soap and water, the way you can the staphylococcus bug. To get Pseudomonas off, you have to scrub and scrub and scrub. And it's also extremely resistant to most disinfectants. But what made the infected-mother theory really attractive was that it seemed to explain what the environmental theory left unexplained. It explained why the infected babies were concentrated in Room 227 and Room 229. Both of those rooms were intensive-care rooms, and there was very little traffic between them and the other rooms. However, it was just a theory. It was based on the supposition that the diarrheal baby's mother was a Pseudomonas carrier. So the next thing to do was find out. The first thing I found was that the baby had been discharged on Saturday, and that he and his mother were now at home. I got the address, and Dr. Taylor and the baby's pediatrician gave me the necessary permission. I went out to the house and introduced myself, and the mother was nice and cooperative, and I got the specimens I needed and took them back to the Health Department lab.

"She wasn't a carrier. The preliminary laboratory report on my specimens was negative for Pseudomonas. That was the following day—Wednesday afternoon. But by then it didn't matter. We had something much more interesting to think about. The way it happened was this. I was in the nursery that afternoon, and one of the residents came over and told me they had another infected baby. New babies had been coming along every day, of course, and this one was a term baby born on Monday and sent up to the premature nursery for special care. He had had trouble breathing at birth, and had required extensive resuscitation in the delivery room. Well, a routine nasopharyngeal culture taken when he was admitted to the nursery had just been found to be positive for Pseudomonas. I looked at the resident and the resident looked at me. This was real news. That baby could not possibly have been infected in the nursery. The laboratory samples had been taken before he was even settled there. He could have been infected only in the delivery room. And there were just two possible sources of infection there. His mother was one, and the other was some piece of contaminated equipment. My guess was naturally the mother. I found out what room the new baby's mother was in, and made the necessary arrangements, and went up and took the standard nose, throat, and stool samples, and arranged for the hospital laboratory to culture them. When I got back to the nursery, Dr. Gezon was there talking to the resident, and I could tell from the look on his face that he had heard the news.

"The three of us went down to the delivery suite. We found the nurse in charge and told her what we were doing. She was terribly upset. It was most distressing to her to have us arrive in her domain on such a mission. But she was a good nurse and she cooperated perfectly. The room where the baby had been born was not in use, and she took us in and showed us around. There was a delivery table in the middle of the room, and a row of scrub sinks along the left-hand wall. On the opposite wall was a resuscitator of the type described to me the day before. It consisted of a rubber face mask and a rubber Emerson bag enclosed in a cellophane casing, and it looked very neat and clean. But something made me take it down for a closer look. There was a little dribble of water in the bottom of the Emerson bag. I showed it to Dr. Gezon, and he raised his eyebrows and passed it on to the resident. The resident took a sample of the water. There were five other rooms in the delivery suite, and luckily none of them were in use. We checked the resuscitator in every room, and every one was wet. The question was: How come? The nurse explained the delivery-suite cleaning procedure. There was one central wash sink, where all delivery-room equipment was washed. Everything was washed after every use, and then sterilized by steaming in an autoclave. Including the resuscitators? No—of course not. They were made of rubber, and rubber can't stand that kind of heat. The resuscitators were washed with a detergent, rinsed with tap water, and left on the drainboard to dry. It was possible, the nurse said, that they were sometimes returned to the delivery rooms before they were completely dry. We asked to see the wash sink. We were all beginning to feel sort of elated. I know I was. And when we saw the sink, that just about finished us. The faucet was equipped with an aerator—a standard five-screen water-bug heaven.

"It *was* a water-bug heaven. The laboratory cultured Pseudomonas from the swab samples we took from the aerator. It also cultured Pseudomonas from five of the six resuscitator samples. You can imagine how the delivery-suite nurses felt when those reports came down. They were crushed. Dr. Gezon was able to reassure them, though. He didn't consider them guilty of negligence. He considered them guilty of ignorance. They assumed, like almost everybody else, that city drinking water is safe. It is and it isn't. It's perfectly safe to drink, but it isn't absolutely pure. This is something that has only recently been recognized. There are water bugs in even the best city water. The concentrations are much too low in ordinary circumstances to cause any trouble, but a dangerous concentration can occur in any situation—like that provided by an aerator—that enables the bugs to accumulate and breed. An aerator is a handy device, but you'd probably be better off letting the water splash. It's certainly a device that a hospital can do without.

"The laboratory gave us three reports in all. The third was on the mother of the new baby. They found her negative for Pseudomonas, and that was welcome news. A positive culture from her would have been an awkward complication. Because everything else was very satisfactory. The contaminated resuscitators seemed to explain the concentration of infected babies in Room 227 and Room 229. Seven of the ten infected babies— inc.uding the diarrheal baby and the two that died—had—received at least some resuscitation in the delivery room, and it was reasonable to suppose that the three others had got their infection from the resuscitated babies by way of the nurses. There's plenty of evidence for that in the literature. I remember one report that showed that nurses' hands were contaminated simply by changing the bedding of an infected patient. It was Dr. Shulman, however, who finally pinned it down. His pyocine typing confirmed the circumstantial evidence at every point. Shulman did two groups of studies for us—one on the original material I sent him, and then another on the new infected baby and the delivery-suite material. The results of his studies were doubly instructive. They identified the delivery-room resuscitators as the source of the epidemic, and they eliminated the contaminated sinks and bassinets in the nursery. The different pyocine types of Pseudomonas aeruginosa are indicated by numbers. The Pseudomonas strain cultured from the delivery-suite aerator was identified as Pyocine Type 4-6-8. So were the isolates from the resuscitator bags. And so were those of all of the infected babies. Type 4-6-8 was also recovered from two pieces of equipment in the nursery, but I think we can safely assume that they had been contaminated indirectly from the same delivery-room source. They were a bassinet used by an infected baby and the sink in Room 227. The other contaminations in the nursery were a wild variety of types—6, 4-6, 6-8, 1-2-3-4-6-7-8, and 1-3-4-6-7-8. And where they came from wasn't much of a mystery. There was only one possible explanation. They came out of the water faucets, too."

That was the end of the formal investigation. It wasn't, however, the end of the trouble. The pockets of contamination in the sinks and elsewhere in the nursery and in the delivery suite were eliminated (and a system of ethylene-oxide sterilization set up for all resuscitation equipment), but the epidemic continued. In spite of the most sophisticated treatment (first with colistimethate and colistin sulphate, and then with colistin in combination with polymyxin B), the infected babies remained infected. Moreover, in the course of the next few weeks twelve new infections were discovered in the nursery. In two of the new victims, the infection developed into serious clinical illness. It was not until the middle of September, when the remaining infected babies were moved to an isolated ward in another part of the hospital, that the epidemic was finally brought under control.

Hospital infections of any kind are seldom easily cured. Pseudomonas aeruginosa is only somewhat more stubborn than such other institutional pathogens as Staphylococcus aureus and the many Salmonellae. These confined and yet all but unextinguishable conflagrations are, in fact, the despair of modern medicine. They are also, as it happens, one of its own creations. The sullen phenomenon of hospital infection is a product equally of medical progress and of medical presumption. It has its roots in the chemotherapeutic revolution that began with the development of the sulfonamides during the middle nineteen-thirties, and in the elaboration of new life-saving and life-sustaining techniques (open-heart surgery, catheterization, intravenous feeding) that the new antibacterial drugs made possible, and it came into being with the failure of these drugs (largely through the development of resistance in once susceptible germs) to realize their original millennial promise. Its continuation reflects a drug-inspired persuasion that prevention is no longer superior to cure. "In the midst of the development of modern antibacterial agents, infection has flourished with a vigor that rivals the days of Semmelweis," Dr. Sol Haberman, director (until his death last April) of the microbiology laboratories at Baylor University Medical Center, once noted. "It would appear that the long sad history of disease transmission by attendants to the sick has been forgotten again."

[1968]

CHAPTER **14**

The West Branch Study

DR. STEPHEN C. SCHOENBAUM arrived at West Branch (pop.
2,025), the seat of Ogemaw County, in northeastern Michigan, in
a rented Ford sedan at about four o'clock on the afternoon of May
19, 1968. It was a Sunday afternoon, and raining. His first stop was
at the Tri-Terrace Motel, on the outskirts of town. That part of
Michigan is resort country, and the motel was built around a trout
pond. He booked a room and left his luggage and asked the
manager where he could get something to eat. He was directed to
the Model Restaurant, near the traffic light on Houghton Avenue
(or Route 76), the main street of town. He drove to the restaurant
and had his supper, and then walked around the corner to the
office of the District Health Department. Dr. Schoenbaum was an
Epidemic Intelligence Service officer. He was twenty-six years old,
he had just completed a year of intensive epidemiological training
at the National Communicable Disease Center, in Atlanta, and
this was his first field assignment. He had come up to West Branch
(by plane that morning from Atlanta to Saginaw), in response to
a request for help from the Michigan Department of Public Health

in its investigation of an outbreak of infectious hepatitis that in just two weeks had grown from two to thirty-two cases.

Hepatitis is an inflammation of the liver. Its name defines its nature: *hepat* derives from the Greek for "liver," and *itis* is a Greek suffix meaning "inflammation." Many different agents, including drugs and chemicals, have the power to inflame the liver. Infectious hepatitis is one of two closely related inflammatory diseases of the liver that are confidently assumed to be of viral origin. The other is called serum hepatitis. These two varieties of hepatitis are by far the most common forms of the disease, and they constitute a public-health problem of ever-increasing concern throughout the Western world. In 1969, the most recent year of record, some fifty-five hundred cases of serum hepatitis and some forty-eight thousand cases of infectious hepatitis were reported in the United States. This would seem to present a total of around fifty-four thousand cases of viral hepatitis for that year, but public-health authorities read the record differently. It is their conviction that (for reasons of indifference or carelessness or misdiagnosis or subclinical infection) only a fraction of all cases of viral hepatitis that occur in this country are actually reported—probably only one in ten. They therefore read the record not as fifty thousand cases but as upward of five hundred thousand.

Until the middle nineteen-forties, when their differences were demonstrated by an international round of experiments involving human volunteers, infectious hepatitis and serum hepatitis were everywhere regarded as one and the same disease. It is not hard to understand why. They come on with equal abruptness, they produce identical signs and symptoms (chills, fever, lassitude, headache, loss of appetite, nausea, vomiting, abdominal pain, diarrhea, jaundice, prostration), and their impact on the liver is pathologically indistinguishable. They are also equally unresponsive to any specific treatment, and although full recovery (within two or three months) is the reassuring rule, both of them can permanently debilitate and sometimes even kill. They are not, however, identical. They spring from different viruses, and they are differently disseminated. Infectious hepatitis is usually transmitted by direct person-to-person contact or by food or water contaminated with the excreta of an earlier victim. Serum hepatitis is less conventionally, and less easily, spread. The virus of serum hepatitis lodges not in the gastrointestinal system but in the blood stream, and it can be conveyed only by an inoculation of infected blood. In addition to (and perhaps because of) these different modes of transmission, the different viruses establish themselves and multiply at different rates of speed. Infectious hepatitis generally manifests its presence about thirty days after exposure. In serum hepatitis, the incubation period ranges from three weeks to three or four months. An attack of hepatitis, like an attack of any viral disease (poliomyelitis, smallpox, measles, whooping cough) produces in its surviving victims an immunity to future infection. But the immunities conferred by infectious hepatitis and serum hepatitis are separate and distinct. An attack of infectious hepatitis protects against a second attack of infectious hepatitis but not against an attack of serum hepatitis, and serum hepatitis also protects only against itself. It is this fact—the fruit of a hundred painful human experiments—that makes it certain that infectious hepatitis and serum hepatitis are, at bottom, different diseases.

The avoidance of the viral hepatitides, despite their immunological potential, is very largely a matter of chance. Medical science has not yet succeeded (as it has with smallpox and poliomyelitis and measles and so many other viral diseases) in developing a protective vaccine. It is far from certain that it ever will. There is a formidable reason for this. The development of a vaccine begins with the laboratory cultivation of the relevant pathogen, and to the best of current knowledge the viruses of human hepatitis are parasites so specialized that they can grow and proliferate only in a living human cell. A reliable passive protection against infectious hepatitis is offered by gamma globulin. Gamma, or immune, globulin is a blood protein extracted from the pooled blood of at least ten thousand donors that (because many of these donors will have had infectious hepatitis) contains infectious hepatitis antibodies. The force of an attack of infectious hepatitis can be bridled and held to a subclinical, or unapparent, infection by an injection of gamma globulin given soon after probable exposure. The effectiveness of gamma globulin, however, is limited to this prophylactic role, for antibodies thus acquired are quickly (in four or five weeks) metabolized and dismembered. The effectiveness of gamma globulin is limited also to infectious hepatitis. Serum hepatitis does not seem to lend itself to gamma-globulin prophylaxis. Why this should be is uncertain. One suggested reason is that serum hepatitis is insufficiently common for its antibodies to be powerfully present in the ten-thousand-donor pools.

Infectious hepatitis has a natural history as long as that of any of the many epidemic diseases. There is no reason to doubt that its origins go back at least five thousand years, to the first coming together of men in city, or community, life. It is thought to have been glimpsed as an entity as early as the Hippocratic era, and it has certainly long been known (under such names as "field jaundice" and "camp jaundice") as a major military scourge. It has been recorded (by army surgeons) as a serious source of misery in every war since the Napoleonic Wars. Serum hepatitis has no such lineage, and its history is anything but natural. It is one of a growing number of upstart diseases that have been brought into being by a triumphant medical technology. Although it probably first appeared when man first learned to puncture his skin for religious or cosmetic purposes, serum hepatitis came into prominent medical view only with the development of smallpox vaccination, and it has mounted to its present eminence with the extension of that procedure to other diseases, and with the invention of the blood transfusion and the hypodermic syringe. Serum hepatitis very seldom occurs in epidemic form. Its rigid epidemiology inhibits its rapid spread. The few large outbreaks of serum hepatitis on record have chiefly stemmed from either blood-bank blood contaminated by infected donors or from slipshod immunization programs, and most of these have involved military or institutional personnel. Serum hepatitis most commonly occurs in occasional cases of insidious origin (its generally long incubation period makes retrospective explication difficult), and its typical appearances are almost always the result of a disregard of elementary hygiene. In 1961, a New Jersey osteopath infected some forty patients by not bothering to autoclave or otherwise sterilize the syringe he used for drug injections. More recently, in 1969, a girl was infected in a Larchmont jewelry store when she had her ears pierced with an unwashed communal punch. And the promiscuous tattoo needle has over the years produced almost as many cases of serum hepatitis as it has epidermal declarations of patriotism and filial love. But these are not typical victims. The typical victim of serum hepatitis in the United States these days is a phenomenon as grimly specialized as the disease itself. He is a young heroin (or amphetamine) user who shares a syringe with a friend. Serum hepatitis would thus seem to be, as Dr. Michael B. Gregg, chief of the Viral Diseases Section of the National Communicable Disease Center, has put it, "a public-health problem compounded by a sociological problem."

The typical victim of infectious hepatitis is a more prosaic person. He can be anyone confronted by a victim or who eats a contaminated meal or takes a drink of contaminated water. Not much (in the absence of a scrupulous universal cleanliness) can be done to protect him. The only defensive procedure possible is to block the spread (or a recurrence) of a discovered outbreak by finding and stopping its source and immunizing potential victims. That, however, is well worth doing, and that was Dr. Schoenbaum's job in West Branch.

Dr. Schoenbaum was received at the West Branch office of the District Health Department by the District Health Officer. She was Dr. Ophelia Baker. Her husband, Dr. Thomas Baker, one of five physicians then practicing in West Branch, was with her. The Bakers led Dr. Schoenbaum into Dr. Ophelia's office and gave him a cup of coffee and brought him up to date on the epidemic.

"It was still going," Dr. Schoenbaum says. "Seven new case reports had come in since Saturday. They raised the total from thirty-two to thirty-nine. There had been no deaths, and none was expected, but there were a couple of patients sick enough to be hospitalized. They were in the local hospital—the Tolfree Memorial Hospital. A gamma-globulin program had been started five days ago, on May 14, and the contacts of all the known cases were being inoculated. The epidemic was definitely an epidemic. Dr. Ophelia got out her records. Only seven cases of hepatitis had been reported in the preceding twelve months in the whole of Ogemaw County. There was as yet no epidemic curve—no chronological pattern. The data needed to draw it—the dates when the victims first became sick—were lacking. Dr. Ophelia had been too busy with the gamma-globulin program. It was quite a program. It ran almost two weeks, and they ended up immunizing something over seven thousand people. The information on the reported cases was the standard minimum—name, address, age. The cases were of both sexes, they seemed to live all over the county, and they ranged in age from five to thirty-nine. There was one moderately interesting fact. A majority of the cases were teen-agers, and most of them were boys. I didn't know what that signified, and I didn't try to guess. I simply copied off the names and addresses. All of them would have to be seen and interviewed. The results of the interviews would set the course of the investigation. I hoped they would, anyway. I knew there weren't any shortcuts. I could only hope that the truth would emerge from the evidence.

"I met with Dr. Ophelia again on Monday morning. The rain was over, and it was a fine spring day. We met by prearrangement at the hospital, and she introduced me at rounds to the other West Branch physicians. They gave me a very warm welcome. They were overworked and tired, and apparently under some pressure from the community. The point appeared to be that Ogemaw County is a summer-resort area, and a continuing epidemic could be bad for business. Also, I guess, a lot of the working population was laid up sick. We left the hospital, and I suggested to Dr. Ophelia that maybe I ought to have a look at the schools. There were two of them, and she took me around and I met the principals. One of the schools was the public school—a consolidated, all-grades school with an enrollment of around fifteen hundred. The other was a small Roman Catholic school. Both schools were within a block or two of the Houghton Avenue business district, and both were served by the same school buses. We then went on to Dr. Ophelia's office. It was almost ten o'clock and time for her to get back to the gamma-globulin program. The District sanitarian was waiting for us at the office. He's a very nice fellow named James Hasty. Mr. Hasty knows everybody in West Branch and almost everybody in Ogemaw County, and Dr. Ophelia had arranged for him to guide me around on my interviews and generally smooth my way.

"We took right off in my car. The family I wanted to call on first was a West Branch family I'll call Simpson. It was the largest family on the list, and it had the largest number of cases. I thought that might make for an instructive start. There were eight in the family—the parents, four boys (including twelve-year-old twins), and two girls. The children ranged in age from five to fifteen. Five of the children and Mrs. Simpson were sick. Only Mr. Simpson and one of the twins had escaped the epidemic. The Simpsons lived on the edge of town. Mrs. Simpson was in bed, too sick to see me, but Mr. Simpson happened to be at home. He and the five sick children told me what they could. There was municipal water in West Branch, but the Simpsons pumped their water from a private well. They bought milk from one or another of the Houghton Avenue stores, usually the A & P. They did their marketing at the same stores. They occasionally bought something at the West Branch Bakery, and they occasionally had a meal or a snack at one of the restaurants or one of the Dairy Queen drive-ins. There were two Dairy Queens in Ogemaw County—one in West Branch and one about twelve miles out of town on the Rose City road. None of the family had recently attended any large gathering at which food or drinks were served. Hepatitis was the only recent illness in the family. The children all were pupils at the public school. I took a sample of water for laboratory analysis, and then got down to the individual cases. What I wanted now were the onset dates of illness—when the first symptoms appeared. This was the essence of the interview. The onset date minus thirty days would give me the approximate date of exposure. We take thirty days as the average incubation period in infectious hepatitis. Mr. Simpson went upstairs and talked to his wife and came back with a calendar, and he and the children worked out the dates. The onset dates for the two girls were May 3 and May 4. The boys— all three of them—took sick on May 5. Mrs. Simpson's first symptoms appeared four days later, on May 9. Well, that was a start. Those six dates made a kind of pattern. If the Simpsons were in any way typical, the epidemic had its beginning in early April.

"They seemed to be entirely typical. Mr. Hasty and I made three more calls that day, and I interviewed four more patients. One was a boy of eleven whose history almost exactly paralleled that of the three Simpson boys. His illness, like theirs, had begun on May 5. He, too, attended the public school. He occasionally had a meal with his parents at one or another of the Houghton Avenue restaurants, he sometimes had a snack at one of the Dairy Queens, and his parents traded at the A & P and most of the other food stores, including the West Branch Bakery. His history differed from that of the Simpsons in only one respect. The family water was municipal water. Two of the three other patients were a woman of thirty-nine and her eight-year-old daughter. She had become sick on May 11 and her daughter on May 10. The daughter attended the Catholic school. Their histories otherwise were substantially the same as the rest. The last patient I interviewed that afternoon was a man—a bachelor of thirty. He was a salesman. His illness had begun on May 13. That suggested that his exposure had occurred on April 13, but his records suggested an earlier date. They showed that he had been out of town between April 11 and April 13, and from April 15 to April 20. April 14, the one day in that period when he had been in West Branch, was a Sunday, and he had spent it at home. His general history, with one exception, was much like that of the others. It included municipal water, the A & P and other markets, the restaurants, and the two Dairy Queens. The exception was the West Branch Bakery. He said he never set foot in the bakery. He and the owner had quarreled. I thought about that on the way back to the office. I had practically decided that water and milk could be eliminated as likely vehicles of infection here. And now I thought maybe we could drop the bakery, too. I mentioned this to Mr. Hasty. He agreed about water and milk. The community was about equally divided between municipal water and private wells, and all the milk came from sources outside the county. But about the bakery —the fact that the salesman never traded there didn't really mean very much. The bakery was more than just a store. It also supplied bread and pastries to most of the markets in town, and to all of the restaurants.

"I talked to Atlanta that night. I didn't have much to report, but they had some news for me. They were sending me some help. He was a fellow EIS officer named James M. Gardner. Gardner was between assignments. He had just finished an assignment at the Michigan State Public Health Department, down in Lansing, and he wasn't due to report for his next assignment—in California —until the following Monday, so they had asked him to spend the interval up here with me. He arrived on Tuesday morning. I hadn't asked for help, but I was glad enough to have it. I got gladder as the week went on. Eight more cases were reported on Tuesday, four more came in on Wednesday, and on Thursday there were six. That brought the total number of cases in West Branch and the rest of Ogemaw County up to fifty-seven. It would have taken me two weeks to see that many people, but with Gardner to help it only took three days. By Thursday night, we had seen and interviewed all the West Branch cases and most of those in the outlying county. The results were all we had hoped for. They were even more than that. We got a good epidemic curve. It was virtually the same as our final curve. It showed that the epidemic began with a single case on April 28, built to a peak of thirteen cases on May 12, and then dropped to a final case on May 20. All but five of the cases had had their onset between May 2 and May 14. This clustering was very instructive. It indicated the time of exposure—around the middle of the first half of April. It established what we had only assumed before—that all of the cases were infected by the same source. They were too close to-gether to be related in any other way. And it told us that the epidemic was either ending or already over. But that wasn't all. Those were merely the formal results. We learned some even more interesting things in our travels around the county. One of these was that the epidemic might be wider in scope than any of us had supposed. Several people we talked to had friends or relatives living elsewhere who had been in West Branch recently and who now were sick with hepatitis. Or so they said. We also heard and confirmed that there had been some hepatitis in the community about a month before the epidemic began. There were two cases in particular. One was a girl who worked at the Rose City Dairy Queen. She left work sick on April 4. The other was a man I'll call John Rush. Rush was a baker at the West Branch Bakery, and he took sick on April 6.

"We sat down with Dr. Ophelia on Thursday night and had a conference. The question was what to do next. There was an embarrassment of promising possibilities. Talk to Rush? Talk to the girl? Look into those outside cases? And we still had a few more county cases to see. We decided to let the county cases wait. They couldn't tell us much that we didn't already know. Rush and the girl were a different matter. They were very interesting news. Either one of them could be the index case—the source case of the epidemic. They both were sick at the right time, and they both had jobs that had to do with food. The girl's job was especially provocative. She made us think of a recent classic case—an epidemic in Morris County, New Jersey, in 1965 that was traced to contaminated strawberry sauce in an ice-cream drive-in. But, of course, that wasn't enough. Our histories showed that everybody seemed to patronize the Dairy Queens, but they also showed that everybody seemed to patronize the bakery, too. We needed something stronger than suspicion. We needed something that would narrow the field a bit. That brought us around to the outside cases. A visitor to West Branch would probably have fewer contacts than a resident. There was one presumably outside case within reasonable reach. She was a schoolteacher I'll call Miss Brown, and she lived in the village of Au Gres, on Saginaw Bay, about thirty miles southeast of West Branch. We decided she was worth an immediate visit. Our information was that she had been in West Branch recently. It could be very helpful to find out when and where.

"The three of us drove down to Au Gres the next morning. That was Friday, May 24. Dr. Ophelia was able to come along because the gamma-globulin program was just about over. She was determined to be in on Miss Brown. She had the same feeling that Gardner and I had that this could give us a lead. You may wonder why we didn't just pick up the telephone and give Miss Brown a call. The reason is this. The epidemiological experience has been that telephone interviews are seldom satisfactory. A telephone conversation is too abrupt, too remote. People make more of an effort when they're talking face-to-face. So we paid Miss Brown a visit. She was sick, but not too sick to see us. There was no doubt about its being hepatitis. She was still icteric—still jaundiced. Dr. Ophelia handled the introductions. Her presence was invaluable. She helped Miss Brown to relax and think back and remember. Miss Brown's memory was excellent. As a matter of fact, it was exhilarating. The chronology of her illness placed her squarely in the epidemic. Her onset date was May 5. It was true that she had been in West Branch recently. She had stopped there twice in the past three months—on March 20 and again on April 5. But only very briefly. On the first occasion she had a cup of coffee in one of the restaurants. That was all—just a cup of coffee. She was sure of the date because her mother had died the day before and she was on her way across the state to Petoskey for the funeral. Petoskey was her home town. She drove back to Au Gres on March 23, but this time she didn't stop at West Branch. She was also sure about her second West Branch visit. April 5 was a Friday and the beginning of spring vacation, and she was driving to Petoskey again to spend the holiday with her father. It was late afternoon, and she was hungry. She stopped in West Branch and went into the bakery there and got something to eat on the way. She bought some pastry—a piece of coffee cake and three cupcakes with yellow Easter-bunny icing. That definitely was on her way to Petoskey. It couldn't have been on her return. She returned on April 14, and April 14 was a Sunday—Easter Sunday. We thanked Miss Brown and marched out to the car and headed back to West Branch. We sang all the way."

Miss Brown's emphatic testimony marked the end of the floundering phase of the West Branch study. It also marked the end of Dr. Gardner's participation in the investigation. He left on Saturday for his new assignment in California. Dr. Schoenbaum was alone again, but again for only a day. He spent that day on a final round of interviews. Four more cases had been reported since Thursday. That brought the epidemic total up to sixty-one. By Sunday night, however, he had histories on them all. His new reinforcement arrived on Monday. He was an EIS officer on loan from the Shelby County Public Health Department, in Memphis, named E. Eugene Page, Jr. Dr. Schoenbaum and Dr. Page spent Monday afternoon and evening at the Tri-Terrace Motel reviewing the collected case histories. They spent Monday night at the West Branch Bakery.

"We got there around midnight," Dr. Schoenbaum says. "That was when their baking day began. They baked six nights a week —Sunday through Friday. I had already talked to the owner and made the necessary arrangements, and he was there to meet us and show us around. I had explained the nature of hepatitis and told him what we knew about the epidemic. I told him that it unquestionably stemmed from a common source, and that we had some evidence that the source might be his baked goods. One of his people, as he knew, had been sick with hepatitis back in April, and he could be what we called the source case. Our evidence wasn't proof, but it was enough to require an investigation of the bakery. We wanted to see how the baking was done. We wanted to see if bakery goods could be a source of infection. The owner was extremely cooperative. He could very easily have been hostile and difficult. In return, I assured him that we wouldn't publicize our visit. We would keep it a secret until the end of the investigation. There was no chance that anybody would see us arrive. West Branch is asleep by midnight.

"He let us in the back door. The back room was the baking room. There were two bakers—the head baker and Rush. The owner was also a baker by trade, but he didn't do much baking anymore. Only when they were shorthanded. He worked days in the shop up front, and after introducing us, he left and went home. Page and I stood around and watched the bakers work. They were mixing dough. There was a bread dough and a pastry dough. They mixed them separately, but they mixed them both by hand. They did everything by hand—mixing, kneading, shaping. Even when they used a mechanical mixer, they scooped the dough out by hand. Rush was older than the head baker, and he was nice and friendly, but he didn't look very bright. But he seemed to know his job. We watched them, and it was interesting for a while, and then it began to get boring. It also began to seem like a waste of time. Rush and the bakery was our one big hope. The Dairy Queen girl had just about dropped out of sight. The case histories made her very unlikely. When we checked them over carefully, it turned out that her Dairy Queen wasn't as popular as we had thought. It was the other Dairy Queen that most of the people went to. She might have been the source of a case or two, but she couldn't be the epidemic index case. That was Rush. Or so we thought until we watched the baking. He had every opportunity to contaminate the dough, but the dough was only the beginning. It then went into the oven and baked for half an hour or more at a temperature of at least three hundred and fifty degrees. Thirty minutes at anything much over a hundred and thirty degrees will kill the virus of infectious hepatitis.

"But we stayed and watched. They made some doughnuts—fried cakes, they called them—and cooked them in boiling oil. Our hopes got dimmer and dimmer. They offered us some of the fried cakes, and Page ate a couple. I declined. I don't know why exactly. There wasn't any risk. Rush was fully recovered, and the head baker had never even been sick. I said I wasn't hungry. I really wasn't—I couldn't have eaten anything anywhere just then. The time dragged on. Around three-thirty, they finished baking and emptied the ovens and stacked the bread and rolls for sale in the shop up front or for delivery to the store and restaurant customers. The next step was icing and glazing the pastry. Some of the frostings were already cooked and ready to use, and some they made without cooking. They started in by glazing the fried cakes. The glaze was in a five-inch pan about five feet long. What they did was take a double handful of cakes and dip them in the glaze and turn them over and around and then lift them out and line them up on a rack. That was glazing. That was the entire process. But it was a revelation to us. Page and I were wide awake in a minute. The process was much the same with the other kinds of pastry. I remember at one point Rush was up to his elbows in glaze. He piled a load of fried cakes on the rack, and then turned and walked across to another counter and started icing cupcakes. He didn't bother with a pastry tube. He used his hands—the same bare hands he had used on the fried cakes. He scooped up a handful of icing and squeezed it out through his fingers. He was an expert. He did a beautiful job. It was horrifying. Page and I could hardly keep from shouting. They called the West Branch Bakery a home bakery. What they should have called it was a hand bakery.

"Rush was almost certainly the index case. We watched and asked questions, and by five o'clock, when we finally called it a night, we had everything we needed but proof. Rush had had every opportunity to contaminate the pastry, and also had had the capacity. His understanding of hygiene was very primitive. Baking was the only protective process. Anything that was contaminated after that would stay contaminated forever. Watching Rush work, we began to wonder when the contamination might have occurred. Did it happen on a succession of nights? Or on a single night? We found that a single contamination could even account for people being exposed over a period of several days. The head baker told us that unused glazes and icings were carried over from one night to the next. He said they often used leftover glaze to start a new batch. We also learned that the bakery regularly sold day-old pastry as well as day-old bread. And not only that. Unsold pastry was often frozen and sold a few days later. The hepatitis virus can be killed by freezing, but not that kind of freezing. A few days isn't long enough. It takes a year or more. All of those points were persuasive. We began to incline toward the idea that the contaminated pastry was contaminated in the course of a single night. It actually seemed more probable than a series of accidents —even for a man like Rush. The night that seemed most probable was the early morning of Friday, April 5. It was supported by the clustering of onset dates, and by Rush's own experience. April 6 was the day he consulted a doctor—the apparent date of onset. But he had obviously been infectious at work that Thursday night and Friday morning.

"There was only one way to pin it down. We would have to interview the cases again. If we were right about Rush and the bakery, the proof would be forthcoming. We would find that all of the victims had consumed some iced or glazed pastry from the West Branch Bakery between Friday, April 5 and the end of the following week. The period could be a bit open-ended, but it had to begin on April 5. That was the day—the only day—that Miss Brown had visited the bakery. Page and I didn't do much work on Tuesday. We slept until early afternoon. At some point, I had a call from Atlanta. They were sending up a distinguished foreign visitor to observe our methods. His name was Zdenek Jezek, and he was a senior WHO medical officer stationed at Ulan Bator, in Outer Mongolia. He arrived that evening. I remember his greeting. 'I am Jezek of Czechoslovakia,' he said. So he was a Czech. That was a little disappointing. We had been expecting a Mongolian. As a matter of fact, he was a thorough Czech. We briefed him on the epidemic, and when we came to the age and sex distribution of the cases, he shook his head. There was something wrong. Teen-age boys didn't eat pastry. Not in Czechoslovakia. Bakery customers were women and young children. He couldn't believe that things were different in Michigan. He thought it over, and then announced that he would get the facts by conducting an investigation of his own. All we had to do was arrange for him to sit somewhere out of sight in the bakery and record exactly who came in. I thought that might be interesting, and I said I would make the arrangements, but maybe he ought to wait a day or two and get his bearings first. I suggested that he sit in with Page and me on our new round of interviews.

"We began on Wednesday morning, and we began with the out-of-county cases. They would be the quickest way to determine if we were moving in the right direction or completely wrong. This time we worked by telephone. We had to. There were thirteen substantiated cases, including Miss Brown, and all of them were out of easy reach. Two were even out of state—a man in Indianapolis and another in Wethersfield, Connecticut. The first case we called was a man in Mount Pleasant, about fifty miles to the south. He confirmed that he had hepatitis. The onset date was May 3. Yes, he had been in West Branch, and in early April—April 5 through April 7. He and his wife had driven up to attend a wedding. We knew about the wedding, and none of the principals or other guests was on our epidemic list. He and his wife had eaten all of their meals with the other wedding guests. He stopped and thought a minute. Except on the day they arrived. That was April 5, and he dropped in the West Branch Bakery and bought four iced cupcakes. Which he ate. He wasn't a teen-ager—he was twenty-two—but he ate them all. I don't know what Jezek made of that. But I do know that Page and I began to feel very good.

"The other out-of-county cases told the same fascinating story. It was fantastic. There were two married sisters from Detroit with four children who spent three days in April—April 8 through April 10—in a cottage in the woods about ten miles out of town. On the morning of April 9, one of the women drove into West Branch alone to mail some letters, and brought back half a dozen fried cakes from the West Branch Bakery for lunch. Three of the cakes were glazed and three were plain. The two women and one of the children ate the glazed cakes, and they were all three sick with hepatitis. Two became sick on May 7 and the other on May 11. The three other children were well. I talked to the man in Connecticut. I ran him down in a hospital. Yes, he had hepatitis. Yes, he had been in West Branch. It was his home town and he had been there to visit his mother. He arrived on April 4 and left on April 6. On April 5 or April 6, he wasn't sure which, he had dropped into the West Branch Bakery to see the owner. They were old friends. The owner gave him a couple of glazed doughnuts, and he ate them while they talked. The man in Indianapolis was like the group from Detroit. He spent three days—April 7 through April 9—at a cottage down at Clear Lake, and on Monday, April 8, he came into town to get his shoes repaired. While he was waiting, he walked into the West Branch Bakery and bought a piece of iced Danish pastry. A thirteen-year-old boy from Roscommon County was in town on April 5—and so on. Every one of the out-of-county cases confirmed the place, the date, and the vehicle.

"We still had the community cases to question. There was no real doubt in our minds about the answers we would get, but we had to make the effort. We had to be thorough and sure. Most of what is known about viral hepatitis has come from epidemiological studies, and every careful study can be a contribution. There was also Jezek and his bakery survey. I introduced him to the bakery owner, and they worked out the details—the logistics. Jezek is a small man and they found a little corner where he could sit unnoticed behind a showcase and have a view between the shelves of all the comings and goings. He spent a full day there —from seven in the morning until the bakery closed at six o'clock that night. His findings were useful. They combined with the information that Page and I were getting to explain the big teen-ager clientele. He found that adults—mostly women—were in and out of the store all day. Very few young children—children under ten—ever came into the bakery, and then only in the late afternoon. The teen-agers showed up in a body at noon and again after three o'clock. The reason, as our interviews showed, was very simple. The teen-agers were, of course, high-school kids, and the high-school kids were allowed to leave the school grounds for lunch. The younger kids—the grade-school children and all of the children at the parochial school—had lunch at school. Jezek's tabulations showed that the bakery was a favorite teen-ager place for lunch. Almost half of the customers he counted in his survey —ninety-six out of a total of two hundred and fourteen—were teen-agers. Moreover—and this again confirmed our information —they all bought pastry: glazed fried cakes, iced cupcakes, chocolate éclairs, frosted twists, iced brownies. That and a Coke or something seemed to be the popular lunch. The only difference between Jezek's findings and our epidemic data was that his teen-agers were about evenly divided between girls and boys. I have no idea what that means—if it means anything. Maybe the girls were dieting back in April. But it seemed to satisfy Jezek.

"The second series of community interviews took about a week. We finished the job on Monday, June 3. We tidied up and said our good-byes on Tuesday, and on Wednesday we left for home. The total number of cases in the epidemic turned out to be seventy-six —sixty-three in West Branch and Ogemaw County, and thirteen in Detroit, Indianapolis, Wethersfield, and elsewhere. We managed to interview all but one of the local cases. The only one we missed was a man who was away on a trip of some kind. We ended up on Monday night, and the place we ended at was exactly where I had started on that Monday morning just two weeks before—with the Simpson family. With Mrs. Simpson, actually. She was now fully recovered and able to see us. It was very strange. I mean, if I had known the right questions to ask, I might almost have solved the whole problem on that first visit. This is what Mrs. Simpson told me. On Saturday morning, April 6, she went marketing, and on the way home, she stopped at the bakery and bought some pastry—an assortment of glazed fried cakes and iced cupcakes. She was sure of the date because she remembered hurrying home to watch the news on television. Martin Luther King had been assassinated on Thursday, and there had been riots in Memphis on Friday and she wanted to see if the trouble was still going on. Mr. Simpson was working, but the two daughters were at home. Mrs. Simpson and the girls sat down at the television and they each ate a cupcake while they watched the program. After a while, the two oldest boys came in, and they helped themselves to some pastry. Then the twins arrived. There was only one piece of pastry left—a fried cake. They saw it and made a grab, and the one I'll call Jerry lost out. He didn't get any pastry. And he was also the twin who didn't get hepatitis."

[1971]

CHAPTER 15

The Huckleby Hogs

THE TELEPHONE RANG, and Dr. Likosky—William H. Likosky, an Epidemic Intelligence Service officer attached to the Neurotropic Viral Diseases Activity of the Center for Disease Control, in Atlanta—reached across his desk and answered it, and heard the voice of a friend and fellow-EIS officer named Paul Edward Pierce. Dr. Pierce was attached to the New Mexico Department of Health and Social Services and he was calling from his office in Santa Fe. His call was a call for help. Three cases of unusual illness had just been reported to him by a district health officer. The victims were two girls and a boy, members of a family of nine, and they lived in Alamogordo, a town of around twenty thousand, just north of El Paso, Texas. The report noted that the family raised hogs, and that several months earlier some of the hogs had become sick and died. The children were seriously ill. Their symptoms included decreased vision, difficulty in walking, bizarre behavior, apathy, and coma. This complex of central-nervous-system aberrations had immediately suggested a ural encephalitis, but that, on reflection, seemed hardly possible. The encephalitides

236

that he had in mind were spread by ticks or mosquitoes, and the season was wrong for insects. This was winter. It was, in fact, midwinter. It was January—January 7, 1970.

"I'm afraid I wasn't much help," Dr. Likosky says. "I could only listen and agree. It was just as Ed Pierce said. The clinical picture was characteristic of the kind of brain inflammation that distinguishes the viral encephalitides. More or less. But there were certainly some confusing elements. It wasn't only that the time of year was odd for arthropod-borne disease. The attack rate was odd, too. There were too many cases. A cluster like that is unusual in an arbovirus disease. Also, the report to Ed had made no mention of fever. That was odd in a serious virus infection. It was odd, but not necessarily conclusive. The report was just a preliminary report. It was very possibly incomplete. At any rate, Ed was driving down to Alamogordo the following day and he would see for himself. Things might look different on the scene. Meanwhile, about all I could suggest was the obvious. Verify the facts. Check into the possibility of a wintering mosquito population. Check up on the hogs. This might be a zoonosis—an animal disease. It might be a disease of hogs that the three children had somehow contracted. Review the signs and symptoms. Not only for fever but also for stiff neck. Stiff neck is particularly characteristic of encephalitis.

"That was a Wednesday. Ed called me from Alamogordo on Friday. He and one of his colleagues—Jon Thompson, supervisor of the Food Protection Unit of the Consumer Protection Section of the state health department—had been out to the house, and he had some more information. The victims were children of a couple named Huckleby. They were Ernestine, eight years old; Amos Charles, thirteen; and Dorothy, eighteen. Ernestine was in the hospital—Providence Memorial Hospital, in El Paso. She had been there since just after Christmas. Amos Charles and Dorothy were sick at home. None of the other children—two girls and two boys—was sick, and neither were the parents. Huckleby worked as a janitor at a junior high school in Alamogordo. He only raised hogs on the side. All that was by way of background. The interest-

ing information was about the Huckleby hogs. There were several
items. The hog sickness happened back in October. Huckleby had
a herd of seventeen hogs at the time, and all of a sudden one day
fourteen of them were stricken with a sort of blind staggers—a
stumbling gait and blindness. Twelve of the sick hogs died, and
the two others went blind. That was interesting enough, but it was
really only the beginning. In the course of their talk with Huckleby, Ed and Jon Thompson learned that the feed he gave his hogs
included surplus seed grain. Well, seed grain isn't feed. Seed grain
is chemically treated to resist all manner of diseases, and when it
is past its season and has lost its germination value, it is considered
waste and is supposed to be destroyed. Huckleby said that the
grain—it was a mixture of several grains, apparently—was the
floor sweepings of a seed company upstate. He said a friend had
given it to him and he knew it contained treated grain. Or, rather,
he knew it included treated grain. It was also partly chaff and
culls, but he took care of that by cooking it with water and garbage
in a metal trough before he fed it to his hogs. He said he had been
given about two tons of the grain, and he still had some left. He
also said that he had slaughtered one of his hogs back in September for home use. They had been eating it right along, but there
was still some left in his freezer. Ed told him to leave it right where
it was. There might be nothing wrong with it, but there was no
use taking chances.

"Poison was a tempting possibility. The symptoms had the look
of poison, but it wasn't a look that any of us were familiar with.
That was the trouble. Nothing seemed to fit. Huckleby was even
a little vague about whether the slaughtered hog had been fed the
seed-grain feed. Moreover, he wasn't the only Alamogordan to
feed that grain to his hogs. He said that five of his friends had got
the same grain from the seed company, and none of their hogs had
become ill. Apparently, that was true. Ed and Jon had talked with
the other hog raisers. They weren't quite sure about the grain they
had used, but they were positive that none of their hogs had
sickened or died. The local records confirmed what they said. The
local records also confirmed that the human outbreak was

confined to the Huckleby family. There were no comparable illnesses anywhere in the Alamogordo area. Ed and Jon had visited
the butcher shop where the Huckleby hog had been processed.
The butcher testified that the meat had appeared to be in prime
condition. An examination of the shop was negative—there was
nothing to suggest that the meat might have become contaminated
during processing. And, finally, there was the fact that others in
the family had eaten quantities of the pork, and only those three
were sick. Nevertheless, Ed was taking the usual steps. He had
samples of grain from Huckleby's storehouse, samples of pork
from the family freezer, and samples of urine from all members
of the family then at home—that is to say, all but Ernestine. He
had arranged for the State Laboratory to examine those for viral
or bacterial contamination, and he was sending specimens to William Barthel, chief of the Toxicology Branch Laboratory of the
Food and Drug Administration, in Atlanta, for toxicological analysis. But even that wasn't all. Ed had still another piece of information. The clinical picture was pretty much as originally reported. Except in one respect. Ernestine's symptoms *did* include
fever. High fever. A local doctor reported that her temperature at
one point got up to 104 degrees. And—oh, yes, there were no
mosquitoes in Alamogordo in January.

"I talked to Ed again on Saturday. He was back in Santa Fe,
and that was the reason for his call. He called to invite me to
participate in the field investigation. Ed had a big rubella-immuni-
zation program coming up at a Navajo reservation the next week
that he couldn't put off, and there was nobody to run it but him.
It was up to him to at least get it started. He had talked to his chief
in New Mexico, Dr. Bruce Storrs, director of the Medical Services
Division of the state health department, and he had talked to my
boss at CDC, Dr. Michael Gregg, chief of the Viral Diseases
Branch. They both approved the proposal. So did I. I was very
eager."

Dr. Likosky left Atlanta by plane the following morning. That was
Sunday, January 11. He flew to Albuquerque, where, by prear-

rangement, he met Dr. Pierce in the early evening. They talked and ate dinner and talked until nearly eleven. Dr. Likosky then rented a car and drove down through the mountains and the desert to Alamogordo. He spent the night (on Dr. Pierce's recommendation) at the Rocket Motel there, and on Monday morning (following Dr. Pierce's directions) he drove out to the Huckleby house.

"I wasn't checking up on Ed or Jon Thompson or the district health officer, or anything like that," Dr. Likosky says. "I simply wanted to see for myself. I wanted to start at the beginning. I got to the Huckleby house at about eight-thirty. Huckleby was at work, at the school, but Mrs. Huckleby was at home, and she received me very nicely. The first thing I learned was that Amos Charles and Dorothy were now also in Providence Memorial Hospital. They were too far gone in coma to be treated at home anymore. They had been taken down to El Paso only the day before. I liked Mrs. Huckleby at once. You could tell she was a good mother. Easygoing, but kind and loving. And she was going to be a mother once again. She was very obviously pregnant. The baby was due, she said, sometime in March. She was a religious woman, too, and that gave her a certain serenity. She believed that everything was in the hands of God. She was also an excellent historian. She remembered every detail of each child's illness. We began with Ernestine. She came home sick from school a little before noon on December 4. She said she had fallen off the monkey bars at recess, and she had a pain in her left lower back. Mrs. Huckleby said she felt hot to the touch. A few days went by, and Ernestine continued to complain of pain and just not feeling right. On December 8, Mrs. Huckleby took her to a neighborhood doctor. It was he who found that she had a temperature of 104, but he found nothing else of any significance. He prescribed aspirin and rest in bed. There was still no improvement, and on December 11, Mrs. Huckleby took her back to the doctor. It was a different doctor this time—the first doctor was off that day—and he did find something. He noticed that Ernestine wasn't walking right— that she was staggering—and he arranged to have her admitted to

Providence Memorial the next day for observation. I got the details later from the doctor. Ernestine's walk and the history of her fall had frightened him a little. It raised the possibility in his mind of a subdural hematoma. A subdural hematoma is a gathering of blood between certain membranes that cover the brain. It is usually caused by a blow or a fall, and it can be extremely dangerous. I saw his point. It was a perfectly reasonable suspicion. But, of course, he was mistaken. It wasn't that. The hospital made the various tests, and ruled it out.

"Mrs. Huckleby told me that the hospital sent Ernestine home for Christmas. She was discharged on December 19. She got steadily worse during her stay at home, and she was readmitted on December 27. It was a pretty dreary Christmas for the Hucklebys. Amos Charles took sick while Ernestine was home—on Christmas Eve. He went to bed with an earache, and when he woke up Christmas morning, he said he couldn't see very well. The next day, Dorothy took sick. It began as a generalized malaise. Then she began to feel 'woobly.' Then she couldn't walk at all, and her speech began to slur. Meanwhile much the same thing was happening to Amos Charles—trouble walking, trouble talking, trouble seeing. Then he went into what Mrs. Huckleby called a 'rage.' It was a good descriptive word. He was wild, uncoordinated, thrashing around on the bed. He and Dorothy both got steadily worse. They sank into coma. And, finally, on Sunday, they had to go into the hospital. I spent the rest of Monday talking to the Alamogordo doctors, and then to Huckleby. He wasn't easy to talk to. He was shy, and he didn't say much. He raised his hogs in a couple of pens he rented at a hog farm on the outskirts of town. He collected garbage from his neighbors, and he fed his hogs a mixture of that and grain and water. He cooked it because the law required that garbage be treated that way for feed. He stored his grain in a shed at home, and he kept the shed locked. I saw the grain—what was left of it. It was a mixture of grain and chaff. Anyone would have wondered about it. Most of the grain was coated with a pink warning dye. Huckleby said he had stopped feeding the grain as soon as he heard it was dangerous. I said I certainly hoped so. I questioned him particularly about the sickness that had afflicted his hogs in October. He told me this. When he saw his hogs one morning, they were well. When he saw them next, around four o'clock that afternoon, they were blind and staggering, sick and dying. That was strange. I hadn't ruled out encephalitis in the children, but this was different. There almost had to be a connection between the sick hogs and that chemically treated grain. And I couldn't connect a sudden illness and sudden death with any chronic poisoning.

"Everybody was still thinking in terms of encephalitis. That was the admitting diagnosis on Amos Charles and Dorothy at Providence Memorial Hospital. I spent Tuesday and Wednesday in El Paso. I read the records and I talked to the doctors involved. The relevant test results were either normal or not very helpful. The children's kind of blindness was identified as tunnel, or gun-barrel, vision—a constriction of the visual fields. Urinalysis was positive for protein in all three patients, and the presence of protein in the urine always indicates some impairment of function, which is unusual in most encephalitides. And I saw the patients. It was terrible. I'll never forget them. It was shattering. Amos Charles was a big, husky, good-looking boy, and he lay there just a vegetable. His brain was gone. Ernestine was much the same. Dorothy was a little more alive. Her arms kept waving back and forth. Pendular ataxia, it's called. The charts on the children spelled everything out.

"I talked to Ed on Tuesday afternoon, and he drove down to El Paso that night. He had his rubella campaign well started, and he was eager to get back into the Huckleby investigation. He brought some confusing news along. One of Jon Thompson's people had run down the source of Huckleby's grain and identified at least one formulation of the material it was treated with. It was a fungicide called Panogen. Panogen is cyano methyl mercury guanidine. And mercury is a classic poison. It's one of the most dangerous of the heavy metals. Well, that should have clarified things a bit, but it didn't. It only added to the confusion. Because the clinical picture the Huckleby children presented looked nothing like classic mercury poisoning. The textbook symptoms of acute mercury poisoning are essentially gastrointestinal—nausea, vomiting, abdominal pain, bloody stools. That and severe kidney damage. Chronic mercury poisoning is entirely different. The kidneys are not seriously involved. Apparently, they can safely handle small amounts of mercury. The features in chronic mercurialism are an inflammation of the mouth, muscular tremors—the famous hatter's shakes—and a characteristic personality change. Shyness, embarrassment, irritability. Like the Mad Hatter in *Alice in Wonderland.* We didn't know what to think.

"We couldn't dismiss the possibility of mercury. But we couldn't quite accept it, either. The only thing we were finally certain about was that we weren't up against an acute viral encephalitis. The clinical picture—particularly the gradual development of symptoms—and the epidemiology made it quite unlikely. It made it incredible. We decided that we were left with two general possibilities. One was a more insidious kind of encephalitis—a slow-moving viral infection of the brain. The other was an encephalopathy. Encephalitis is a disease of the gray matter—or gray nervous tissue—of the brain. Encephalopathy involves the white matter—the conducting nerve fibres. A toxic encephalopathy was the kind we chiefly had in mind. The cause could be one of a variety of substances. The heavy metals, of course. Arsenic. Or numerous drugs on the order of sedatives and tranquilizers. The slow-moving-viral possibilities were more exotic. Rabies is in that class. And kuru, a fatal neurological infection in New Guinea that is perpetuated by cannibalism. And scrapie, a disease of sheep, but a human possibility. And others. The trouble was that Ed and I weren't neurologists. But I had a friend who was—Dr. James Schwartz, at Emory University, in Atlanta. So, just on a chance, I called him up and gave him the clinical picture and asked him what *he* thought it sounded like. He gave us quite a shock. He said it sounded a lot like one of the multiple scleroses—a rapidly progressive demyelinating process called Schilder's disease. Except, he said, three cases would constitute an epidemic, and an epidemic of Schilder's disease had never been reported before. He was inclined to doubt that one ever would be. Another, and more reasonable, possibility, he said, was heavy-metal poisoning. Except that our picture wasn't quite right for lead. Or mercury. Or anything else that readily came to mind. I thanked him just the same. Ed and I decided we had better go back to Alamogordo.

"We went back on Thursday. Jon Thompson joined us there, and we spent most of the day at the Huckleby house, taking it apart. It would have been very exciting to turn up the first epidemic of Schilder's disease in history, but we decided that my talk with Dr. Schwartz had just about narrowed the possibilities down to a toxic encephalopathy, and we were looking for a possible source of poisoning. Ed and Jon had already searched the house, of course, but this time we really left no stone unturned. We went through every room and everything in every room. We went through the medicine cabinet. We checked the cooking utensils. Some pottery clay, for example, is mixed with lead, and there is sometimes lead in old pots and pans. We looked for spoiled food. We examined everything in the family freezer, including what was left of the hog they slaughtered back in September. We didn't find anything new or suspicious, though. We had another useless talk with Huckleby, and ate dinner and went back to the motel, and we were sitting there around eleven o'clock trying to think of what to do next—when the telephone rang. It was a call for me from Dr. Alan Hinman, in Atlanta. Alan was one of my bosses. He's the assistant chief to Mike Gregg in the Viral Diseases Branch at CDC, and he was calling from his office—at one o'clock in the morning! It was fantastic. And then it got more fantastic.

"Here's what Alan had to tell me. That afternoon, he said, just before quitting time, Mike Gregg had gone in to see Alexander Langmuir—Dr. Langmuir was then director of the Epidemiology Program at CDC—about something, and as he was leaving, he mentioned the Huckleby outbreak. Mike said there was this damn disease out in New Mexico with gun-barrel vision and ataxia and coma, and we didn't know what to make of it. Dr. Langmuir said oh? Then he stopped and blinked, and said it sounded to him as if it might be Minamata disease. Mike looked blank. Dr. Langmuir laughed, and picked up a reprint on his desk. He said his secretary had dumped a pile of accumulated reprints on his desk to either discard or keep on file, and this was one he had just finished looking at. It was a paper from *World Neurology* for November, 1960, and it was entitled 'Minamata Disease: The Outbreak of a Neurologic Disorder in Minamata, Japan, and Its Relationship to the Ingestion of Seafood Contaminated by Mercuric Compounds.' Well, Mike took it home and read it, and it struck him just as hard. So he called up Alan, and Alan came over and got the reprint and read it, and he reacted in the same way, only more so. He drove down to the office and got into the library and checked the references, and there wasn't any question in his mind. Our problem was Minamata disease. He read me the original paper. It began like this: 'In 1958, a severe neurologic disorder was first recognized among persons living in the vicinity of Minamata Bay, Japan. Now 83 cases have been recorded, most ending fatally or with permanent severe disability. Epidemiologic investigations . . . helped to establish the relationship of this illness to the consumption of seafood from Minamata Bay. The effluent from a large chemical manufacturing plant which emptied into the bay had been suspected as the source of a toxic material contaminating fish and shellfish. Subsequent work has provided evidence that the responsible toxin is associated with the discharge of organic-mercury-containing effluent from the chemical factory.' The clinical features had a familiar sound. Ataxic gait. Clumsiness of the hands. Dysarthria, or slurred speech. Dysphagia, or difficulty in swallowing. Constriction of the visual fields. Deafness. Spasticity. Agitation. Stupor and coma. And this: 'Intellectual impairment occurs in severely affected patients; children seem particularly liable to serious residual defects.' And, finally, there was this: 'When the first cases were recognized, Japanese B encephalitis was considered as a diagnostic possibility; however, evidence against encephalitis (aside from the cardinal clinical features) includes the following: the onset of symptoms was usually subacute and not accompanied by fever, cases were limited geographically, and they occurred throughout the year.' The Minamata paper would have been enough for me, but Alan had found another that was even more convincing. It had the title 'Epidemiological Study of an Illness in the Guatemala Highlands Believed to Be Encephalitis,' and it had been published in the *Boletín de la Oficina Sanitaria Panamericana* in 1966. The passage that really clinched it was this: 'Possible toxic elements in food, such as edible mushrooms, were investigated. In the course of this investigation it was noted that during the period of the year in which the illness occurred, many families, and especially the poorest, ate part of the wheat given them for seed, which had been treated with a fungicide known commercially as Panogen—an organic-mercury compound.'

"That was the key—that word 'organic.' It instantly clarified everything. Panogen is cyano methyl mercury guanidine, and cyano methyl mercury guanidine is an *organic* mercury compound. Ed and I were perfectly well aware of that, but the significance just hadn't penetrated. When we thought of mercury, we naturally thought of *inorganic* mercury. That's the usual source of mercurialism. But the pathology of the two afflictions is very different. *Organic*-mercury poisoning has always been something of a rarity. Until recently, anyway. At the time of the Minamata report, the total number of cases of organic mercurialism on record was only thirty-nine. And only five of them were American. Well, I finally finished talking to Alan and gave the news to Ed and Jon, and we all immediately agreed. It was a fantastic kind of coincidence—Dr. Langmuir, Minamata, Guatemala. But it fitted to the letter. Or practically. Talking it over, we did see one objection. The behavior of the Huckleby hogs. Chronic poisoning doesn't produce a sudden, fatal illness. It doesn't, and—the way it turned out—it didn't. On Monday, January 19, I went out and had another talk with Huckleby, and I don't know why, but this time we got along much better than we had before, and he finally said that he could have been mistaken—that it probably wasn't all that sudden, that his hogs probably did get sicker and sicker over a period of several weeks.

"The only thing we needed now was proof. It arrived that night—Monday night—in another call from Alan. He had had a call from Mr. Barthel at the Toxicology Laboratory. Mr. Barthel had analyzed Ed's samples of grain and pork and urine, and he had found mercury in everything but two of the urine samples. The two negative samples were those of the two youngest Huckleby children. One of them was two years old and the other was just ten months, and neither of them had eaten any of the pork. The highest mercury levels were found in the patients' urine. That, of course, was as expected. One of the lowest was Mrs. Huckleby's. Which was fortunate. After all, she was pregnant. But the fact that she and the other asymptomatic members of the family showed any mercury at all was unexpected. It raised a question, and it was a question that would have bothered us if we hadn't been doing a little reading in the literature on our own. The question was: Why didn't *all* the pork-eating Hucklebys get sick? They all consumed about the same amount of meat. We found an acceptable answer in a 1940 paper in the British *Quarterly Journal of Medicine.* The paper was entitled 'Poisoning by Methyl Mercury Compounds,' and the relevant passage read, 'The fact that eight men, exposed in a similar way to the four patients, excreted mercury in the urine, yet showed no symptoms or signs of disease, suggests that most of the workers absorbed mercury compounds, but that only four . . . were susceptible to them.' That was the best explanation—Ernestine and Amos Charles and Dorothy were simply more susceptible to mercury than the others. It satisfied me, anyway. And the next day I flew home to Atlanta."

The chain of laboratory evidence that linked the Huckleby children's affliction to the diet of the Huckleby hogs brought the epidemiological investigation of the outbreak to an end. As it happened, however, that was not the end of the case. On Monday, January 19—the day on which Dr. Likosky and Dr. Pierce and Mr. Thompson wound up their joint investigation—the case took a new and ominous turn.

The turn occurred in Clovis, New Mexico, a livestock center some two hundred miles northeast of Alamogordo, and it was set in motion by a state sanitarian on duty there named Cade Lancaster. Mr. Lancaster was the investigator who had identified Panogen as a chemical contaminant in the Huckleby grain. His second contribution to the case stemmed from a monitoring impulse that prompted him to leaf through the recent records of a Clovis hog broker. He didn't have far to look—only back to Friday, January 16. On that day, he was interested to read, the broker had bought a consignment of twenty-four hogs from a grower in Alamogordo. What particularly interested Mr. Lancaster was the grower's name. He knew it: the man was one of the Huckleby group who had included treated grain in their hog feed. He also knew that all the growers in that group had been instructed to withhold their hogs from market until otherwise notified. The feeding of treated grain had been stopped by horrified common consent at the very start of the investigation, but that was not enough to render the hogs safe for immediate use. Organic mercury is eliminated slowly from living tissue. It has a half-life there of two or three months. Mr. Lancaster sought out the broker. The broker stood appalled. He hadn't known. He had had no way of knowing. What was worse, he added, he had already disposed of the hogs. They were part of a shipment of two hundred forty-eight hogs that had gone out on Friday afternoon to a packing plant in Roswell. Mr. Lancaster turned to the telephone.

The recipient of Mr. Lancaster's call was Carl Henderson, chief of the Consumer Protection Section of the New Mexico Health and Social Services Department, and he received it in his office at Santa Fe. He thanked Mr. Lancaster for his enterprise, and sat in thought for a moment. Then he himself put in a call—to the packing plant in Roswell. The manager there heard the news with a groan. Yes, he said, he had the Alamogordo hogs, but that was the most he could say. He didn't know which they were. They were no longer hogs. The big Clovis shipment had been slaugh-

tered on Saturday, and it was simply so many carcasses now. In
that case, Mr. Henderson said, he had no choice but to immobilize
the lot. The plant was therewith informed that all those two
hundred forty-eight carcasses were under state embargo. Mr.
Henderson hung up and moved to reinforce his pronouncement.
His move was at once effective. The following day, January 20, the
United States Department of Agriculture, through the Slaughter
Inspection Division of its Consumer and Marketing Service,
placed the state-embargoed carcasses under the further and
stronger restraint of a federal embargo, and arranged for a sample
of each carcass to be tested for mercury at the Agricultural Research Center, in Beltsville, Maryland. That same day, the state
extended its embargo to the grain-fed hogs still in the possession
of Huckleby and his five companion growers in Alamogordo. The
next day, the hogs were tallied and found to total two hundred and
fifteen head. One hog was selected from each grower's herd and
killed and autopsied, and specimens were sent to the Toxicology
Laboratory of the Food and Drug Administration, in Atlanta, for
definitive examination.

Two days later, still shaken by Mr. Lancaster's discovery, the
New Mexico Health and Social Services Department took another
protective step. A cautionary letter, signed by Dr. Bruce D. Storrs,
director of the Medical Services Division, was distributed to every
physician and veterinarian in the state. The letter read:

A recent outbreak of central-nervous-system disease among three
children of an Alamogordo family has been traced to the ingestion
of pork contaminated with a methyl-mercury compound. The animal involved had been fed over a period of several weeks with grain
treated with this material. At the time it was butchered, the hog
appeared to be in good health, and visual inspection of the carcass
during processing failed to reveal any significant abnormalities.
Laboratory analysis of the meat eaten by the family, however, revealed a high concentration of methyl mercury.

The grain had been obtained by the children's father, free of
charge, in the form of castoff "sweepings" from a seed company

located in eastern New Mexico. This was mixed with garbage and fed to the hogs despite the knowledge that it had been treated with the fungicide. Several weeks after slaughter of the boar, 14 of the family's remaining 17 hogs became ill with symptoms of blindness and staggering gait. Over a three-week period, 12 died and the remaining two were permanently blind.

Because of the potential danger to humans of ingesting meat contaminated with methyl mercury (all three children are comatose and in critical condition), as well as the likelihood that the practice of feeding treated grain may be widespread among indigents raising livestock in the state, this matter is being brought to your attention. The possibility of methyl-mercury poisoning should be considered in any outbreak of . . . unusual central-nervous-system illness. Please notify the Preventive Medicine Section (Phone 505-827-2475) of any such occurrence, and we in turn will be happy to provide technical advice and assistance.

The letters were mailed out on Thursday, January 22. The Preventive Medicine Section spent Friday and the weekend waiting for the ring of the telephone and the urgent voice of a doctor. But nothing happened. Nothing happened the following week. February loomed, and began. Still nothing. The Preventive Medicine Section sat back and relaxed. They let themselves assume—correctly, as it turned out—that the outbreak was confined to the Huckleby family.

Nevertheless—as it also turned out—the danger of an epidemic had been real. Early in February, the results of the hog examinations were announced. The first report was on the embargoed packing-plant carcasses. It seemed to justify Mr. Lancaster's investigatory zeal, but its meaning was otherwise not entirely clear. Of the two hundred forty-eight carcasses examined by the Department of Agriculture, one—just one—was found to contain a high concentration of mercury. The others, curiously, were uncontaminated. The contaminated carcass was ordered destroyed, and the others were released to the market. The Toxicology Laboratory's report on the hogs taken from the Alamogordo pens was very different. Six hogs were examined, and all six were found to be dangerously contaminated with mercury. This unequivocal finding condemned the remaining hogs in the pens, and the state issued an order for their destruction.

Dr. Likosky returned to Atlanta with the Huckleby outbreak still very much on his mind. What particularly disturbed him was its implications. This led him back to the library and he undertook a comprehensive exploration of the literature on organic mercury compounds and organic mercurialism. It was not a reassuring experience. Mercury fungicides, he learned, were developed in Germany around 1914 and came into almost universal use shortly after the First World War. He learned that the Minamata episode was not the only outbreak of organic-mercury poisoning caused by contaminated fish. It was merely the first. In 1965, despite the morbid Minamata example, a similar outbreak occurred on the Japanese island of Honshu. A total of a hundred twenty people were stricken there, and five of them died. He learned that the Guatemalan episode was not the only instance on record of poisoning caused by eating treated grain. A similar outbreak occurred in Iraq in 1961, and another in Pakistan in 1963. A total of almost five hundred people were struck down in the three outbreaks, and the mortality rate was high. He learned that the Huckleby hogs were not the first American hogs to be stricken with chronic mercury poisoning. They were merely the first involved in human mercurialism. He learned that on February 1, 1966, Sweden had revoked the license for the use of certain highly toxic mercury compounds (including methyl mercury) in agriculture. He learned that the 1969 pheasant-hunting season in the Canadian province of Alberta had been canceled after a survey of the pheasant population showed an average mercury level of one part per million— a hazardous concentration. And, closer to home, he learned that a similar survey in Montana that same year revealed a level of mercury contamination that prompted the State Game Commission to advise hunters against eating the birds they shot. In both instances, the pheasants had been exposed in the field to treated seed.

Dr. Likosky arose from his reading with a sense of apprehension. He sought out two of his colleagues, Dr. Hinman and Mr. Barthel, chief of the Toxicology Branch, and found that they shared his concern about the proliferating casual use of a substance as toxic as methyl mercury. They also felt that a broader discussion of the matter was desirable, and Mr. Barthel arranged for a meeting in Washington with interested representatives of the Department of Agriculture and of the Food and Drug Administration. At the meeting, which was held on February 12, Dr. Likosky, Dr. Hinman, and Mr. Barthel reviewed the Huckleby case and its varied antecedents, and suggested that they constituted a looming public-health problem. Their arguments were well received, and on February 19 the Department of Agriculture issued a statement on the subject. It read:

> The U.S. Department of Agriculture has notified pesticide manufacturers that Federal registrations are suspended for products containing cyano methyl mercury guanidine that are labeled for use as seed treatments.
> U.S.D.A.'s Agricultural Research Service suspended cyano methyl mercury guanidine fungicide because its continued use on seeds would constitute an imminent hazard to the public health. Directions for proper use and caution statements on labels of the product have failed to prevent its misuse as a livestock feed. The U.S.D.A.-registered label specifically warns against use of mercury-treated seed for food or feed purposes. The pesticide may cause irreversible damage to both animals and man.
> The action was taken following the hospitalization of three New Mexico children after they ate meat from a hog which had been fed seed grain treated with the now-suspended mercury compound. Subsequently, 12 of the remaining 14 hogs also fed the seed died.

"Other movements of this treated seed that found its way into livestock feed posed a potential for similar incidents," Dr. Harry W. Hays, director of the Pesticides Regulations Division, USDA-ARS, said in announcing the suspension action. "In each case, USDA and state public health officials have taken prompt action to protect the public health." Dr. Hays also announced that the ARS had asked the Advisory Center on Toxicology of the National Research Council to review the uses of other organic-mercury compounds to determine whether similar hazards to human health existed in connection with the use of these compounds.

March came on. Ernestine and Amos Charles Huckleby were discharged from Providence Memorial Hospital, in El Paso, and transferred to the chronic-care facility of a hospital in Alamogordo. They were still comatose. (Ernestine will probably be permanently comatose. She is almost certainly blind. Amos can communicate on a primitive level. He, too is blind.) At the same time, Dorothy Huckleby was removed to a rehabilitation hospital in Roswell. (It is expected that she may in time recover enough to care for herself under general supervision.) Meanwhile, Mrs. Huckleby's term approached. Because of the delicate nature of her pregnancy, state health officials had recommended that her confinement take place in the scientifically sophisticated environment of the University of New Mexico Medical Center, in Albuquerque, and she was admitted to a maternity ward there. Her condition was satisfactory and her course was uneventful. On Monday, March 9, she was delivered of a seven-pound boy. At birth, the baby appeared to be physically normal, but a few hours later he experienced a violent convulsion. He survived the seizure, and recovered. The prognosis, however, was uncertain. It is very possible, in view of his long fetal exposure to mercury, that some physical or mental abnormalities will eventually manifest themselves.

April arrived. Manufacturers affected by the federal suspension order on seed dressings containing cyano methyl mercury guanidine stirred, and suddenly struck. On April 10, Morton International, Inc., and its subsidiary Nor-Am Agricultural Products, Inc., the makers of Panogen, applied in United States District Court in Chicago for an injunction relieving the company of compliance with the suspension order, and the application was granted by Judge Alexander J. Napoli. The grounds for granting the injunction were that the Department of Agriculture had acted without holding a hearing to establish the hazardous nature of the fungicide.

The government appealed, and on July 15 a three-judge panel of the Court of Appeals turned the government down, again citing insufficient evidence. The government then petitioned for a review by all six judges of the Court of Appeals. This request was granted, and the full panel met and reviewed the case. Their decision was announced on November 9. The panel ruled, by a vote of four to two, to uphold the government suspension. Thus, after almost seven months of swinging in the wind of legal technicalities, the door was closed again on methyl mercury guanidine dressings, and this time firmly latched.

[*1970*]

CHAPTER **16**

All I Could Do Was
Stand in the Woods

———————◆———————

ADOLPH (RUDY) CONIGLIO says his trouble began on the night
of July 27, 1969, a Sunday. He was then fifty-three years old and
the proprietor of a restaurant called Rudy's Pizza, in Closter, New
Jersey. He remembers:

"Two of my people, the cleaners, they didn't show up that night
to clean. So I stayed after closing and did the cleaning myself. I
did the kitchen, the counter, the dining room, where I got twenty
tables—I did it all. So I perspire even in that air-conditioned place,
and I got so hot that I didn't sleep good all night. So I got up the
next morning with this heavy cold and fever. I don't feel good. So
my wife goes down to run the place, and I went over to this doctor
—this local, family doctor—and he prescribed some yellow antibiotic pills. I took those pills and stayed three days at home. On
Thursday, I went back to the restaurant. I got there early, and I
felt all right. Not bad. So I start to work in the window up front
and make some pizza dough. Pizza is my speciality, I make it all
myself. O.K. Well, now the dough is rising good, so I get out some
tomatoes and the other stuff for a *pizza alla napoletana.* All of a

255

sudden, I start to smell something funny. Something bad. I'm peeling a tomato, and I smell it. It smells rotten. It smells like garbage. Then I taste it. It has the taste of garbage. I call the wholesaler, and I'm mad. What's the matter, you send me rotten tomatoes? He says I'm crazy—he don't handle rotten tomatoes. Now *he's* mad. I hang up and go out to the kitchen to find some good tomatoes. But—oh, my God! Did you ever when you're a kid burn a plastic comb? Everything in the kitchen smelled like that. I had to get out. I went back to the front, and it was just as bad. That smell—that garbage stink. Some of my people were coming in, and I told them, but they don't smell nothing. So it must be me. It must be that heavy cold.

"I figure I'm still sick. So I got my car and went home. The street was almost as bad as the place—the exhaust fumes. It was the same stink. It was better at home. Some. But not much. My wife went down to the restaurant again. I went outdoors, out in the back yard. It was better there than in the house, but it had a smell. The grass had a smell. It smelled like regular grass—only twenty times stronger than normal. It smelled so strong it smelled bad. I got some woods behind my house, so I tried there. That was better. It was O.K. I tell you, it felt good to get away from that garbage stink. I sat down on a log, and it was nice sitting there, so after a while I lit a cigarette. Oh, my God! The smell was like everything else—and the taste! It burned on my tongue like a hot pepper. So I don't have anything to do but just sit in the woods. I stayed there all day. When it was time for dinner, I couldn't eat. Everything had that stink. I thought I might as well go to bed. I got ready and laid down, but I had to sit up. The pillow—it smelled like dirt. So my wife went out to the store and bought a new pillow. It had a smell, but it wasn't too much. But I had to sleep on my back. I couldn't have my face in the pillow. So the next day I went back to the doctor.

"The doctor was surprised. But he thought like me—it was still that cold, that flu I had. He gave me a new prescription, for some pink antibiotic pills, and said to come back in a couple days. But the pills didn't help. I went back to the doctor. He told me to wait a while. Be patient. So I had to hire somebody to run my place. All I could do was stand in the woods all day. Of course, I had to eat—I didn't want to die. But the regular food—it was all like garbage. I could drink a little cold milk. I could eat a little cold boiled potato. I could eat a white grape. I could eat a little vanilla ice cream. That stuff, it didn't taste good, but it didn't taste bad. It didn't have any taste at all. So I lived on that. No coffee—God forbid. Even a banana—I couldn't go near it. I went back to the doctor again, and he sent me to another doctor—an ear-nose-and-throat man. A specialist. He looked at my nose and he took some X-rays, but he couldn't find anything wrong. He called the family doctor and said, This Rudy, what he needs is psychiatry, his trouble is psychosomatic. He sent me to a psychiatrist. So the psychiatrist started talking about my life history—did I like my work? My God, I love my work. I listened and I went out and I didn't go back. Well, now it was almost September. I wasn't any better, I was losing weight, and the only time I was comfortable was out in the woods. So I decided to go to Italy. My mother is there, and I thought maybe the Italian doctors could help me. I went to Naples and saw a doctor there, and I went to a doctor in Rome. But they were the same as at home. It was all in my mind. Did I like my work? They gave me some tranquillizer pills. Five a day. Ten a day. I stayed a month and came back home in October.

"By this time, I don't know what to think. Am I crazy? Maybe those doctors they're right. They keep saying it's my mind. Or maybe they're making me crazy. I try another doctor. Not in Closter—in another Jersey town. This doctor looks at my nose with a needle, and a little blood comes out. Ah, he says, I know what you got. You need an operation. So he gives me a date. But I don't take his word. I go to another doctor next day. He says he first wants X-rays. But I just had X-rays yesterday. He wants to know who, and I tell him, and he calls that doctor and they talk. Then he looks at my nose again, and he finds where that needle touched me. Ah, he says—that's it. You got a tumor. Very serious. Operate at once. But I went home. I thought I was going to drop dead. Then I thought I better kill myself. I've lost forty pounds of weight and I've spent a hell of a lot of money, and now I've got a tumor in my nose. But, thank God, I've got a neighbor friend. His job is hospital administrator at a big hospital in New York, in Manhattan. I talk with him sometimes, and so one day—this is maybe December—he says, Maybe you better try my hospital. He says they've got good men, and so he arranges everything and I get a date. So I walk in the hospital and they're waiting for me —three doctors. Everything is organized right. But first another X-ray. O.K. And then—my God, good news. The head doctor says no tumor. They don't find anything. But the head doctor is interested. He says he wants to see me in his office. He's got a private office up around Fifty-seventh Street, and I went there the next day. We visited and talked for an hour. He couldn't find what was wrong. He ordered me again some pills, and a spray for my nose. He says to come back in a couple days so he can see the reaction. But he don't find any reaction. No better. No change. So then I tell him what I'm worried about. I say, Listen, Doctor— they send me to all these psychiatrists, and they talk about psychosomatic. I say, I'm willing to go to the crazy house. I don't care anymore. I had enough. But I want to ask one thing. Do crazy people have their food taste like garbage? And the doctor says no. He says they eat like everybody else. He says, I don't know the trouble with you, but I know one thing. You're not crazy.

"That made me feel better, so I was ready to go. But he said wait a minute. He said he couldn't help me but he thought he knew a man that could. He said there was a doctor down in Washington, a specialist at the National Institutes of Health, that was coming up to New York pretty soon. He wanted me to see him. So I went home and waited. At least I wasn't crazy. But I still can't eat or smoke or work or do anything but go out and stand in the woods. It's Christmastime, and I'm living like a dog. Well, I wait a couple weeks, then I get the call. There's a Dr. Gould, with an office on Seventy-seventh Street, and this man from Washington is waiting for me there. So I go over and meet him—Dr. Henkin, of the National Institutes of Health. He takes

me into a little room. I talk and he listens. Then he says, Open
your mouth. So I open. Stick your tongue out. So I do, and he puts
a drop of something on my tongue. He says, 'How does it taste?'
I tell him, 'No taste at all.' 'Good, he says—now I know what you
got. But I can't treat you here. You got to come down to the
Institutes of Health.' My God, I didn't know it then, but there was
a god. That's all I can say about Dr. Henkin. He's a god."

Dr. Henkin is Dr. Robert I. Henkin. He was at that time chief of
the Section of Neuroendocrinology, Experimental Therapeutics
Branch, of the National Heart and Lung Institute of the National
Institutes of Health, in Bethesda, Maryland, an appointment he
had held since July of 1969. Since 1975, he has been director of
the Center for Molecular Nutrition and Sensory Disorders at the
Georgetown University Medical Center, in Washington. He says:
"Well, I *thought* I knew what Rudy's trouble was. I thought he
had what we call idiopathic hypogeusia. It's a condition that
involves a loss of taste acuity—the ability to distinguish tastes—
and it is often accompanied by a distortion of taste and smell. The
cause is unknown, or idiopathic. The reason I suspected it in Rudy
was simply that I had it on my mind. I had seen the same condition twice before—and only a few weeks earlier. Aberrations of
taste and smell are not, of course, at all unusual. They are common
symptoms of the common cold. And hepatitis. And pregnancy.
But my introduction to idiopathic hypogeusia was entirely accidental. I had occasion to mention the loss of taste acuity, with a
related distortion of taste and smell, in a refresher training course
in the basic sciences that I was teaching at the Armed Forces
Institute of Pathology in the fall of 1969. It was sometime in early
September. Well, about a month later I got a call from one of the
officers who had taken the course—an Air Force colonel down in
Texas. He had a patient at his base he thought I might want to
see. Sergeant Mack, I'll call him. Mack's complaint was pretty
much what I had been talking about. Everything he ate or drank
smelled and tasted foul. But he didn't have a cold, he didn't have
hepatitis—he seemed in every other respect to be in normal good

health. I said I was very much interested. I said to send him right up.

"Mack came into the N.I.H. clinical center on October 18. He was forty-six years old, and a big, powerful man. He said he had lost some weight in recent months, but he still weighed a good two hundred pounds. We gave him the usual workup, including neurological and ear-nose-and-throat evaluations, and the results were, as we say, uneventful. He was a little hard of hearing from working around airplanes, but that was all. He was, in general, in excellent health. I sat down with him and we had a long talk. His trouble had come on gradually. Back in June, he said, he began to notice that onions had a peculiar taste. In July, he began to notice that meats of all kind also tasted peculiar—tasted as if they had been cooked in rancid grease. Then other foods began to taste bad. Vegetables had an 'iron and tinny' taste. Cigarettes smelled like Vicks Vapo-Rub. For the past three or four months, he had been living on lettuce, cottage cheese, milk, and ice cream. It was weird—fantastic. I wondered if he might be crazy. I'd never heard of such a thing except as a manifestation of some other and demonstrable disease. I took him down to the hospital kitchen. I got out a cold roast of beef and cut off a beautiful slice and asked him just to taste it. I've never seen such a look of revulsion. He said he wouldn't, he couldn't—he would vomit. Just the smell of it made him feel sick. I took him back to his room. I had to believe him—but I certainly had something to think about. And I thought about it. I knew from a study I'd been involved in a couple of years before that a substance called D-penicillamine can diminish taste acuity, and even distort taste. This has happened a number of times to patients being treated with that drug for various diseases —rheumatoid arthritis, idiopathic pulmonary fibrosis, scleroderma. When D-penicillamine was discontinued, normal taste returned. Our study also showed that the administration of D-penicillamine was usually followed by a drop in the concentration of copper in the blood. We treated those patients with supplemental copper, in the form of copper sulphate, and normal taste returned. That certainly suggested that copper might play a role in the physiology of taste. We then began to wonder about the possibility of a similar role for other metals that are minutely present in the body—zinc, for example. We tried the same experiment again, only this time using zinc sulphate instead of copper. We got the same result. Normal taste returned. Well, with that in mind, I arranged to measure the levels of copper and zinc in Mack's blood serum. Which was done. And the results were normal. There was no decrease at all.

"That was rather a shock. I had been all set to start him on one of those metals—probably zinc. It has a very low toxicity. So now what? But I still wanted to *do* something. And I thought, What's the harm—why not try it anyway? So I wrote out an order for oral zinc sulphate—four hundred milligrams a day. And what happened? It was really quite dramatic. Two days later, he said things didn't seem to smell as bad as they had. The next day, there was more improvement, and the day after that I dragged him down to the kitchen again and offered him another slice of roast beef. He ate a couple of bites, and said it didn't taste too bad. A few more days, and his taste was back to normal. He was well. It was a gratifying recovery, and it raised an interesting point. Mack's serum-zinc level was never less than normal, and yet he had—or seemed to have—a certain zinc deficiency. I wondered if something could happen to deplete the supply of zinc in a specific area —as in the taste buds on the tongue. In any event, the zinc had seemed to work. Mack was discharged on November 4, with a good supply of pills, and that was the end of him as far as I was concerned. I never heard of him again.

"I didn't forget him, though. That wouldn't have been possible. And for several reasons. One was this. I have an old friend, an English colleague, and about a week after Mack went back to Texas my friend—I'll call him Wilson—turned up in Washington on his way out West somewhere. I hadn't seen him for a couple of years, and I picked him up at his hotel and drove him home to dinner. On the way, I happened to mention that my wife was cooking a leg of lamb. I thought Wilson acted a little strange. He cleared his throat and looked embarrassed. Finally, he said some-

thing about not eating much these days. He wondered if he might
have something very light—a little cottage cheese, perhaps, and
some lettuce. The truth was, he said, he couldn't eat meat anymore. He couldn't stand the smell of it. As a matter of fact, there
were very few foods that didn't have a most unpleasant smell. It
was like talking to Sergeant Mack. I asked him how long this had
been going on. A couple of years, he said. It was a nuisance, but
he had learned to live with it. I felt—it's hard to say how I felt.
But I said I thought I might be able to help him. He gave me a
hopeless look. I asked him if he would let me try. Would he come
into the hospital for a week or two? He shrugged. He didn't see
much point in it, but—all right. I made the arrangements, and he
came in, and we worked him up very thoroughly to exclude any
other possibilities—liver disease, a tumor, all the rest. We found
nothing. He was normal in every way but one. His serum-zinc
level was down. I put him on placebos for three days. If he
improved on placebos, which are nothing but sugar pills, that
would strongly suggest that his problem was functional—was
rooted in some emotional disturbance. There was, however, no
change. Then I started him on oral zinc. It wasn't like Sergeant
Mack—it wasn't in the least dramatic. But after four or five days
his serum zinc began to rise. His sense of smell improved. He
could eat a little something—an egg, a slice of meat. But then he
had to leave. He couldn't postpone his engagements any longer.
I discharged him on November 17. I gave him a supply of zinc-
sulphate pills. Zinc sulphate is not a prescription drug, you know.
We were using it in an entirely experimental setting. The official
view is spelled out in the standard text 'The Pharmacological Basis
of Therapeutics,' which says, 'The systemic actions of zinc have
no therapeutic application. . . . All drugs must by law have a
specific use. Zinc is not recognized as having any usefulness inter-
nally except as an emetic.' Well, Wilson went on his way, but he
called me a few weeks later, just before flying back to England. He
had continued to improve. He wasn't all the way back to normal,
but he felt enormously better.

"And that was when I heard about Rudy Coniglio. A friend and colleague named Wilbur James Gould made the arrangements, and I went up to New York and saw him. It sounded like the same sort of thing. I gave him a drop of relatively concentrated hydrochloric acid. This is a compound that normal people find extremely sour. But Rudy didn't bat an eye. He couldn't even taste it. That didn't immediately rank him with Sergeant Mack and Dr. Wilson, but it was more than a little suggestive. I certainly wanted to know a bit more. Those first two cases had merely whetted my curiosity. I was really interested now. So I got Rudy into the Bethesda clinical center as soon as I could—on January 11. We started with a very comprehensive workup—all the routine tests, plus a skull series, EEG, brain scan, sinus series, a neurological examination, an ophthalmological study, copper and zinc evaluations, and a tongue biopsy. An examination of tongue tissue could tell me if there had been any anatomical changes in the area of the taste buds. The test results were essentially normal—all but the trace-metal evaluations and the tongue biopsy. Zinc and copper—serum and urinary both—were low. On the fifth hospital day, I started Rudy on oral zinc sulphate: four hundred milligrams a day. There was some improvement by the end of the first week, and it was marked by the end of the second week. That was when I saw the results of the tongue biopsy. They were quite impressive—startling, even. The normal structure of the taste buds was almost totally absent. I was looking at the pictures—the photographic enlargements of the electron-microscope examination—and you couldn't miss it. The taste buds looked frayed, worn down, moth-eaten. But now it was two weeks later. So I made another tongue biopsy. And waited with some impatience for the results. They were very interesting indeed. They looked like a normal tongue. Very close to it, anyway. Rudy had left by then. I had discharged him on January 31, with the usual supply of pills. I asked him to call me in a couple of months to keep me posted on his progress, and I also arranged for him to come back to the center in July for another checkup. He called me in March. He was taking his four hundred milligrams of zinc sulphate every day, and he was fine—he was all the way back to normal. On July 12, he came into the clinical center again, and he stayed a little over a week, until July 21. His trouble seemed to be completely under control. But I wanted to follow him—I had lost track of Mack, and Wilson was back in England. I wanted to be sure, so I made arrangements to see him again in another three months or so."

Mr. Coniglio says: "I wonder how many people went into the crazy house with this thing I had. I mean, years back—before Dr. Henkin. I bet plenty. I was almost there myself. But Dr. Henkin fixed me up. I came back home that second time from the hospital, and I never felt better in my life. It was even better than the first. I felt better in my mind. For almost a year now, I could eat and drink and smoke and work at my place. I had a normal life. But I think it was around the beginning of October that I noticed something different. It starts very gradual, very slow. At first, I don't believe it. But things are beginning to smell again. They stink. I'm taking the pills, but I'm losing. They don't do any good. Oh, my God—I'm scared. It wasn't as bad as before, thank God, but I know that smell and that garbage taste. I know I'm going backward. I can still work, but my nerves are bad, and I think maybe I better sell out—get rid of the place before I get worse. So I did. I sold Rudy's. I did all right—I got a good price. But the way I'm feeling, I don't really care too much. Because it is getting close to time for me to go back to the hospital. I got to go back, and I don't know what I'm going to say to Dr. Henkin. He thinks I'm cured. He thinks he made me well, and I'm not. I almost don't want to see him. You know, I'm embarrassed. But the time came, and I got the appointment letter and I went down to Bethesda and Dr. Henkin came in and I told him. And he laughed."

Dr. Henkin says: "Laughed? I don't know—maybe I did. But it wasn't because I was amused. If I laughed, it could only have been from relief. I was pleased. Rudy had confirmed an important hypothesis. Maybe *I* was a little embarrassed, too. Rudy's suffering those past few weeks was all my fault. He should, of course, have called me when the first symptoms returned. That's what I told him to do. I realize now why he didn't—he didn't want to disappoint me. Which tells you a lot about Rudy. It was terrible that he didn't call. I could have helped him right away. Because the pills I sent him off with in July were not zinc-sulphate pills. They were placebos. That was something I had to do. This is a research hospital. My work is clinical research. Sergeant Mack had dropped out of sight, and Dr. Wilson was way to hell and gone in England. But I still had Rudy. He gave me an opportunity I had to take. I had to see if the recoveries I had seen were truly in response to medication—if it wasn't a matter of spontaneous remission or a psychosomatic response to a sympathetic hospital and medical situation. And now I was sure—reasonably sure. It was only necessary to put Rudy back on the old regime. Which I did. And then I *was* sure. In Rudy's case, at any rate, oral zinc sulphate, at four hundred milligrams a day, was genuinely therapeutic."

Adolph (Rudy) Coniglio has the distinction of being the first victim of idiopathic hypogeusia whose experience has been recorded in definitive detail. He thus is classically commemorated in the infant annals of the disease. He is, however, no longer alone in affliction. Others—many others—have since followed in his enigmatic footsteps. More than ten thousand cases of taste and smell dysfunctions are now known to Dr. Henkin from correspondence with colleagues throughout the United States, and some fifteen hundred more have been diagnosed, attended, and treated by him and his associates in Bethesda and Georgetown.

"Call it ten thousand," Dr. Henkin says. "But I'm inclined to think that those cases we have on record are only a fraction of the real total. I'll tell you why. Fifty-five of the cases we saw at the clinical center were our own people—National Institutes of Health employees. There were around ten thousand men and women working in one or another of the several institutes, and fifty-five cases gave us an incidence rate of roughly one in one hundred and eighty. Now, let's project that on a national population of over two hundred million. And what do we get? It suggests that more than one million Americans are suffering from hypogeusia or its related disorders. That really is a lot of cases. I think it's a reasonable figure, though. Hypogeusia isn't a new disease, of course. It's been around for a long, long time. It's merely a newly discovered disease. I think loss of taste is probably just as historical an affliction as loss of sight or hearing. But we don't know much about it, because it's only just been recognized. Remember all those doctors who thought Rudy Coniglio's trouble was psychosomatic?

"So we're beginning to get somewhere. I think zinc sulphate is therapeutic. It works—not always, less often than we would like, but often enough to prove its worth. We're beginning to understand that problem. We understand, at any rate, that it is an immensely complicated one. But treatment isn't all. We've also been learning something about the nature of the disease. One thing we know is this: As a group, the victims of hypogeusia are measurably deficient in zinc. There are a number of conditions that can bring about hypogeusia. We know it can happen in pregnancy—perhaps because a portion of the mother's zinc is transferred to the baby. We know it can be brought about by head injury, by a fall or a blow. We know it occurs in cancer—I've seen it in my own experience. About ten per cent of the patients we've studied who were found to have hypogeusia have later been found to have cancer. Which raises an interesting question. Is hypogeusia a disease? Or a symptom? Hypogeusia and its related disorders occur in a variety of infectious diseases. Hepatitis is one. Flu is another. About half the people in whom we have confirmed a loss and distortion of taste and smell developed the disorder following an attack of flu. Like Rudy—Rudy is our model case. Another group of our patients developed aberrations of taste and smell after some surgical procedure. I don't know why. Maybe stress is a factor. That casts some possible light on a rather familiar phenomenon. Postoperative patients are always complaining about the quality of hospital food. Can *all* hospital kitchens be that bad? Or could it be the patients' sense of taste? I wonder. And then we have a third group. These are the patients whose hypogeusia is truly idiopathic. We can't find any precipitating insult. I think we will in time, possibly in the area of nutrition.

"A team of pediatricians at the University of Colorado Medical Center, in Denver, has reported some interesting findings. They examined a group of a hundred and thirty-two boys and girls with ages ranging from four to sixteen. Zinc concentrations were normal for all but nine of the group. Seven of the nine were found to be below normal in both height and weight, and they also reported poor appetite. They were then tested for taste acuity. All were found to be suffering from hypogeusia. They then were treated with oral zinc sulphate. Normal taste acuity returned. Other studies, here and elsewhere, have confirmed our conviction that zinc can be a crucial factor. One of our studies involved a group of normal people—volunteers. We biochemically depleted their body zinc. We used an essential amino acid called L-histidine, which has the power to strip zinc from its binding proteins and excrete it through the urine. The immediate effect of depletion was appetite loss. That was followed by a loss of taste and smell. Then distortions of taste and smell appeared. Oral zinc therapy reversed the condition, even in the face of continued administration of L-histidine. So once more zinc emerged as a major marker for normal taste acuity.

"Much of what we feel we have learned has come from a double-blind test that we staged in 1973 at the N.I.H. The true effectiveness of a drug can best be established in a double-blind test. The test we did on Rudy was a single-blind: the patient didn't know the nature of the drug he was given, but the doctor did. In a double-blind test, the true drug and the placebo are delivered to the clinician in containers labelled in a code known only to the investigator who prepared the medication. The doctor is as blind as the patient—he can't anticipate results. Our double-blind test involved a group of one hundred and six patients. Half of them were given placebos, and the other half got zinc sulphate. The results were a shock. Some of the people on placebos got better. Some didn't. Some people on zinc got better. And some didn't.

Statistically, we could not distinguish between the group that got placebos and the group that got the zinc. We couldn't believe our eyes. What was going on here? We thought we had been put back to zero. Then we calmed down and took a closer look at the data and we got a clue. In a single-blind test, we give the patient the placebo first, and if there is no response we then try zinc. If, however, he responds to the placebo, we drop him from the study. A placebo response suggests that other factors are involved— maybe a functional debility, maybe a psychosomatic problem. In any event, when we dropped the placebo responders from the double-blind study we ended up with results much like those of our single-blind tests. So we had another confirmation. But the study actually gave us something more than that. It encouraged us to reconsider, to rethink. We realized that there is a great diversity of taste and smell disorders, with a great variety of causes, and we were forced to conclude that zinc is no magic bullet. That is to say, it is a drug like other drugs. Magic bullets are rarer than people like to think. There are, for example, insulin-resistant diabetics. And the famous antibiotics are far from being comprehensively effective.

"We were also encouraged by the study to reconsider some of our criteria. We had been using serum-zinc levels—the concentration of zinc in the blood—to measure both zinc deficiency and the response to zinc therapy. We decided that there might be a more definitive approach. We had plenty of room to maneuver in. The phenomenon of taste has never been of very great interest to medicine. The reason probably is that its pathology had never seemed to have much diagnostic value. Not like vision, for example, or hearing. Consequently, very little is known about the mechanics of taste. The conventional explanation supposes it to be entirely a neural process. Our studies have suggested a different explanation. We now think the perception of taste is initiated chemically. We don't know just how this chemical information is then translated into electric signals to the brain, or how those signals are coded to define the basic qualities of salt, sweet, sour, and bitter. But we do know that taste is initiated at the taste buds, of which there are more than a thousand in the oral cavity. We thought that zinc must be present in the bud. But then the question arose: How does the zinc get there? Well, the first possibility that came to mind was the obvious one—the saliva. Saliva serves the oral cavity as blood serves the rest of the body. We decided that our next move would be a study of the saliva.

"That took a little time. We had to develop a technique for measuring zinc in the saliva. What we did finally was adapt an atomic-absorption-spectrophotometric technique to measure the tiny amounts of zinc involved. We isolated the zinc-containing protein in saliva, and named it gustin. Gustin is responsible for the growth and maintenance of the taste buds, and in hypogeusia victims it is diminished in concentration. The result of our work was to establish that zinc is one essential of the taste process. And realizing that enabled us to revise our clinical thinking. We had made the mistake of using the serum-zinc concentration as a diagnostic measure. But now we know better. We now measure salivary zinc—and with notably accurate results. We find in an impressive number of cases that victims of hypogeusia have a low level of salivary zinc. And we find that an impressive number of such cases respond to zinc therapy. The salivary-zinc level goes up. And taste acuity returns. We're beginning to know who we can help—and who we can't. We've still got a long way to go. But I think we're over the hump.

"I mentioned that well over a million Americans are suffering from some form of taste or smell dysfunction. That's a lot of people, and problems of taste are very disagreeable problems. Eating is a basic pleasure. I would say that eating and sex are the two basic pleasures in life. We're being told now that the pleasures of sex may well be enjoyed into very ripe old age, and I hope it is true. But it is certain that eating is with us *all* the years of our life. A life without pleasure in eating—a life in which nothing tastes good and much tastes awful—would be hard to endure. We hope that we're on the way to bringing help to those unhappy people who are forced to endure it. Oh, there's so much ahead of us. I'm an activist and an enthusiast, and the potential in our work could hardly be more exciting. We're on the threshold of a new technology. I'm thinking beyond the treatment of hypogeusia— important as that is. I'm thinking, for example, of the day when we can order the chemistry of saliva in such a way as to prevent dental caries and gum disease. I'm also thinking of what a fuller knowledge of the physiology of taste could mean to the food industry. We know now that the taste receptors can be manipulated. Sour or bitter can be made to taste sweet. Think what the refinement and application of that phenomenon would mean to the diabetic, the obese, the hypertensive, the heart sufferer. Sweetness without sugar, sweetness without sweeteners. No more worries about saccharin or the cyclamates. Think—just think—how happy the U.S. Food and Drug Administration would be!"

[*1977*]

CHAPTER 17

As Empty as Eve

NATALIE PARKER, as I'll call her, is an attractive woman of medium height, with large gray eyes and light gray hair, but thin —still painfully thin. She is in her early fifties, and is an economist by profession. Her husband, Alan, is an artist and illustrator. The Parkers live in Washington, in an apartment on J Street that also includes Mr. Parker's studio. They have no children. Until September of 1973, when she retired for reasons of health, Mrs. Parker was employed—and had been for more than twenty years —at the Department of Commerce. Her work there involved certain aspects of the computation and analysis of the gross national product.

The misfortunes that led to Mrs. Parker's premature retirement began in the spring of 1972, when she learned from her dentist that she had a serious gum problem. He referred her to a periodontist of his acquaintance. In June, the periodontist, after a long and interested examination, referred her to an orthodontist for the realignment of several teeth affected by the condition of her gums. The orthodontic work was not a success. Indeed, as Mrs. Parker

271

subsequently declared in an application for disability retirement, the results were both mechanically and cosmetically "disastrous." They were also emotionally daunting. As the summer passed and autumn came on, she began to despair, and fell into a deep depression. She was tired all day, she couldn't sleep at night, her teeth hurt, her appetite vanished. Her family physician prescribed a conventional tranquillizer. She continued to work, but with increasing difficulty. Her weight dropped from a normal hundred and eighteen pounds to a hundred and ten, and then to ninety-eight. By Christmas, it was down to eighty-nine. Her physician referred her to a consulting psychiatrist. The psychiatrist's prognosis was guardedly hopeful. He thought a period of rest in a relaxed environment would be a sufficient restorative. The environment he had in mind was that of a psychiatric hospital. Mrs. Parker was at first appalled. The idea was socially unacceptable. In time, however, her resistance weakened, and at last, too discouraged to care, she allowed herself to be admitted for observation to a well-appointed hospital with which the psychiatrist was associated. That, according to Mr. Parker, was on February 8, 1973, and his recollection is confirmed by the hospital records. Mrs. Parker herself has no recollection of that decisive event. She has, in fact, no recollection of any part of her hospital stay. Mrs. Parker's stay in the hospital lasted about nine weeks. She was discharged around the middle of April. Nothing of that time remains in her memory but an occasional shifting shadow, a half-heard sound, an indefinable feeling. She has had to recover the nature of the experience from sources outside herself. One of these, of course, is her husband. Mr. Parker came to the hospital every day for a leisurely visit, often joining her for lunch or dinner, and he sometimes took her out for a drive or for an afternoon of shopping. Another source—an almost eerily definitive source—is a series of letters that she wrote to her parents, in Boonville, Missouri, nearly all of which they fortunately preserved. The first of these letters was written on February 16, the ninth day of her hospitalization. It reads:

Well, here I find myself in a totally new experience for me, and one I never expected to have—residing in a mental hospital. This place is a nice new building. The atmosphere and decor are those of a hotel. The patients, nurses, and other staff members all dress in casual clothes—no uniforms. There are two patients to a room. Most of the patients here are perfectly lucid, though some are kind of mixed up. One woman laments that she knows that she is going to live forever, whereas she would rather die. There is a cute young colored girl who worries because she thinks the Communists are taking over the world etc. But all are perfectly harmless.

[Alan] has been coming over to have lunch and dinner with me. I am a unique patient here. Some are on drugs, some are on electric shock, some are on individual psychotherapy, and some are on group therapy. I seem to be on a sort of do-it-yourself therapy—in other words a sort of rest cure. I had been assigned to the head doctor (I mean the chief doctor; I guess they are all "head doctors"), but I felt dubious about him. Also, one of the patients told me that he is noted for wanting to give all of his patients electric shock. I told him that I had the same kind of intuition about him that I had had about the orthodontist—namely: This man doesn't understand my case. I said that after being turned into a monster by the orthodontist, I didn't want to take any chances on being turned into a hopeless lunatic by him. So I lounged around for several days in nobody's charge, and then was assigned to a young fellow (thirtyish) named Dr. [Smith]. I told him my problem was that I have to get adjusted to life as a damned ugly woman. He said, "You certainly have a gutsy way of putting it." Actually, the name of what ails me is depression. After fighting the battle of the hopeless dental work for so many months, I was so worn down that I lost all appetite for food, work, or anything else. Every little thing—even putting my clothes on—seemed as difficult as climbing Mt. Everest. I tried resting at home, and tried going back to work, but nothing got me into gear again. The medical doctor recommended that I talk to a "consulting psychiatrist," and the latter recommended that some time in the hospital—change of environment and no need to push myself—might get me going again. So here I am. I look at it philosophically, and when I get out I will tell people that it is "mod" to be "mad."

Don't worry about my being out of my head. It is not like that at all. I pulled myself together a few days before I came here and made out all the income-tax forms—federal and D.C. and estimated tax. I didn't want to leave poor [Alan] with that job because I have always done the household paperwork and he wouldn't know how.

The next letter in the series was written about a week later, on February 25. It reads:

I continue taking it easy here—something like living in a college dorm. My main pals seem to be the girls in their twenties. Perhaps the teeth braces automatically identify me with the younger group. I had thought that I might get to tell my dreams and interesting life experiences. But instead I get a diet supplement and a laxative pill. There is quite a lot of humor around here of the type [Alan] calls gallows humor. One little girl is determined to commit suicide but can't figure out any way to do it here. I offered to save up a hundred laxative pills for her. She got the giggles. My case has been so mixed up that they only got around to giving me my entry exam a couple of days ago—two weeks after I had come here. Apparently I am O.K. except for being weak and run down. A nurse was doing something at her desk with rubber bands, and I told her in case I got violent, she could use one of those to restrain me.

A following letter in the series, written early in March, appears to have been lost. The next, and penultimate, letter is dated April 5. It reads:

I certainly am in a strange state. Early last week I suddenly came to—so to speak—and wondered where I was and how I got here. I learned that I had had something called "electric-shock treatments" that had caused me to lose my memory. Now I know how Eve must have felt, having been created full grown out of somebody's rib without any past history. I feel as empty as Eve.

I can remember a few things. I know my phone number and who my relatives are. However, the letters from you all that I found in my dresser were completely new to me. I reread them without any recollection of having read them before. There were a number of get-well cards, and some were from names I didn't recognize. There were a couple from a [Margaret Davis], who [Alan] says lives on our floor, but I don't remember her. Also some from people in the office whose names sound familiar but whom I couldn't visualize.

A cute-looking young fellow with a turned-up nose came in to see me and was asking how I felt. I told him I felt all right except that I couldn't remember much of anything. I asked him who he was and he said he was my doctor. Then I asked him what kind of doctor he was. He looked surprised, and said he was a psychiatrist. I said:

"A psychiatrist! Then I must be crazy." He said: "Oh, no, no!" However, I presume I must be off balance to some extent or I wouldn't be in what turns out to be a mental institution.

Actually, the so-called patients here all seem to be in command of their wits except for one—namely, the woman who has been my roommate for the past couple of days. She is a nice-looking woman and about my age, but she talks a blue streak, pure nonsense—on and on about the Pope and sex and euthanasia and syphilis and strangers' toothbrushes and God knows what all.

This morning I went to the orthodontist and this afternoon I have an appointment with the periodontist, Dr. [Brown]. [Alan] has kept track of my appointments—otherwise I wouldn't know about them. Oddly, I can remember Dr. [Brown's] attractive red-and-black waiting room, but I can't remember him—don't know whether he is young or old, short or tall. I haven't the slightest idea what kind of dental work he is doing to me.

Mrs. Parker wrote her final hospital letter to her parents on the following day, April 6. She had learned by then that she had been given a total of eight treatments. Her letter reads:

For the first several days after those electric-shock treatments were over—possibly as long as a week—I felt just fine, perfectly relaxed and comfortable and also very hungry, as if I were making up for lost time. However, beginning Monday night (today is Friday) I began feeling all churned up and nervous and jittery and tense for no reason, and I have felt that way ever since. [Alan] said that what I was describing was the way I have felt for a long time. Then, beginning last night, in addition to feeling tense and agitated, I also felt scared, also for no reason.

Somebody the other day asked me to make a fourth at bridge and I refused, saying I hadn't played bridge for at least fifteen years. Then he said I had been playing right along for the two months I had been at the hospital. Also I was surprised to find a very pretty black-and-white checked raincoat in my closet. [Alan] told me I had bought it one day soon after I came here when we went out to get my hair fixed. They tell me that gradually I will get my memory back. I hope so.

Electroconvulsive-Shock therapy is a relatively recent refinement of a primitive procedure that was first employed around the turn of the eighteenth century. Johann Christian Reil (1759–1813), a German neurologist and anatomist whose name distinguishes several structures of the human body, is generally regarded as its pioneer proponent. Reil's curious contribution to psychiatry was a product of his interest in the then just forming humanitarian opposition to the traditional chain-and-shackle treatment of the mentally ill. Reil went further than most of his associates in the movement. It was not enough, in his opinion, that the inmates of the madhouse be merely freed of their fetters. They should also, he proposed, be given some sort of restorative treatment, and after consideration he came up with a plausible psychotherapeutic program. Its aim was to frighten, or shock, the patient into rationality. Reil's regime could be administered in many different ways. The unsuspecting patient might be suddenly seized and flung into a pond. Or a cannon might be shot off behind him. Or he might be wrenched from sleep to face a hovering ghost. More heroic measures were prescribed for stubborn cases. Medical historians of the period have reported elaborate tableaux (with large casts drawn from the madhouse staff) that depicted such salubrious horrors as a resurrection of the dead, the Last Judgment, the yawning gates of Hell. Reil called his treatment "noninjurious torture."

If Reil was the first to attempt an instant psychotherapy, he was also (except for the perennial hypnotist) the last. Most subsequent attempts to achieve an immediate emotional rehabilitation by means of a cathartic shock have employed a chemotherapy. The renowned eighteenth-century American clinician Benjamin Rush was responsible for one of the earliest of these. Rush treated mental patients in his Philadelphia practice with a shock therapy that involved the induction of suppuration at the back of the neck to excite a tonic discharge "from the neighborhood of the brain." The triumphant confirmation in the late nineteenth century of the germ theory of disease provided a more convenient method of producing a chemotherapeutic shock. In 1890, the Austrian neuropsychiatrist Julius Wagner von Jauregg used an extract of the tubercle bacillus to ignite what he hoped would be an explosively curative fever in an insane patient. This early effort was not a success, but many years later, in 1914, he tried again, with the malaria organism, and this time achieved a distinct improvement in the condition of a group of men suffering from general paresis. The discovery of insulin, in 1921, made possible a variety of chemotherapeutic shock that remains the awesome ultimate in the pharmacopoeia of psychiatry. Another Austrian, a clinical investigator at Berlin's Lichterfelde Hospital named Manfred Sakel (1900–57), is recognized as the discoverer of insulin as a psychiatric tool. Insulin is distinguished for its power to reduce the sugar content of the blood, and it is this power, of course, that makes it a salvational drug in the treatment of diabetes. Its impact on a normal person is very different. A large injection will produce confusion, deep sleep, and, finally, coma. It was this capacity that interested Sakel. He first experimented with the induction of insulin shock, or hypoglycemic coma, as a means of calming morphine addicts during a withdrawal period. The results (when he learned to reverse the action with a timely dose of glucose) were gratifying enough to encourage him to try the same treatment in other excited states, and in 1933 he reported its usefulness in the treatment of schizophrenia. Sakel's estimation of the value of his work was soon confirmed by other investigators, and insulin-shock therapy—though not without drawbacks, and even dangers—is still in widespread use. It is, however, most highly esteemed as the inspirational prototype of electroconvulsive therapy.

The principles of electroconvulsive therapy were developed by the Italian clinicians Ugo Cerletti and Lucio Bini, and first described by them in a report (entitled "L'Elettroshock") to the journal *Archivio Generale di Neurologìa, Psichiatrìa, e Psicanalisi* in 1938. Electric shock differs from other forms of shock therapy in that it involves the direct mechanical manipulation of the brain to produce a generalized convulsion, or epileptiform seizure. It is thus a physiotherapeutic treatment. Cerletti and Bini conceived electric-shock therapy as a treatment for schizophrenia—as an improvement on Sakel's insulin shock—but subsequent investigators have found it most effective in treating the depression of old age (involutional melancholia) and the depressive phase of manic-depressive psychosis. The procedure currently in vogue is as simple as it is direct. Electroconvulsive therapy is usually given in the morning, and the patient is prepared for it as if for surgery: he is allowed no breakfast, and false teeth, if any, are removed. He is positioned comfortably in bed on his back, and given an intravenous injection of a muscle-relaxant drug and a complementing injection of a hypnotic to maintain normal respiration. An electrode is then placed on each temple, and an alternating current of (usually) eighty or ninety volts is passed between the electrodes for a fraction of a fraction of a second. In that stupendous instant, the brain is so raced that the mind cannot function, and it is from this eerie quietus that the beneficial results of the treatment appear to spring. Just why a halt in cerebral function should be therapeutic is not, however, known. Most cases of depression require several such treatments, and the usual course is between eight and twelve convulsions. In the early days of electroconvulsive therapy, before the development of a satisfactory muscle relaxant, fractures or dislocations were frequent in the moment of violent seizure, but they are now relatively rare, and the patient passes from a brief (four or five minutes) unconsciousness into a peaceful sleep.

The states of mind of most patients emerging from the postconvulsion sleep are similar. There is a harrowing sense of confusion, and then a full awakening in the midnight dark of total amnesia. The patient has no idea who he is or where he is or what has happened to him. He is often also weak, unsteady, and dizzy. Nausea, sometimes with vomiting, and headache are not uncommon. Some sense of identity soon and spontaneously returns, and from the attending doctors and nurses the patient learns his whereabouts and the nature of his situation. At that point, reorientation slows, and the deepest amnesia remains. The distant past —the past of childhood and adolescence—is the first to gradually reappear. The middle past is more difficult to recover, and the immediate past—the weeks or months just preceding treatment— is almost always irretrievable.

Psychiatrists are generally inclined to regard electroconvulsive therapy as a useful psychiatric tool. Some are more enthusiastic than others. The late Arthur P. Noyes, director of psychiatric education at the Pennsylvania Department of Public Welfare, and Lawrence C. Kolb, chairman of the Department of Psychiatry at the Columbia University College of Physicians and Surgeons, who together wrote the standard text "Modern Clinical Psychiatry," observe, "In the depressions of involutional melancholia and of manic-depressive psychosis, the improvement following electroconvulsive shock therapy is striking. In eighty per cent or more of these disorders, five to ten treatments are followed by full or social recovery." Justin Hope, clinical professor of psychiatry at the Tufts University Medical School, and Raymond D. Adams, Bullard Professor of Neuropathology at Harvard Medical School, take a somewhat guarded position. In a collaborative contribution to "Principles of Internal Medicine" they note, "Although carefully controlled experiments cast some doubt upon the efficacy of electric-shock therapy in terminating an individual depressive episode or preventing recurrences, nevertheless it is the authors' clinical impression that it does indeed favorably influence the course of the individual depressive episode." Aubrey Lewis, professor of psychiatry at the Institute of Psychiatry of the University of London, has expressed what seems to be the opinion of a majority of clinicians. "Electric-convulsive therapy," he suggests in a current monograph on the psychoses, "has been over-used in the last twenty years, being applied in unsuitable cases or when less severe methods would have sufficed; but this reproach has been taken from it since the new drugs have superseded it as the easiest acceptable form of somatic treatment."

Most psychiatrists are satisfied that electroconvulsive therapy is as benign as it is beneficial. They are generally agreed (on the basis of numerous psychometric tests and other objective studies) that the patient undergoing such treatment runs no risk of basic intellectual impairment. There is less agreement on the question of memory impairment. Some investigators have recently suggested that the more or less permanent amnesia resulting from repeated electric-shock treatments may not be confined to the period immediately preceding treatment. Larry R. Squire, assistant professor of psychiatry at the University of California School of Medicine at La Jolla, reported to the third annual meeting of the Society for Neuroscience, in San Diego in 1973, that controlled tests involving memory of long-past events indicated that repeated electroconvulsive stimulation "apparently produces a defect in recall which can extend to memories that are some twenty years old, [but] it is not yet known how long this amnesia . . . persists." And J.-O. Ottoson, a participant in the 1967 International Congress of the Academy of Psychosomatic Medicine, observed in a paper entitled "Memory Disturbance After E.C.T.—A Major or a Minor Side Effect?" that while "in most cases memory impairment soon vanishes . . . some patients have prolonged, perhaps irreversible, disturbances." These, however, are minority cautions. The majority view would seem to be the one proclaimed by the editors of the 1973 edition of "The New Home Medical Encyclopedia." They conclude, "Memory may be somewhat impaired as a result of the treatment, but it returns when the full course of treatment is terminated." This is the reassurance that most patients are given as they prepare to leave the hospital after treatment. It is precisely the reassurance that Mrs. Parker received at the end of her hospital stay.

Mrs. Parker's return, on April 13, to the apartment that had been her home for many years was something of a *déjà-vu* experience. She had an uncertain feeling that she had been there before. It was strange, and yet familiar.

"It was all very peculiar," Mrs. Parker told me shortly after her retirement. "I was puzzled—but only vaguely. I really felt too vague to care. Nothing really bothered me. Not at first. I felt physically very well. I felt vegetablized and calm. I didn't have enough memory to think, or even worry, with. And then, because the apartment was so familiar, my mind seemed to open up a little, and my memory began to come back. I mean my memory for where I was—for simple, household things. Although there were odd gaps even there. I remember my first morning at home. I thought of breakfast, and my mind was a blank. I turned to Alan: 'What do I usually have for breakfast?' He looked a little startled, but he told me—an egg and a cookie. Oh, yes. I remembered. I was full of questions. It was like beginning life all over again. I said something one day about the hospital, about the bills, and Alan said Blue Cross was taking care of it. Blue Cross? I didn't know what he was talking about. I'd never heard of it. And, of course, Watergate. I kept hearing 'Watergate' on the radio, and seeing it in the paper. It meant nothing to me. So Alan had to explain. That was the way I remembered, the way things came back, the way I relearned.

"The hospital had told me to take it easy, to rest at home for a few weeks, not to even think about my job. I had a general memory of my job. I knew where I worked, and that I was an economist and analyst. But it was no problem not to think about my work. Work was just something that drifted across my mind from time to time. It didn't interest me. I was too comfortable doing nothing. I've always been a great reader, but even reading didn't interest me now. I read a couple of novels, and the minute I put them down I forgot everything about them. I read a book called *Zelda,* but I don't remember a single thing about it. Any more serious reading—a book that required any background of general knowledge—I simply couldn't read. I couldn't understand it. So I gave up trying and just let myself be comfortable. And I was comfortable. I got to know our friends again. We went out to dinner now and then. We went to the movies. I kept house. I functioned very well.

"I went back to work in July. The rest at home and some sessions with a sympathetic psychotherapist had done me good, and I felt almost like my old self again. My memory seemed to be coming back the way the hospital had told me it would. I was eager to work, eager to put my mind to work again. And I was curious. I wanted to find out what I had been working on before I took my sick leave. So I went back to work one Monday morning and up to my office and sat down at my desk, and my old associates flocked around. Most of them looked familiar, and I was able to remember some of the names. I was still feeling pretty good. Then I started going through my desk—all the current papers and pamphlets and so on. I gathered that I'd been working on the income of securities dealers—relating their earnings to the gross national product. The papers were full of professional terms that seemed familiar. I knew what they were, but I didn't know what they meant. 'Over-the-counter,' for example. It was a familiar term, but I didn't know—I couldn't remember—exactly what it referred to. 'Mutual funds' was another. And 'odd-lot dealers.' All blanks. But I had a vague idea that there was something that might help, that might get me reoriented. I went to one of the girls. I hemmed and hawed, and said I'd forgotten but wasn't there some particular book that I had been using? 'Oh, sure,' she said. 'You mean that book you got at the National Association of Securities Dealers meeting in December.' I said I guessed that was it, but had I been at the meeting? She almost laughed. She said, 'Were you there? Why, Natalie, you practically ran it. It was you who asked most of the questions. It was you who got most of the information we needed.' It was a terrible moment. I thought I was losing my mind. I had no recollection of it at all. And then, like a shadowy film, I got a dim sense of a man sitting on my left in a meeting room. But that was all. Just the shadow of a presence. My friend just stared at me.

"I came home from the office that first day feeling panicky. I didn't know where to turn. I didn't know what to do, I was terrified. I've never been a crying person, but all my beloved knowledge, everything I had learned in my field during twenty years or more, was gone. I'd lost the body of knowledge that constituted my professional skill. I'd lost everything that professionals take for granted. I'd lost my experience, my knowing. But it was worse than that. I felt that I'd lost my self. I fell on the bed and cried and cried and cried.

"But you know how it is. One always hopes, or tries to hope. I told myself that maybe it was only a matter of time. If I was patient, maybe in time everything would come back to me. So I went back to the office determined to try. I was going to start all over again. I was going to relearn. I started looking everything up and making elaborate notes. It was like learning to walk—I started out taking little baby steps. The days and weeks went by, and everybody at the office was good and patient and helpful. Every now and then, I'd get a little glimmer. But mostly it was discouraging. There weren't just gaps in my memory. There were oceans and oceans of blankness. And yet there seemed to be a kind of pattern. My childhood recollections were as strong as ever. That, I've gathered from my reading about electric shock, is quite typical. The fog of amnesia increased as I came forward in time. The events of the past several years were the blurriest and the blankest. Another area that didn't seem to be affected was ingrained habits—repetitive acts and procedures. I mean, I hadn't lost my command of the English language, I still knew the multiplication tables, I could still do double-entry bookkeeping. And then there was an area that I call emotional. I could still remember experiences from any period of my life that had had a big emotional impact. Good *and* bad. I could remember pains and hurts. And I could also remember a trip we took to Spain only a few years ago. A wonderful trip.

"But the worst of all my problems was that I couldn't seem to retain. I couldn't hang on to my relearning. Or only a part of it. The rest kept sliding away again. I think there was—and is—another factor involved in that. I mean my teeth. My ordeal at the orthodontist goes on and on. And it's a constant worry, a constant distraction. I think that stands partly in the way of my relearning. Anyway, sometime in August I was so discouraged that I had an idea. I was still seeing my psychiatrist—the psychotherapist. Well, one day I asked him about the possibility of recovering my memory through hypnosis. He said he didn't know but he would try to find out. The next time I saw him, he gave me the name of a professional hypnotist. I got in touch with the man and made an appointment. I told him my troubles, and I told him what I had in mind. My idea was to be put to sleep and then asked where I had bought the dress I had on. That was to sort of start my memory working. Then, if I remembered that, he was to ask me the meaning of a term we use at the office—a certain labeling of a concept on which I've written a dozen little treatises. But that wasn't what *he* wanted to do. He was a Freudian psychoanalyst at heart. He got me talking—to blubbering out a sort of intellectual life history. It began with how I could hardly wait to go to school when I was a little girl, and then on to how I never cared about amassing money but only about amassing mental capital— and now it's gone and I want it back. That was on my second visit. I saw him three times, but we never got together. All I wanted was a kind of parlor trick. I wanted him to pull my memory back. All he wanted was to analyze me. That was the end of it. He said he could help recover an emotional memory loss—but not a loss from brain damage. He didn't seem to know anything about electric-shock memory loss. He said he couldn't do much about that. So we both gave up.

"I believe the electric-shock literature is right in one regard. My brain may be damaged insofar as part of my memory has been erased, but my mentality is certainly not impaired. I can still use my mind. And, except for that period of vegetating at home, I've never lost my intellectual curiosity. I was curious almost from the beginning to learn more about what had happened to me—about the whole idea of electric-shock therapy. So I began to look into the literature. I got a list of references from the National Institute of Mental Health. That was around midsummer. I went through the list at one of the medical libraries—the George Washington University library. The result was almost nothing. The authorities all seemed to be parroting each other. I couldn't find a single study that tested the permanence of memory loss. Then, almost by accident, I got started on a little investigation of my own. Soon after I went back to work, I devised a routine to handle my inability to recognize the names and faces of people around the office. I would say, 'I'm sorry. I haven't any memory. You will have to tell me who you are and what you do.' I was going through this one day with a man from one of our coordinating sections when he stopped me. He said, 'You don't have to apologize to me. My wife had shock treatments a couple of years ago, and she hasn't any memory either.' Well, you can imagine my interest in that. I got him to sit down and tell me all about it, and I made notes on it later. He told me, 'Within a few weeks after her discharge, my wife got reoriented to the main outlines of her life. After that, there was very little further spontaneous memory return. I'm a statistician, so I'll put it this way: maybe she improves three per cent a year. She gets by. For the life of a suburban housewife, she doesn't need much memory—the kaffeeklatsches and all that.'

"That conversation was the beginning. I started bringing up the subject of electric shock whenever I met new people, and it's absolutely astonishing how many people have a relative or a friend or somebody who has had the treatment. I met a museum friend of Alan's at an art exhibition in Baltimore whose aunt had had the shock experience. He said her memory for the year that preceded treatment was a blank, but she could function. She has money, he said, and doesn't work—just lives quietly at home. I remembered an older woman from my home town who had had shock treatments maybe twenty years ago. She was a professional woman— a dietitian. I wrote and asked her about her experience. She told me, 'There was a lot of memory I never got back. But I did manage to relearn my work—all those recipes and things.' Another person I questioned was a man who had been at the hospital with me. He was a political analyst for the C.I.A., but he hadn't gone back to work. I talked to him at his home. He said he could remember the general type of work he had been doing but not the specifics. Then there was a lawyer I met. He had had shock therapy about four months before. He told me, 'I haven't any memory, but I have a book that I look things up in.' I questioned about a dozen people in all, and there really wasn't much difference in the answers. Their experience was pretty much mine. Oh, yes—I even wrote to Senator Thomas Eagleton. I thought his experience would be interesting. After all, it cost him the Democratic Vice-Presidential nomination. But he never acknowledged my letter.

"All that while, of course, I was trying to work—desperately trying to relearn my job. But it was heartbreaking. It was so slow.

I was relearning, but only a little, only after a fashion. It was like tunnel vision. I couldn't seem to see the whole panorama anymore. As far as my actual job went—the job I was being very well paid for—I was doing nothing. I was totally unproductive. I wasn't worth my salary, and I didn't see how I ever would be again. No one was pushing me. Everyone was wonderful. Still and all, the office isn't running a home for incurables. So I did what seemed to me the only sensible thing. I applied for disability retirement. I asked for one concession. I asked to be allowed to stay on—without pay—as what's called a 'guest employee.' They were kind enough to grant both of my requests. I have my retirement, and I also have a desk at the office. I go there almost every day. I can type. I can do low-level clerical work. And I'm trying, still trying, to rebuild my mental capital."

Mrs. Parker and I had our first conversations in the fall of 1973. We met again, by prearrangement, some four months later—early in 1974. She told me at once that there had been no appreciable progress in her efforts to recover her professional past.

"But I don't want to sound like a pill," she said. "I mean, I mustn't give the impression that my experience with electric shock was a total disaster. There have been some beneficial results. For one thing, my physical health has improved. I'm beginning to eat again, my digestion is much improved, and I have no trouble with sleep. I also feel emotionally relaxed. And I've lost a lot of bothersome inhibitions. I don't shrink the way I used to. I got in a cab the other day that had a big 'No Smoking' sign, and the driver was one of those know-it-all non-stop talkers. But I interrupted him. I said, 'I see you have a rule against smoking. Well, I've got a rule against talking.' He gave me a look, but he shut up.

"I'm thankful for those little blessings. I'm thankful that I got something for the price I paid. Because my memory is still as blank in those certain areas as it was when I went back to work in July. I walked out of the office one evening last week with a man I'd worked with very closely for a number of years. He was saying something about his children, and I asked how many children he had. He looked surprised. 'Why, six,' he said. I said, 'Well, that's a statistic I would certainly think I'd have remembered.' He said, 'Yes—I would have thought so, too. You were always telling me that six was too many.' That's just one example. I could give you dozens more. It happens all the time, and it makes me feel so stupid. It keeps reminding me of how much of myself I've lost. There are times when I almost wish I were back in those weeks of resting and vegetating at home. When I didn't know what I know now. But that's a little frightening. If I hadn't been a professional woman—if I hadn't been a woman with a highly specialized and demanding job—I might never have realized the extent of my amnesia. I would have thought that I was still perfectly whole and complete."

[*1974*]

CHAPTER 18

Two Blue Hands

DEAN M. BERGER is a big, tall, smiling man, and on the night of January 10, 1974—a cold night with a light snow blowing, a night he still remembers with a shudder—he had just turned fifty-two. Berger is a paint chemist. He is associated professionally with Gilbert Associates, Inc., a firm of engineering consultants in Reading, Pennsylvania. His avocation is bridge, a diversion in which he has achieved the gratifying rank of Life Master, and whenever possible, whenever he is not on one of his many out-of-town consultations, he plays in the regular Thursday-night games of an American Contract Bridge League club in the nearby town of Lancaster. His usual partner, until she moved to Michigan with her family, was his daughter Cheryl—Mrs. R. Douglas Crews—also a talented amateur. The Crewses lived in Palmyra, twenty-five miles from Lancaster, and on his occasional bridge nights, Berger would dine with them at their home. January 10, 1974, was a Thursday, and it was such a night.

The event that evening was a tournament, and it was held in a private room at the Distelfink Restaurant, on the outskirts of

Lancaster. Berger and his daughter arrived there, in Berger's car, at around seven-thirty. The meeting was well attended, with thirteen or fourteen tables in play, but the room was cold. Everybody complained about it. Berger felt it in his hands, in his fingers. He was warmly dressed, in heavy socks and a flannel sports shirt, but his fingers were like ice. He sat for a while, between games, between rubbers, whenever he was dummy, with his hands tucked in his armpits. Then he tried keeping them warm in his trouser pockets. He went back to warming them under his arms. The room got colder. Finally, a little before eleven, the meeting broke up. Berger and his daughter placed second. They were sitting at their last table recalling the more decisive plays of the evening when Mrs. Crews gave a sudden gasp.

"Daddy!" she said. "Your hands! Look at your hands!"

Berger looked. He stared. His hands were blue—a gray, slaty blue, about the color of a ten-cent postage stamp.

Mrs. Crews said, "Are you all right?"

"I don't know," Berger said. He stared at his hands. They looked dead. "I feel a little woozy."

"I think we'd better go," Mrs. Crews said.

When they got out to the car, Berger gave Mrs. Crews the keys. He didn't feel like driving. It had just occurred to him that his hands looked like the hands of a corpse. He didn't feel he could trust them on the wheel. They drove in silence. Berger was too worried, too bewildered to talk. Mrs. Crews was trying to think what to do. A fork in the highway loomed ahead. She abruptly made up her mind, and took the turn to the left. Berger sat up.

"Cher—where are you going?"

"I'm taking you to Hershey Medical Center," she said. "I'm worried about you, Daddy. I want a doctor to look at your hands."

The Hershey Medical Center, in the chocolate town of Hershey, is no ordinary hospital. It is the Milton S. Hershey Medical Center of the Pennsylvania State University, and it consists of a college of medicine and a teaching hospital—a three-hundred-and-forty-bed hospital—of the first rank. Just before midnight, Berger, with

Mrs. Crews at his side, walked into the emergency room there. He
was received by a nurse, who took one look at his hands and called
the physician on emergency duty that night. The physician was a
first-year resident named Robert D. Gordon. Dr. Gordon and the
nurse both stood and stared at Berger's hands. They had never
seen a case of cyanosis so chromatically arresting.

Cyanosis (which takes its name from the Greek *kyanos,* meaning "blue") is a discoloration of the skin that reflects an insufficient
concentration of oxygen in the blood. The presence of oxygen in
the protein known as hemoglobin is what gives normal blood its
rich red color, and when for any reason the normal oxygen supply
is reduced, the color fades and dulls. This phenomenon most
commonly manifests itself in the face (especially in the thin skin
of the lips) and in the extremities—the toes and fingers. The
appearance of cyanosis, if at all pronounced, is a serious sign of
warning. It signals either a circulatory problem or some ailment
affecting the lungs. The number of diseases in which cyanosis is
an early symptom is very considerable. Even those in which the
cyanosis involves the hands are numerous. They passed in baleful
procession through Dr. Gordon's mind—congestive heart failure,
Raynaud's disease, polyarteritis nodosa, Buerger's disease, scleroderma, dermatomyositis, systemic lupus erythematosus, arterio-
sclerosis, polycythemia, obliterative vascular disease, syringomy-
elia, congenital heart disease, arteriovenous aneurysm, myx-
edema, and several forms of poisoning, including the gangrenous
agony induced by the ingestion of fungus-infected rye grain, which
has been known since medieval times as ergotism. The parade of
possibilities halted only once: at the phonetic coincidence of
Buerger's disease. It then moved evenly, and uninstructively, on
to the end. Dr. Gordon had, of course, expected nothing else.
Cyanosis of even the deepest lividity is merely an indication of the
presence of disease. It is not—like certain rashes, like certain
lesions, like certain neural responses—a definite diagnostic sign.
It is a symptom to be explained.

Dr. Gordon left Mrs. Crews to the hospitality of the emergency-
room nurse, and led Berger down the hall and into an examination

cubicle. Berger seated himself on the examination table. Dr. Gordon leaned against the wall and guided him through the usual interrogation. Berger's past was uneventful. There was nothing in his medical history (scarlet fever as a child, an appendectomy in 1941, a hemorrhoidectomy in 1961) that cast any light on his present problem. Dr. Gordon brought him up to the present. Berger didn't smoke or drink, but he had that night drunk several —maybe four—cups of coffee. His hands were neither numb nor painful. They were sensitive to the cold. He had no pain of any kind, but that evening he had had a couple of moments of light-headedness. Dr. Gordon interrupted only once. Berger had mentioned eating rye bread at dinner. Did he often eat rye bread? Yes, he did—quite often. He much preferred it to white. Dr. Gordon mentioned ergotism, and explained it. Berger nodded. As a matter of fact, he said, he had found some mold on a rye-bread sandwich the other day, and had simply scraped it off. That sort of thing didn't bother him. In the Second World War, in the Navy in the South Pacific, he had thought nothing of eating bread full of embedded Oriental beetles. Dr. Gordon reflected, looked again at Berger's hands, and put the thought of ergotism aside. Cyanosis alone was not enough. He moved methodically on. Berger continued fully cooperative. He said he suffered occasionally from headaches. He had been told that he had high blood pressure, but had never done much about it. He had an occasional high-pitched ringing in both ears. He had been afflicted off and on for several months with a dry cough. He had no abdominal pain, no nausea or vomiting, no urinary-tract symptoms. He suffered occasionally from constipation. That, to Dr. Gordon's satisfaction, completed the standard review of systems. He made a perfunctory note: "Tinnitus, cough."

The standard physical examination came next. Dr. Gordon took a sample of blood and a sample of urine, and sent them off for laboratory analysis. He then proceeded with the examination. He noted down his findings in the standard mode. He found Berger to be "a well-developed, well-nourished white man, in no acute distress, with dusky hands." Berger's blood pressure, in his right arm, was 212/152. His pulse rate was 80, respirations 12, temperature 36.8° C. (or 98.2° F.). There was no evidence of any intradermal, submucosal, or subcutaneous hemorrhaging. There were normal tympanic membranes. Berger's teeth were "in poor repair," but there was no mucosal cyanosis. There was no neck-vein distension, no back tenderness. Lungs were clear to percussion and auscultation. Abdomen was soft and non-tender. There were normal bowel sounds. Genitalia were "within normal limits." Rectal examination showed normal sphincter tone. Prostate was normal. Examination of the extremities "revealed the absence of peripheral edema." There was "no pedal cyanosis, but the patient's hands appeared definitely dusky in color." There was no calf tenderness. Chest X-ray was normal. Electrocardiogram was normal. That completed the general physical examination. Only one finding was of any pathological interest. Berger's blood pressure was elevated. Even allowing for his history of high readings, and for the usual rise in moments of stress and tension, it was high —ominously high. Dr. Gordon added "hypertension" to his diagnostic notes. The source of Berger's trouble was still anything but clear. There was, however, sufficient evidence to consider his admission to the hospital. Dr. Gordon excused himself and went in search of the admitting resident.

Berger sat huddled on the edge of the examining table, and watched Dr. Gordon go. Hershey Medical Center felt as cold as the Distelfink Restaurant. He tucked his cold blue hands in his armpits. He sat alone and waited. "I was scared," he says. "I mean I was getting really frightened. What the heck was the matter with me? What was going on?"

The admitting resident on duty that night was a second-year resident named John M. Field. Dr. Field accompanied Dr. Gordon back to Berger's cubicle. He greeted Berger and looked at his hands—his dusky, cyanotic hands. He was uncomfortably struck by the fact that the cyanosis included the nail beds. He talked with Berger for a moment. He reviewed Dr. Gordon's notes and findings. He rechecked Berger's blood pressure, and confirmed Dr. Gordon's reading. Berger was indeed seriously hypertensive. That, however, did nothing to explain his dusky hands. An explanation of that, at this point, could come only from the laboratory. Dr. Gordon reported that the results of the blood and urine studies were on the way. While they waited, he shared with Dr. Field his lingering suspicion of ergotism. Dr. Field was interested in so exotic a thought, then skeptical. He finally shook his head. The results arrived from the laboratory. Berger's urinalysis was normal. So were certain aspects of his blood chemistry. These included the white-cell count, hematocrit, prothrombin time, blood urea nitrogen, calcium, phosphorus, creatinine, and bilirubin. Certain other aspects, however, were not. These were in the category known to hematology as "blood gases." Berger's blood pH was 7.24, or excessively acid. His blood-carbon-dioxide pressure was 38, compared to a normal pressure in the middle thirties. His blood-oxygen pressure was 78, compared to a normal of at least 80. And his blood-oxygen saturation was 92 per cent. Normal is upward of 94.

Dr. Field reread the blood-gases report. He found it disconcerting. On the basis of the laboratory findings, Berger was suffering from both hypoxemia (low blood oxygen) and acidosis. That would seem to account for his cyanosis. But that would also indicate that he was seriously ill. And he didn't seem to be. Except for his dusky hands (and the hypertension, irrelevant in this connection), he showed no clinical signs of illness. Moreover, and inexplicably, his chest X-ray was normal. "Not only that," Dr. Field says. "There was also a conflict that I didn't understand. I didn't think his high carbon-dioxide pressure was entirely compatible with his low pH. Well, we have a rule here on admissions. It requires that we either admit or discharge after two hours. I didn't do either. I couldn't possibly discharge a man with that degree of cyanosis. But I wasn't quite ready to admit him. I wanted some more information. I decided to keep him where he was for the time being. It was now around two o'clock. I wanted his blood pressure taken every hour, and at six o'clock I was going to run another

check on his blood gases. Then I would know. At least, I hoped
I would."

Berger was left alone again. Dr. Field and Dr. Gordon had
gone. His daughter had been in to say goodbye. The doctors had
advised her to leave; she had talked to her mother; she would be
back again in the morning; she knew everything was going to be
all right. The nurse appeared and took the first of the hourly
blood-pressure readings. Berger tried to make himself comfortable
on the narrow examining table. The room was dimly lighted, and
cold. He tried not to worry. He tried not to think. "I wanted to
leave with Cher," he says. "They told me they couldn't let me. It
would be against their better judgment. But they couldn't tell me
what was wrong. I had high blood pressure. There was also something wrong with my blood chemistry. They weren't sure what.
But I got the impression it was bad news. I dozed off. Then the
nurse came in for another blood pressure. I don't think I slept at
all after that."

But, of course, he did. He was asleep, hunched up on the table with
his feet hanging over the end, when Dr. Gordon looked in a little
after six. Dr. Gordon awakened him gently. He took a sample of
arterial blood from Berger's wrist. He gave him a reassuring word
or two. He then went along to the laboratory and arranged with
the night technician for another blood-gases analysis. The results
were ready by seven o'clock. Dr. Gordon and Dr. Field read them
together. The second report more than confirmed the first.
Berger's blood pH was much the same as before. It was now 7.25,
compared to the earlier 7.24. But his blood-carbon-dioxide pressure was up from 38 to 41. His blood-oxygen pressure was down
from 78 to 62. And his blood-oxygen saturation was down six
points, to 86 per cent.

Berger lay awake after Dr. Gordon had gone. He and his boss at
Gilbert Associates were scheduled to fly to a business meeting in
New Orleans tomorrow—today. The nurse came in for the seven-
o'clock blood-pressure reading. When she left, he got up and

found his way along the hall to the men's room. He straightened his clothes and combed his hair and washed his cold blue hands. "Dr. Gordon came in while I was drying my hands," Berger says. "He watched me for a moment. Then he said he'd like to try something. He took me back to the washbasin and had me hold my right hand under the cold water. And was that water *cold!* It couldn't have been more than forty degrees. He kept it there for a full two minutes. And then—my God! It was the damnedest thing anybody ever saw. My hand began to change like a damn chameleon. First it was blue. Then it was red. Then it was blue again. There was something about it that really scared me. That's when I knew I was a goner."

The attending physician on call from the medical service to the emergency room on Friday, January 11, was an assistant professor of medicine named Joseph J. Trautlein. Dr. Trautlein arrived at the medical center that morning at his usual time, a minute or two before eight. He parked, as he always did, in the staff parking lot, and came in through the emergency-room entrance. Dr. Field knew Dr. Trautlein's habits. He was waiting for him in the foyer of the emergency room. He greeted him there and walked with him while he put away his overcoat and got into his long white coat. He told him about Berger.

"He said he hadn't wanted to bother me during the night," Dr. Trautlein says. "But he had this man with two blue hands and abnormal blood gases and a negative chest X-ray. And high blood pressure. He said he didn't know what the hell was going on. Would I take a look at him? Just a curbside consultation—that's all. Just an opinion. I said O.K.—be glad to. It was entirely appropriate. I was a physician of record. And it sounded interesting. So I went along to the examining room and met Mr. Berger and looked at the resident's notes and talked for a while—and I was just as puzzled as John Field was. That acute onset, those two grossly cyanotic hands, those severely altered blood gases. It was strange. I thought, for one thing, that a lung scan was indicated, and a lung scan is an in-patient procedure. Dr. Field had come to the same conclusion. So it was agreed. I admitted him to my service. I then tactfully withdrew. I was the attending, but this is a teaching hospital here, and our residents are given responsibility. It's an observed responsibility—we're always there within an arm's reach. But it's still responsibility. Mr. Berger was Dr. Field's patient."

Berger was admitted to the hospital under a tentative diagnosis of hypoxemia and acidosis, and shown to a private room on the fourth floor. After a night on an examining table, he was more than ready for bed. "But first I had some telephone calls to make," he says. "I wanted to talk to Marilyn—my wife. She would have heard by now from Cher, but that wasn't enough. I wanted her to hear the worst from me. I arranged for some clean clothes and toilet articles and all. Then I called Cher and brought her up to date. Then I called my boss, and we cancelled the trip to New Orleans. I can't tell you how hopeless I felt. When I talked to Marilyn, I told her to go out first thing and increase my life insurance. I was joking—but not really. Then I undressed to my shorts and got into bed. At least I had things organized. That made me feel a little better. And I was warm and comfortable for once. I began to feel hungry. But nothing was said about breakfast. Around nine o'clock, a nurse came in with some apparatus, and the next thing I knew I had a tube up my nose and they were giving me supplementary oxygen. Then a beautiful woman doctor came in and took some more blood samples. I hardly felt a thing. Then I was left alone again, and I knew I wasn't going to get any breakfast. Or any lunch. Or anything. I was too far gone to waste food on."

By nine o'clock, the news of Berger and his perplexing case had spread through his floor and beyond. Dr. Graham H. Jeffries, the chairman of medicine, heard the news as he stepped out on his regular morning rounds, and he at once added Berger to his itinerary. There were eight physicians in his entourage—Dr. Trautlein and two other staff members, and five house officers, including Dr. Gordon and Dr. Field. The group arrived at Berger's bedside at a little past nine-thirty. Dr. Gordon made the presentation. Berger was asked to display his hands. Dr. Jeffries examined them with interest. He had never seen anything quite like them. They made, however, an unusually challenging teaching exposure. Questions were asked and answered. Opinions were solicited. Ergotism was mentioned, and again discarded. Berger's work as a paint chemist was discussed. It was established that he was exposed almost daily to paint and solvent fumes. The order for a lung scan to determine the presence of a possible obstruction was approved. It was scheduled for one o'clock. There was an evaluation of Berger's blood pressure. It had dropped significantly in recent hours—to 160/100. There was still a degree of hypertension, but it was agreed that his condition could satisfactorily be treated on an out-patient basis. It was also agreed that his hypertension was unrelated to his other, more urgent troubles. Dr. Jeffries endorsed the admitting diagnosis, and the reservations of the two residents about it. He thanked Berger for his patience and cooperation. He and his entourage moved on.

Berger watched them go. "Nine doctors," he says. "Four head doctors and five of the brightest young doctors in a big institution like Hershey Medical Center—and there wasn't one of them could tell me what was my trouble. And it worried them. I could tell they were really worried. So now I knew I'd really had it."

Dr. Jeffries completed his teaching rounds at about ten-thirty and dismissed his entourage. Dr. Field had a moment of leisure. He took the elevator down to the second floor—down to the laboratory. He sought out the technician to whom Berger's latest blood sample had been delivered for still another blood-gases assessment. The results were ready and waiting. They were very different from the two earlier sets of findings. Berger's blood pH was 7.41, or normal. His blood-carbon-dioxide pressure was 36.4, or normal. His blood-oxygen pressure was 109, or normal. And his blood-oxygen saturation was 98 per cent. Also normal. "That shouldn't have been surprising," Dr. Field says. "It was exactly what we would expect after a course of oxygen therapy. The oxygen balance in Berger's blood had been very nicely restored. But it *was* surprising. It didn't make sense. If Berger's blood gases were back to normal, how come he was still cyanotic? How come his hands were still blue? I had seen them less than half an hour before. There was something wrong somewhere."

Dr. Field went back up to the fourth floor. On the way, he decided to discontinue Berger's oxygen therapy. He wanted him breathing room air again. Then he would take another blood sample for another blood-gases study. Then he would see. He stopped in the corridor for a drink of water, and was joined by one of his colleagues, a resident named James E. Meyer. Dr. Meyer had also been a member of Dr. Jeffries' entourage. He said it had been an interesting round—particularly the Berger case. He said the look of Berger's hands had stuck in his mind. Those hands reminded him of something. And he had just realized what.

"What?" Dr. Field said.

"I don't know if I can explain it," Dr. Meyer said. "But you know how your hands look after you've been shovelling snow? I mean the way they look when you come in and take off your gloves. They have a certain look."

"Mmmm," Dr. Field said. "Yeah—I know what you mean."

But he didn't. He and Dr. Meyer walked together down the corridor. He had no idea what Dr. Meyer meant. But the thought of gloves stayed with him—gloves and skin discoloration. It hung on in the back of his mind. And suddenly it came alive. He stopped.

He said, "Wait a minute."

He cut off to the stockroom. He rummaged around and found a surgical acetone swab. He rejoined Dr. Meyer. Dr. Meyer looked at the swab.

"What's that for?"

"I just had an idea," Dr. Field said. "It was what you said. You gave me a brainstorm."

They came to Berger's room. Berger was alone and dozing. Dr.

Field turned off the oxygen supply. He removed the nasal cannula. He asked Berger to hold out a hand. He took the hand by the wrist and gave the back a vigorous rub with the acetone swab. A pinkish pale streak appeared on the back of Berger's hand. The face of the swab turned blue.

Berger had been almost half asleep. Now he was fully, galvanically awake. He stared at the swab. Then he stared at his striped hand. "I didn't understand it," he says. "I just looked. My head was spinning. It was just too much too fast. Then I calmed down a little and saw something. The blue on the swab looked different from the blue on my hand. My hand looked dead. But the swab was simply blue, an attractive shade of blue. And blue is one of my favorite colors. I raised my head and looked across the room. My flannel shirt was hanging there on the back of a chair. It was the exact same shade of blue. The shirt was new—Cher had given it to me for Christmas. I looked at my shirt, and I didn't even have to think. I knew what had happened. I remembered sitting in the Distelfink and warming my hands in my armpits. And later on, during the night in the examining room. It was as simple as that. The dye had come off on my hands. But how? That stopped me for a moment. But I'm a chemist, not just a paint chemist. The answer was that the dye was fast by ordinary standards. It hadn't come off when I washed my hands. But it was soluble in what Dr. Field was using—in acetone. And in sweat. Sweat isn't just water and a little sodium chloride. It also contains certain solvents and acids. Acetic acid, for example. And butyric acid. And valeric acid.

"Dr. Field was reaching for my other hand. I gave it to him. And while he was scrubbing it back to normal, I told him what I thought had happened."

Dr. Field believed him. "It was weird," he says. "The whole thing was staggering. But I knew it was true. It had to have happened that way. The only trouble was that that only explained his hands —his cyanosis. It didn't explain those altered blood gases. Was it possible that there actually was something wrong with his blood chemistry? Well, the way to make sure was to do what I had planned to do anyway. When he had been breathing room air for a sufficient length of time, I drew a sample of blood and took it down to the lab. The technician went to work. Blood gases don't take too long. Maybe ten or fifteen minutes. Then we went over the results. They weren't precisely the same as the previous study, but the differences were minimal. Insignificant. Berger's blood gases were normal. That was the answer I wanted, of course, but still—what about last night? What about those two other studies? No answer. There isn't any absolute explanation. My feeling— everybody's feeling—is that the lab had made a mistake. And that's the way it stands in the record: 'It was felt that evidently some error must have been made in the determination of the blood gases during the night, and it was felt that the patient's apparent hypoxemia was a lab error.' "

Dr. Trautlein says: "They told me that Dr. Field was looking for me. I found him, and he told me the story. Gulp!"

Berger was discharged from the hospital at five o'clock that afternoon. "I was ready to leave at noon," he says. "But that isn't the way they do things at Hershey. They wouldn't let me go until they had done a lung scan. It turned out normal. They expected it to be normal, but I guess they wanted to be absolutely certain. I didn't mind the wait too much. I was thankful just to be alive."

[*1975*]

* * *

AUTHOR'S NOTE: Several readers of this piece found Mr. Berger's seeming cyanosis a more mysterious mystery than I intended it to be. The question that occurred to these readers was put with some succinctness by a gentleman from Kentucky. "Weren't

Mr. Berger's armpits also blue?" he wrote. "If not, why not?"

Well, Mr. Berger's armpits were *not* blue. Nor, for that matter, were the underarms of his undershirt. And the reason for this, though chemically complex, is simply explained. The sweat evolved by the glands in the axilla, or armpit, was not the sweat that leached the dye from Mr. Berger's shirt. The sweat that drew the dye from his shirt and so alarmingly stained his hands was sweat evolved from his palms. These two glandular secretions are differently constituted. Dye—at least the dye of the sort contained in Mr. Berger's shirt—is fast to the sweat of the armpit. It yields, however, to that produced by the palms.

I am indebted to Dr. C. A. Hilgartner of the University of Rochester School of Medicine for a useful amplification of the sweating phenomenon. "Palmar sweating," he noted, "occurs more as a function of excitement of one kind or another than as a function of temperature—and probably the bridge tournament provided enough excitement to induce relatively profuse palmar sweating, at least at times."

CHAPTER 19

Antipathies

A YOUNG WOMAN I'll call Sara Strong is sitting in the waiting room of the Tulsa Dermatology Clinic—a private group practice —on East Twenty-first Street in Tulsa, Oklahoma, and she is, of course, uncomfortable. The cause of Mrs. Strong's discomfort is an itchy rash on her elbows, on the backs of her legs, and here and there on her chest. She has had this rash for almost a month. When it first appeared, around the middle of April, her husband, an interne at St. John's Hospital, just down the street from the clinic, identified it as an eczematous allergic contact dermatitis, and treated it in the conventional manner—with an application of cortisone cream. The rash was then on her elbows only, and the cream at first seemed effective. But a few days later, the rash had spread and worsened. It erupted first on her left leg, in the sensitive pocket behind the knee, and then on the other thigh. Mrs. Strong continued her husband's prescription of cortisone cream, and an added soothing lotion, but the rash continued to spread. It appeared on her chest, and Dr. Strong gave up. The proper treatment of her trouble was plainly beyond his professional powers. She needed more experienced and more specialized help. He put in a call to the Tulsa Dermatology Clinic and spoke to the appointments nurse. That was on the afternoon of May 15, a Saturday. The nurse responded with the sincerest form of professional courtesy: she gave Mrs. Strong the earliest possible appointment—Monday morning at eight o'clock.

The allergic condition—with its multiplicity of pains, gasps, fevers, sneezes, itches, nauseas, swellings, diarrheas, and strangulations—is at once the oldest and the newest of man's afflictions. There is no disease more ancient than asthma, and none more newly arrived than a flaming response to the latest creation of the cosmetics laboratory. Almost everything in the human environment (or, as the New York University dermatologist Alexander A. Fisher has put it, "everything under the sun, and even the sun itself") has the power to produce in some unfortunate person some form of allergic reaction. The list of allergens includes most common foods (wheat, corn, eggs, fish, milk, nuts, beans, chocolate), many popular drinks (cola, beer, orange juice), the most useful drugs (aspirin, quinine, codeine, the sulfonamides), the inescapable inhalations (dusts, spores, pollens, vapors, feathers, hair), the many kinds of injections (vaccines, penicillin, mosquito bites, bee stings, insulin), the infinite physical agents (poison ivy, chemicals, metals, soaps, dyes, paints, cosmetics, fabrics, furs, leather, plastics, rubber), and light and heat and cold.

The existence of the phenomenon now known as allergy was recognized long before there was any scientific understanding of its nature. It was too peculiar to be easily overlooked. The allergic phenomenon was probably first observed in what is still its most dramatic expression. Indeed, this manifestation forms the subject of the most celebrated aphorism of the pre-Christian Roman poet Lucretius: "What is food to one is to others bitter poison." Increase Mather, one of the more enlightened of the early Massachusetts clergymen, recorded it in larger scope. "Some men," he wrote in his treatise "Remarkable Providences," in 1684, "have strange antipathies in their natures against that sort of food which others love and live upon. I have read of one that could not endure to eat either bread or flesh; of another that fell into a swoonding fit at the smell of a rose. . . . There are some who, if a cat accidentally come into the room, though they neither see it, nor are told of it, will presently be in a sweat, and ready to die away." Hay fever, which wrung from Sydney Smith the anguished cry "The membrane is so irritable that light, dust, contradiction, an absurd remark, the sight of a Dissenter—anything—sets me sneezing," was accurately described in the sixteenth century (by an Italian anatomist named Leonardo Botallo) and was given its misleading name in 1829 (by an English physician named Gordon), but it was not until 1873 that Charles H. Blackley, another English physician, demonstrated that the causative irritant was pollen. Blackley's finding was accepted, after the usual exposure to derision and indifference, around 1906.

Allergy takes its name from the German *Allergie*—from the transliterated Greek *allo ergon,* which means, roughly, "altered reaction." The term was invented in 1906 by the Austrian investigator Clemens von Pirquet (1874–1929), who is also remembered as the inventor of the standard diagnostic skin test for tuberculosis. Von Pirquet developed the tuberculin test in the course of his explorations of allergy, and he came into that all but empty room through the then just opening door of immunology. This door (which legend recalls was first tried by Mithridates VI of Pontus when he attempted to gradually habituate himself to poison and thus confound his enemies) had suddenly yielded to the efforts of such men as Pasteur, von Behring, Kitazato, and Ehrlich. Standing on the shoulders of these and other giants (most notably those of Theobald Smith, of Harvard, and the subsequent Nobel laureate Charles Richet), von Pirquet observed that patients given a second injection of an immunizing serum a few days after a first injection sometimes developed outbreaks of hives, and even asthmatic attacks. Something had changed a protective serum into a hazard. He decided that what he had seen was an alteration in the immune reaction, and gave to the phenomenon the descriptive name by which it is now universally known.

The nature of allergy, though it has been the object of fifty years or more of increasingly sophisticated scrutiny, is still far from fully understood. There are, however, certain areas of at least translucent enlightenment. It is generally accepted that the phenomenon involves a perversion of the antigen-antibody reaction. The fundamental mechanics of immunity and allergy are much, if not entirely, the same. The effect of immunization—in its active, and most satisfactory, form—is to stimulate the natural defensive powers of the body. This is accomplished by first exposing the body to a safely bridled brush with a specific infecting agent, or antigen. The body instinctively reacts to the presence of this antigen by producing in certain tissues a specific chemical counteragent, or antibody. The body thus armed is immune to that particular antagonist. In the event of a reexposure to that antigen, the sentinel antibody mobilizes in the bloodstream and destroys it. Allergy appears to be a blundering version of this beneficent process. The phenomenon has its beginning in a confusion of identity. A substance that is harmless to most people is identified as inimical by the body of a susceptible person, and (as in the development of the immune state) the body defensively elaborates a chemical weapon against its reappearance. This antibodylike material compounds the initial confusion into a kind of pandemonium. Through some constitutional derangement, it attacks not the allergen (as the antigen in allergy is called) but, in effect, the cells of the body itself. It is at this climactic moment that the scientific visualization of allergy begins to blur. It is not known what renders a person sensitive to one allergen or another—or to none at all. Nor is it known just how the damage in allergy comes about. It is accepted that the collision of allergen and antibody releases an irritating substance, but questions continue to loom. Is this substance totally toxic? Or are the affected tissues congenitally defective? About the most that can be said with any assurance is that susceptibility to allergy seems to be an inherited characteristic, and that the irritant involved may consist of one or more of several pharmacologically active substances. These substances include histamine, 5-hydroxytryptamine, and acetylcholine, all of which are natural and (within reason) necessary constituents of the body. "In this connection," E. B. French, Reader in Medicine at the University of Edinburgh, has observed, "it is worthy of note that the juice of the stinging nettle *(Urtica dioica),* from which is derived the name urticaria [hives], contains histamine, 5-HT, and acetylcholine."

Allergy differs from most other diseases in that its victims (including even most asthmatics) seldom die and almost as seldom recover. In this bittersweet respect, as well as in the fortitude with which its long embrace must be borne, it much resembles gout. But gout, though no longer considered a rarity, is nevertheless a disease of the unfortunate few. Allergy is a commonplace disaster. It is, in fact, the most common of all chronic complaints throughout the industrially developed world. The National Institutes of Health report that around eighteen million Americans are afflicted with one or another of the many chronic digestive diseases. Some twenty-two million, its records show, are afflicted with one or another of the several forms of arthritis, and about thirty-five million with hypertension. The allergic population of the United States is on the way to forty million. Almost fifteen million of these are hay-fever sufferers. Nine million are asthmatic, and another several million are allergic to some food or drink or drug. The rest, in whose company Sara Strong unhappily found herself on that spring morning, are victims of allergic eczematous contact dermatitis.

The Tulsa Dermatology Clinic is an association of four young dermatologists. One member of the group is a former Mississippian named Vincent P. Barranco, and it was he who had arranged to see Mrs. Strong on that Monday morning in May. He found her to be a small, trim, dark-haired woman of twenty-six with the characteristic expression of a dermatitis patient. She looked dejected.

"And plain," Dr. Barranco says. "I could see that she was actually very attractive, but her trouble was wearing her down. There's nothing like a raw, itchy rash for taking the life out of a

person. I've seen enough to know. And her history, when we got
down to that, was disturbing. She suffered from hay fever. There
were certain foods that she was allergic to. She had a sister with
asthma. The sister was part of the record, because allergy is
strongly familial. Now, all of this immediately suggested a diagnosis of atopic dermatitis. That's an eczema that occurs without any
external or discernible allergic cause. It's an inherited constitutional state—very strange. But not at all uncommon. It differs
from an allergic contact dermatitis in the matter of cause. In
contact dermatitis, if you look hard enough, there is always a
contactant—a cause. Well, Mrs. Strong's history and everything
else she had to tell me seemed to indicate an atopic state. Except
for one thing. This eczematous attack was her first, and atopic
dermatitis usually begins in childhood and continues on through
the years. But it still looked atopic to me. However, I took the
usual precautions to eliminate any possible contactant. I gave her
the usual instructions: no rings, no earrings, no bracelets, no
jewelry of any kind. No cosmetics except those on this list of
hypoallergenic products. No washing dishes or such without protective polyethylene gloves. And so on. I treated her rash as one
would treat any allergic dermatitis—with topical cortisone cream.
The same thing her husband had used, only a new and more
potent type. Then, because her lesions were quite distinct and
comparatively few, I injected each one of them with cortisone.
Intralesional cortisone is a profound treatment. As a matter of
fact, I thought twice before I decided to go ahead. I thought it
might be more profound than the case called for. But she was so
miserable. And she'd been miserable for almost a month—that
made up my mind.

"I honestly thought I'd never see her again. By the end of the
week, I was sure I wouldn't. And then on Monday—the following
Monday, May 24—the phone rang, and it was Mrs. Strong. She
hated to bother me. But she wasn't any better. In fact, she was
worse—much worse. The itch was driving her crazy. I was staggered. If she had called to say there was still some rash—if she had
said she was only slightly improved—well, I would have been a

little surprised. I'd have accepted it, though. But *worse!* I couldn't believe it. Except, of course, I should have known. Her husband was a doctor. That's what always happens with a damn doctor's wife. Something always goes wrong when you've got a colleague looking on. Well, I swallowed my embarrassment, and did the best I could. I told her to continue the cortisone cream, but this time cover it with Saran Wrap—with some plastic wrap—to increase absorption. I also prescribed an antihistamine—fifty milligrams of Benadryl every six hours. My hope was that that would stop the itching. Even if it didn't, it would at least give her the relief of a good night's sleep. Benadryl has about the highest incidence of sedation of any antihistamine. But I was counting on it to end the itching, and I was counting on the continued cortisone to finally wind things up. This had to be the end of the case. I couldn't imagine anything more.

"So another week went by. Eight days, to be exact. Then, on Tuesday morning, June 1, I got a call from Dr. Strong. They had followed my instructions, but the treatment didn't seem to work. She was worse—even worse than the week before. The rash had continued to spread. It now covered every inch of her body. Every inch of her was lobster red except the palms of her hands, the soles of her feet, and her scalp. The itch was unrelenting. She was just about climbing the walls. He sounded frightened. His voice was pure anxiety. And I was frightened, too. I said he'd better bring her right over to the clinic. Or, better still, we'd get her into the hospital. But he said no to both. He said he'd rather handle it in the emergency room at St. John's. He was on that service there. He was firm on that, so I said O.K. Then we talked treatment. We settled on an injection of one cc. of betamethasone. Betamethasone is an extremely potent corticosteroid, and the usual dose is one cc. or less at intervals of six or seven days. So he took her off for treatment, and I waited another week. It was now the seventh of June. Another Monday. And Dr. Strong called. It was the same terrible story. There had seemed to be some improvement for a day or two after the betamethasone injection. But then it turned around again, and, if possible, she was worse. I said I wanted to see her. I wanted to see them both. I asked him to bring her over right away.

"I couldn't understand it. An atopic dermatitis should have cleared up by now under this really rather heroic treatment. So I had to reconsider—I had to wonder if maybe it was a contact dermatitis after all. That presupposed the continued presence of an undetected allergen. But what? There had been nothing in her recent experience that sounded even remotely like an allergenic possibility. And I had eliminated all the obvious possibilities— cosmetics and jewelry and such. But when she and her husband arrived at my office I went back to the beginning and started all over again about anything out of the ordinary that she might have come in contact with around the middle of April, around the time of onset. Nothing. It was all completely negative. And yet I was now pretty well convinced that there was a contactant somewhere. The two conditions aren't mutually exclusive—an allergic contact dermatitis can occur in an atopic state. I've never felt so frustrated. I wondered if I should have risked a patch test at the very start. Patch testing is a means of determining sensitivity to various suspect substances, and a positive reaction is manifested by a rash. But it can be risky when the patient already has an acute and spreading rash. It isn't always accurate then, and it can also exacerbate the trouble. It's best to wait until the rash is under control. And now it was too late. Mrs. Strong was all rash. There was no place on her body to test. So I was left once more with the question of treatment. I prescribed oral corticosteroids—four tablets every other day for ten days. And a tranquillizer to try and relieve her anxiety. But it was no use. She came back to the office by prearrangement ten days later, and she was much the same, or worse. I led her back through her history again. Nothing. And I was finally at the end of my rope. I had tried every treatment I knew. I'd done everything I could. I had to start all over again. I gave her another injection of betamethasone and another potent antihistamine. She went home just about in tears. And she looked at least twice her age.

"I had a call the following day from Dr. Strong. There was, of course, no encouraging news. He was calling because he was getting a little frantic. He was finishing up his interneship in a couple of weeks, and then he would be moving to Dallas for his residency. Could I refer him to a dermatologist there? Should we have another consultation? Was there absolutely nothing more I could do? He didn't mean it as a challenge, but it was challenging. It was threatening. I began grabbing at straws. I dug into their life as deep as I could. I began throwing out questions just on a chance. One of them was, What did they do for birth control. He said she had an IUD—an intrauterine contraceptive device. Was that what she had always used? No—she had started on the pill, but the pill had made her sick. So she had been fitted with an IUD around the middle of April. The middle of April! A bell began to ring. I wondered if there was any metal in her IUD—any copper. There is sometimes metal in an IUD, and when there is it is almost always copper, because copper has a certain anti-spermatozoon action. And copper is a known allergen. It's not common, but it happens. But Dr. Strong said he had no idea. I'd have to ask the gynecologist who inserted it. He turned out to be one of my friends, and I called him about two minutes later. He said yes. The device was one that contained a small amount of copper. I told him what I had in mind, and asked him to pull it out. He thought I was crazy. But, as I said, he's a friend, and I sent Mrs. Strong down to his office the next morning, and he took it out. That was on Friday, June 18. I saw her four days later, on Tuesday. It was fantastic. It was unbelievable. It was the most dramatic clearing of a dermatitis that I've almost ever seen. But, of course, I had to make sure. I waited a couple of days for the last of the rash to subside, and then called her into the office for a patch test. I tested her for copper, for nickel, for potassium, for cobalt, and for mercury. There was one positive reaction—to copper. It gave her a beautiful, definitive, four-plus flare."

The Strongs moved down to Dallas at the end of June. The dermatologist there to whom Mrs. Strong had been referred reported back to Dr. Barranco that she was now entirely recovered. Dr. Barranco enjoyed his diagnostic triumph. It was also, he came to realize, a most provocative one. This realization led him to the library, and his sense of triumph increased. His experience, he found, had been a rare one—almost disconcertingly rare. He was able to find on medical record only six other cases of allergic contact dermatitis involving copper. In three of these, the exposure to copper had occurred in a New England industrial setting. One, reported by a French physician, involved exposure in a chemical-products warehouse, and another, reported from the Netherlands, involved a telephone lineman. The sixth involved an American woman for whom the contactant was copper jewelry. What disconcerted him was the discovery that his was the only case in which the exposure to copper was internal rather than external. There was, however, he was relieved to learn, a considerable record of eczematous dermatitis resulting from internal exposure to metals other than copper. That was reassurance enough. His experience with Mrs. Strong had been rare—notably rare—but not beyond belief. What he wanted now was an audience.

"I knew I had something worthwhile to report," Dr. Barranco says. "I got all my notes together—my clinical notes and my review of the literature—and began to draft a little paper. I thought I might get it published somewhere. Maybe in the *Southern Medical Journal.* Or maybe even in the *Archives of Dermatology.* But first of all I wanted to tell Buzz Solomon about it. Buzz is Dr. Herman Solomon. He's in a small dermatological group practice up in Wichita that's very much like ours, and he is also my best friend. We roomed together at medical school, at Mississippi, and I was best man at his Jewish wedding and he was best man at my Catholic wedding. Buzz and I go to all the same meetings—we take our wives and have a big reunion—and the next meeting on the schedule was a dermal-pathology course in Houston in August. We met as planned, and the first chance we got we sat down and I told him all about Mrs. Strong. He was fascinated. One thing I didn't know—one thing I couldn't precisely explain—was just how her sensitivity to copper originally came about, and we talked that over. Our best guess was that the sensitizing exposure was probably to a piece of costume jewelry. It might have been a copper bracelet—one of those arthritis amulets—or something made of brass. It could have happened weeks ago, or months, or even years. Well, as I say, Buzz was fascinated—really fascinated—by Mrs. Strong, and when I finished he told me why. He had a case that sounded a whole lot like it.

"His patient was a young woman, too. I'll call her Janet Walker. She was now twenty years old, but Buzz had seen her for the first time a little over a year earlier, when she was referred to him by a colleague. She was suffering from a generalized eczematous dermatitis, which the referring doctor had diagnosed as pityriasis rosea. That's an eczema that some people think may be caused by a virus. She told Buzz that her rash had been coming on little by little for a couple of months. He treated her with topical cortisone, and set up an appointment for two weeks later. When he saw her then, she was worse. He hadn't expected that—just like me and Mrs. Strong. So he reconsidered the diagnosis. Her history was a little bit suggestive. Her father was an asthmatic of long standing. That at least hinted at the possibility of allergy. So did the fact that her ears had been pierced for earrings about two years before. A pierced ear provides a very good setting for a sensitization to metal. He suggested the possibility of allergy to Janet, and told her to take her earrings off and keep them off, to wear no jewelry of any kind for the next two weeks, and then come in and he would see how things looked. I gathered that she wasn't much impressed by the allergy idea. For some reason or other, she seemed to resist it. But she said O.K., and Buzz sent her off with an injection of cortisone.

"But she didn't show up for that next appointment. She simply disappeared, the way some patients do. A year went by. Apparently, she had been wandering from doctor to doctor, and nobody had done her much good. And now he had seen her again. Buzz thought she looked pretty bad. He was now convinced that her trouble was an allergic eczematous contact dermatitis. And although he hadn't been able to test it out, he thought the allergen was most likely nickel. Nickel is by far the most common metal allergen. She admitted that she had gone back to wearing jewelry, and there was a very pronounced dermatitis on her earlobes and around her ring fingers. The trouble was, Buzz said, he hadn't been able to help her. He had even had her in the hospital. No kind of treatment helped. He was still convinced that it was a nickel dermatitis, but he couldn't find the nickel. He had eliminated every conceivable source. And that is no small job. The sources of nickel allergen—particularly for women—are everywhere. Hairpins. Hair curlers. Bobby pins. Eyelash curlers. Earrings, of course, and rings. Spectacle frames —particularly those metal granny glasses. Coins. Medallions. Dental instruments. Necklace clasps. Zippers. I.D. tags. The wire in bras. Garter clasps. Handbag handles. Thimbles. Pens. Scissors. Needles. Watchbands. Lipstick cases. The metallic eyelets on boots and shoes. As a matter of fact, nickel turns up as an element in a whole range of alloys, including sterling silver and fourteen-carat gold. Well, I was beginning to see why Buzz had been so fascinated by my Mrs. Strong. He had separated Janet Walker from every possible source of nickel—every *external* source. He hadn't considered the possibility of an internal source. He didn't know if it was a possibility. But he went back to Wichita with that little glimmer of hope.

"A few days after the Houston meeting, I got a telephone call from Buzz. He had a lead. He had learned from Janet that she had had some surgery done about a year before their first meeting. There was some problem with her knees. She had chronically dislocated patellar tendons. They were corrected by a Hauser procedure, and the tendons were secured in the proper position with stainless-steel screws. Buzz had talked to the orthopedist, and the screws were still there. The problem now was to persuade the orthopod to go in and remove the screws. It wasn't that Janet needed the screws. The orthopod was satisfied that they had done their job, and that the tendons were now naturally secured. But he thought Buzz was out of his head. He thought the idea was preposterous. Those screws were *stainless steel*. He had never heard of an allergic reaction to stainless steel. But Buzz had finally managed to persuade him. Janet was going into the hospital at the end of the week. He'd let me know. Which he did—on Saturday. The screws had been removed the day before, and there was already, he thought, some improvement. The erythema—the redness—had very definitely subsided. He would do a patch test as soon as he could. Meanwhile, he had the screws and the name of the maker, and he was writing the company for information about their composition. It was a week before I heard from Buzz again. And it was all over. Janet's dermatitis had completely cleared in just three days. Two days later, he did a patch test with a tray of six substances—pure nickel, nickel sulfate, one of the stainless-steel screws, potassium dichromate, cobalt sulfate, and mercuric chloride. The last three were negative, and the first three were all four-plus positive. Buzz said the orthopod was with him and watching but still had his doubts. So he did a test on his own. He took the screw and taped it to her back. Four hours later, she began to flare. The same generalized pruritus and erythema. And it took a couple of days to clear with topical cortisone. But by then the surgeon was doubly convinced. Buzz had heard from the manufacturer—the company that made the screws. Their stainless-steel screws were steel, all right, but the steel had a nickel content of up to fourteen per cent. That seemed to be a conventional formulation.

"That wasn't the end of the case, however. Not quite. Buzz saw Janet Walker one more time. She came out of the hospital and disappeared from followup. Several months went by, and then one day she walked into his office with the same old rash. It was easy to imagine how he felt. She told him that she had gone back to school for a while and had then quit to go to work, and that she had been working about a week in a dress factory. He questioned her about her job, and the answer soon came out. She did a lot of cutting in her job, and the scissors she used had a ring of some sort that kept them hanging on her thumb. Always in contact, all day long. And, of course, all scissors contain nickel. Buzz advised her to get another job. She was much too exquisitely sensitive to even think about handling nickel. It was the same with Mrs. Strong. Her sensitivity to copper was also extraordinarily pronounced. Most dermatologists are satisfied that an atopic constitution increases sensitivity, but I wonder if an internal exposure may not further heighten sensitivity. It's an interesting thought.

"I never saw Mrs. Strong again. But, even so, she gave me a little scare. I got a telephone call one day from her dermatologist down in Dallas. I had told him that I was planning a report on her case. His call was to tell me that I'd better slow down. Mrs. Strong had just had another bad attack. And it couldn't be related to an IUD this time. Her gynecologist had told him that her present IUD was one that contained no copper. Didn't that seem to suggest some other source of her trouble? I didn't think so. I didn't see how I could have been that wrong. But I did begin to wonder. He left me that way for about a week, and then he called me again. He had taken the precaution of having Mrs. Strong's new IUD removed, and she was much improved, but the picture was still confused. He also had been in touch with the manufacturer, and they had given him a full report on their IUD. It *did* contain copper. But the amount was so infinitesimal —it was thirteen ten-thousandths of one per cent. I had to admit that that wasn't much, but I thought it was enough. It had to be enough. He didn't think so. He thought it was most unlikely. But he did agree to have the IUD replaced. That was sometime in late March or early April. I didn't hear from him again until August. This time, he wrote me a letter. I'll read you what he said:

" 'This is a brief note to catch you up on Mrs. [Sara Strong]. I saw her again this morning for a wart on her hand, and it is interesting that she said that she had absolutely no breaking out since the last time I had seen her, but after three or four months of being clear she felt it peculiar that the IUD device would cause it, and tried it one more time, and had an almost immediate reflare on the wrists of the lichen planus-like eruption. She used her Lidex cream again, and it cleared very promptly, and she has had no difficulty since that time. I am sure that you will find this as fascinating as I did.'

"I did. I did indeed. And I still do."

[*1978*]

CHAPTER 20

Sandy

———◆·◆———

DR. JOEL L. NITZKIN, chief of the Office of Consumer Protection, a section of the Dade County, Florida, Department of Public Health, sat crouched (he is six feet nine) at his desk in the Civic Center complex in downtown Miami, stirring a mug of coffee that his secretary had just brought in. It was around half past ten on a sunny Monday morning in May—May 13, 1974. His telephone rang. He put down his coffee and picked up the phone and heard the voice of a colleague, Martha Sonderegger, the department's assistant nursing director. Miss Sonderegger was calling to report that her Miami Beach unit had just received a call for help—for the services of a team of public-health nurses—from the Bay Harbor Elementary School. There had been a pipe break or a leak of some kind, Miss Sonderegger had been told, and the school was engulfed in a pall of poison gas. Many of the children were ill, and some had been taken to a neighborhood hospital by the rescue squad of the municipal fire department. Dr. Nitzkin listened, considered.

He said, "What do you think, Martha?"

"It sounds a little strange."

"I think so, too."

"But I'm sending a team of nurses."

"Yes," Dr. Nitzkin said. "Of course. And I think I'd better drive out to the school and take a look myself."

He thanked her and hung up—and then picked up the phone again. He made two quick calls. One was to an industrial hygienist named Carl DiSalvo, in the Division of Environmental Health. The other was to a staff physician named Myriam Enriquez, in the Disease Control Section. He asked Dr. Enriquez to meet him at once at his car; as for Mr. DiSalvo, he was already on his way to the school. Dr. Nitzkin untangled his legs and got up. He was out of his office in two easy, five-foot strides. His coffee cooled on his desk, untasted and forgotten.

Dr. Nitzkin is no longer associated with the Dade County Department of Public Health. He has moved up, both professionally and geographically, to Rochester, New York, where he now serves as director of the Monroe County Department of Health, and it was there, on a winter day, that I talked with him about the summons to the Bay Harbor Elementary School. His recollection was undimmed, indelible.

"I remember it was hot," he told me, standing at his office window and gazing down through the palm trees in his memory at the bare maples and last night's foot of new snow. "Warm, anyway—warm enough to make me think that the 'poison gas' at the school might have something to do with the air-conditioning system. And I remember my first sight of the school. The scene was complete pandemonium. It had the *look* of a disaster. We had to park half a block away, because the school parking lot was full of trucks and vans and cars of all kinds—all parked every which way. Ambulances. Fire equipment. Police cars. All with their flashers flashing. And the media—they were swarming. Newspaper reporters and photographers. Radio people with microphones. Television cameras from four local stations. And even—good God!—local dignitaries. Members of the Dade County School

Board. Members of the Bay Harbor Town Council. And neighbors and passersby and parents all rushing around. I had never seen anything like it, and I had to wonder how come. But the explanation, it turned out, was simple enough. The school had called the fire department, and the fire department had called the rescue squad—and the media all monitor the fire department's radio frequency. There was one oasis of calm and order. That was the children. They had been marched out of the building in fire-drill formation and were lined up quietly in the shade of some trees at the far end of the school grounds. There were a lot of them—several hundred, it looked like. Which was reassuring. I had got the impression that most of the school had been stricken by whatever the trouble was. Dr. Enriquez and I cut through the mob, looking for someone in charge. It turned out that the school principal was away somewhere at a meeting. We asked around and were finally directed to the head secretary. She was the person nominally in charge, but you couldn't say she was in control. Nobody was in control.

"She and Dr. Enriquez and I talked for a moment at the entrance to the building. The building was standard design for contemporary Florida schools. The entrance hall ran back to a cross corridor that led to the classrooms. The other school facilities were off the entrance hall. The offices, the clinic, and the library were on the right-hand side. On the left were the teachers' lounge, the cafetorium, and the kitchen. A cafetorium is a room that doubles as an auditorium and a cafeteria. The secretary gave us all the information she had. It was her understanding that there had been a gas leak of some kind. That was what she had heard. But she had seen the first victim with her own eyes. The first victim was an eleven-year-old girl in the fifth grade. I'll call her Sandy. Sandy was a member of a chorus of around a hundred and seventy-five fourth, fifth, and sixth graders who had assembled with the music teacher in the cafetorium at nine o'clock to rehearse for a schoolwide musical program. Halfway through the hour—this, I should say, was constructed later—she began to feel sick. She slipped out of the cafetorium. She was seen by some of the students but not by the teacher. She went across the hall to the clinic and went in and collapsed on a couch. The clinic staff was off duty at the moment, but the secretary happened to catch sight of her, and went in and found her lying there unconscious. She tried to revive her—with smelling salts!"

"My mother used to carry smelling salts," I said.

"Yes. It was rather sweet, I thought. Well, anyway, Sandy didn't respond, and that very naturally alarmed the secretary. And so she very naturally called for help. She called the fire department. Sandy was still unconscious when the fire-rescue squad arrived, and they didn't waste any time. They put her on a stretcher and took her off to the hospital—North Shore Hospital. Then another child got sick, and another, and another. That's when our nursing unit was called. Seven children were sick enough to also be rushed to the hospital after Sandy went. Around twenty-five others were sick enough to be sent home. The school called their parents, and they came and picked them up. Another forty or so were being treated here at the school. They were in the cafetorium." Dr. Nitzkin raised his eyebrows. "That's what the secretary said—in the *cafetorium!* Myriam Enriquez and I exchanged a look. Wasn't the cafetorium where Sandy became ill, I asked. Where the poison gas must have first appeared? The secretary looked baffled. She said she didn't know anything about that. She had first seen Sandy in the clinic. All she knew was that the sick children still at the school were being treated in the cafetorium.

"We left the secretary and went on into the school. I think we were both in the same uncomfortable state of mind. The situation still felt the way it had to Martha Sonderegger. It felt strange. There was also a strange smell in the place. We smelled it the minute we stepped into the hall. It wasn't unpleasant—just strong. We couldn't place it. Well, that was what Carl DiSalvo was here for. He would work it out. I hadn't seen him, but I knew he was somewhere in the building. We went on to the cafetorium. There was the sound of many voices. It sounded like a mammoth cocktail party. We went into a big room full of people, full of uniforms.

Nurses. Police. Fire-rescue workers, in their white coveralls. And a lot of other people. The sick children were stretched out here and there. I could still smell the strange smell, but it was fainter—much fainter—here. Dr. Enriquez and I separated. She had her clinical tests to make. I was the epidemiologist. I walked around the room and looked, and talked to some of the children. The clinical picture was rather curious. There was an unusual variety of signs and symptoms. Headache. Dizziness. Chills. Abdominal pain. Shortness of breath. Weakness. I noticed two kids who were obviously hyperventilating, breathing very fast and very deep. That was an interesting symptom.

"I stood and thought for a moment. I began to get a glimmer of a glimmer. I went across the hall to an office and found a telephone and called the emergency room at North Shore Hospital and talked to the doctor on duty there. He knew about the children from the Bay Harbor school. He said they were in satisfactory condition. He said they seemed to be feeling better. He said he didn't have results on all of the lab tests yet, but the findings he *had* seen seemed to be essentially normal. My glimmer still glimmered. I started back to the cafetorium, and ran into DiSalvo. He had been looking for me. He had made a quick inspection of the physical environment of the building and he hadn't turned up any tangible factors—any gases or fumes or allergens—that could have caused any kind of illness. I mentioned the funny smell. He laughed. He had checked it out. It came from an adhesive used to secure a new carpet in the library. The adhesive was in no way toxic. Anyway, the carpet had been laid a good two weeks earlier. DiSalvo was satisfied with his preliminary findings, but he was going to settle down and do the usual full-scale comprehensive survey. I was satisfied, too. I was more than willing to drop the idea of a toxic gas. I had never really believed it. And I was also satisfied that we could rule out a bacterial or viral cause of the trouble. The incubation period—the interval between exposure and the onset of illness—was much too short. And the symptoms were also wrong.

"I left DiSalvo and went back to the cafetorium, and I remember looking at my watch. It was eleven-thirty. I had been at the school a scant twenty minutes. It felt like forever. But then, all of a sudden, things began to move. I entered the cafetorium this time by a side door at the kitchen end of the room, and there was a woman standing there—one of the kitchen staff. The dietitian, maybe. An authoritative woman, anyway. She called me over. And—Was it some look in my eye? I don't know. But she said, 'Aren't you a doctor?' I said I was. 'Well,' she said, 'then why don't you do something? Why don't you straighten out this mess? This is all perfectly ridiculous. You know as well as I do that there's nothing the matter with these kids. Get them up on their feet! Get them out of here! They're in the way! I have to start setting up for lunch.'

"I must have stood and gaped at her. I'd had a funny feeling —a deep-down, gut suspicion—from the very beginning of the case that there was something not quite right about it. I'd got a glimmer when I saw those two kids hyperventilating. Hyperventilation is a classic psychosomatic anxiety reaction. And now the truth finally hit me. A memory rose up in my mind. I knew what I was seeing here. Something very like this had happened just a year before in an elementary school in a little town in Alabama —Berry, Alabama. The dietician was right. But she was also wrong. She was right about there being nothing fundamentally the matter with the kids. But she was wrong in thinking that all those aches and pains and chills and nausea were illusory. They were real, all right. And this was a real epidemic. It was an epidemic of mass hysteria."

The word "hysteria" derives from the Greek *hystera,* meaning "uterus." This curious name reflects Hippocrates' notion of the point of origin of the disturbance. "For hysterical maidens," he wrote, "I prescribe marriage, for they are cured by pregnancy." His view prevailed in medicine until well into the nineteenth century, and is perhaps still prevalent in the lingering lay association of women and hysteria. The term "mass hysteria" is also a lay survival. The phenomenon is now preferably known to science as "collective obsessional behavior." Collective obsessions occur throughout the animal world (the cattle stampede, the flocking of starlings on the courthouse roof), and the human animal, despite —or maybe because of—its more finely tuned mentality, seems exquisitely susceptible to them. Manifestations among the human race take many forms. These range in social seriousness from the transient tyranny of the fad (skate-boards, Farrah Fawcett-Majors, jogging, Perrier with a twist) and the eager lockstep of fashion (blue jeans, hoopskirts, stomping boots, white kid gloves, the beard, the wig) to the delirium of the My Lai massacre and the frenzy of the race riot and the witch hunt. Epidemic obsessional behavior differs from its companion compulsions in one prominent respect. It is not, as Alan C. Kerckhoff and Kurt W. Back, both of Duke University, have noted in "The June Bug: A Study of Hysterical Contagion" (1968), "an active response to some element in the situation; it is a passive experience. The actors do not *do* something so much as something happens to them."

History is rich in eruptions of mass hysteria. An outbreak was reported toward the end of the first Christian century by Plutarch. In "Mulierum Virtutes," one of his several philosophical works, he refers to a mass "mental upset and frenzy" among the young women of Miletus (a then important port on the Aegean coast of what is now Turkey), in which "there fell suddenly upon all of them a desire for death, and a mad impulse toward hanging." The toll, if any, is not recorded. More recently, in 1936, a similar desire for death fell upon the citizens of Budapest. Its victims, eighteen in all, were mesmerized admirers of a popular song called "Gloomy Sunday." Lugubrious in words and music, it is a cry of despair from a lover whose loved one has died. The lyrics, in English translation, read in part:

> *Gloomy is Sunday, with shadows I spend it all.*
> *My heart and I have decided to end it all.*
> *Soon there'll be candles and pray'rs that are sad, I know.*
> *Let them not weep, let them know that I'm glad to go.*

Suicide is not a common component of morbid mass behavior. In fact, death of any kind is a rarity in such outbreaks. The maniacal speculation in tulip bulbs that swept ruinously through Holland in the seventeenth century was a phenomenon of more classic construction. So were the Children's Crusade of the early thirteenth century and the successive waves of the dancing mania which broke over most of Europe a century and a half later. The Children's Crusade had its beginning in the summer of 1212, when some twenty thousand German and some thirty thousand French children, exalted by a sudden and contagious conviction that Christian love would succeed where Christian arms had failed in the recovery of the Holy Land, left their homes and set out in mile-long swarms for the East. It ended about a year later in Marseilles and several Italian ports, where differently inspired Christians rounded up most of the children and turned them into cash. The girls were thrust into brothels, and the boys were sold to agents of the Egyptian slave trade. The dancing mania is usually taken to have been an overreaction to the abatement of the protracted terrors of the Black Death in the mid-fourteenth century. "As early as 1374," Ralph H. Major, in his "History of Medicine," notes, "large crowds of men and women, obsessed by a strange mania, appeared on the streets of Aachen. Forming circles, hand in hand, they danced around in wild delirium for hours and hours, quite oblivious to the jeers and taunts of the onlookers. From Aachen the malady spread to Liège, then to Utrecht, Cologne, and Metz. The bands of dancers moved from town to town, finding everywhere new recruits to swell their numbers." The nineteenth-century historian J. F. K. Hecker, in his "Epidemics of the Middle Ages," adds, "Where the disease was fully developed, the attacks were ushered in with epileptiform seizures. The afflicted fell to the ground unconscious, foaming from the mouth and struggling for breath, but after a time of rest they got up and began dancing with still greater impetus and renewed vigor. . . . Music appeared to be the sole means of combatting this strange epidemic. . . . Soft, calm harmonies, graduated from fast to slow, proved efficacious as a cure." The frenetic folk dance of southern Italy called the tarantella is widely thought to be a relic of that terpsichorean marathon.

Eruptions of epidemic hysteria in the modern world, though numerous, have tended to be more modest in size. They have also, like the 1973 outbreak in Berry, Alabama, which led Dr. Nitzkin to suspect the nature of the episode at the Bay Harbor school, tended to occur in schools or other closed communities, and to conceal their functional origin behind a mask of organic illness.

Dr. Nitzkin turned away from the winter window. He sat on the edge of his desk and crossed his legs. He reached down and scratched a distant ankle. He raised his head and smiled. "The trouble at Berry was a little unusual," he said. His smile widened. "It was a pruritus—an itchy rash." His smile faded. "But it was otherwise very seriously typical. It was a small disaster. It went on and on. The outbreak began on a Friday, there was a recurrence on Tuesday, and another the following Friday. At that point, the school board closed the school for the remainder of the term. A diagnosis of mass hysteria is largely a matter of exclusion. The investigators at Berry seem to have suspected hysteria pretty early, but they weren't able to convince the community that they had excluded all possibility of an organic cause. Hysteria is self-perpetuating. It doesn't just run its course. It must be promptly recognized and acted upon, or it will go on and on and get worse and worse. Well, I recognized it here. I was sure of that. I trusted DiSalvo's professionalism and plain good sense, and I was sure of my own clinical judgment. So it was up to me to act. But I was critically aware that I was taking a chance. The decision to close the school or allow it to stay open was no small thing. I looked around the room, and spotted a little group at the far end. I recognized the head secretary and two or three school-board people and somebody from the town council. That was the place to start. I went over, and they saw me coming. My height has its advantages. And I guess I looked decisive. I said I had an announcement to make, and they gave me their attention. I said that our Health Department investigation had eliminated the possibility of any toxic gas. I said there was also no evidence of any infectious disease. I called their attention to the hyperventilating kids. By that time, our little circle had grown. Some of the teachers drifted over, and some of the fire-rescue squad, and some of the police, and some parents, and some of the kids themselves. And, of course, the press. All of a sudden, I was talking into microphones, and flashbulbs were popping, and the television cameras were zooming in. I tell you, it was chilling. I had to believe I knew what I was doing. I took a deep breath, and said that it wasn't an outbreak of gas poisoning or any other kind of poisoning. It was an outbreak of hysteria—mass hysteria. I said I didn't know just how it had started, but I knew how to stop it. I said the only way to bring it under control was to get things back to normal. I asked that the cafetorium be cleared, so the kitchen staff could set it up for lunch. I asked that the children out on the school grounds be brought back into the building and sent to their classrooms. And the time to do it, I said, was *now*—right this minute.

"Then I held my breath. Nobody tried to challenge me or contradict me. Nobody said a word. But you should have seen their faces. The public-health nurses looked stunned. I saw Dr. Enriquez smiling and nodding. The parents of the sick children looked horrified and insulted—I was telling them their children were crazy. But most of the others—the teachers and the school-board people and the firemen and the head secretary—just stood there looking thoughtful. The truth was dawning. I think that maybe some of them had half suspected the truth all along. Well, I started in all over again. I gave them my reasons and my reasoning. I could see heads beginning to nod and faces starting to relax. And then, all of a sudden, the tension dissolved. The firemen and the police just sort of disappeared. People began to turn to each other and talk. The sick kids stopped looking so sick. The head secretary went out to the kitchen, and the teachers began to clear the room. It was all over.

"I mean, the *emergency* was over. There was still plenty of work to do. I still had to justify my decision—and not only to the school and the public. I had to explain it to the satisfaction of my office. I had made a bold move, and bold moves are not encouraged in the bureaucratic world. I had some very angry calls from parents. One mother demanded that I apologize to her child. I was finally able to convince her that being suggestible at the age of eleven was not a sign of insanity. But, thank God, I was lucky. The proof was forthcoming. I acquired one useful piece of information even before I left the cafetorium. A teacher came up to me at the end of my speech and said she was sure I was right. She said there could not have been any poison gas or fumes in the cafetorium, because when the first rehearsal class let out, and the first wave of illness broke, a second rehearsal class had marched into the cafetorium and spent the hour rehearsing, and none of the children in that class had been taken ill. So, gas or no gas, the cafetorium could not have been the site of the trouble. But it was in the cafetorium that the first victim—the girl I'm calling Sandy—took sick. How come?

"I hung around the school, trying to figure it out. It was amazing how quickly everything got back to normal. I talked to more teachers. I talked to some of the children. I talked with a neighborhood doctor who had been called when the trouble first began. They gave me a brainful of bits and pieces. I could see what had happened, but I couldn't see how it had happened. Sandy was what we call in epidemiology the index case. Everything that happened stemmed from her. But I couldn't find a clear connection. The timetable didn't seem to tell the right time. Sandy had slipped out of the cafetorium and collapsed in the clinic at about nine-thirty. But it wasn't until ten o'clock—until the nine-o'clock classes let out—that the wave of illness broke. Why did it take so long? And why did it happen when it did? I wandered around talking and listening, and trying to fit the pieces together. It's hard to remember now just how I did it. Did somebody tell me something? Or did I simply take a different look at something I already knew? Anyway, it suddenly all came together. The whole thing hung on coincidence—on two coincidences, actually. One of them had to do with Sandy and the fire-rescue people. They arrived at the school on the head secretary's summons a couple of minutes before ten. Sandy was still in her faint, or whatever it was. They lifted her onto a stretcher and carried her out and down the hall —at the very moment that the nine-o'clock classes, including the group rehearsing in the cafetorium, let out. The kids flowed out of the cafetorium. They saw Sandy passed out on a stretcher. They had seen her slip out of the rehearsal, and now they knew why. Something was wrong. And somebody reacted—I'm reconstructing now. Somebody complained of feeling sick. That triggered it. That's all it takes in these cases. Hysteria took over. And then the second coincidence happened. That was the neighborhood doctor. He arrived at the school, he saw the fire-rescue squad, he saw the children reacting—and he smelled the funny smell of the adhesive on the new library carpet. He thought it was the smell of something toxic. He said something to that effect to somebody. And the word spread. That pulled the trigger on the second barrel.

"When I went back to the office, I knew what had happened, and why, and how. The rest was documentation. I was confident that it would bear me out. And it did. DiSalvo's thorough investigation confirmed his preliminary study. The physical environment of the building was safe and clean. We also did a comprehensive clinical study. The heart of it was a student questionnaire. We found that a total of seventy-three children had reported at least some symptoms of illness—seventy-three out of a total enrollment of around four hundred and fifty. Most of them—sixty-three—were in either the fifth or the sixth grade. Most of the chief reactors were girls. Don't ask me why, but girls—young girls—seem to be more susceptible than boys. In the Berry outbreak, girls outnumbered boys by more than two to one. Anyway, we did a special study of the seven children who were hospitalized along with Sandy. All of them had been in the first rehearsal class, and they all knew Sandy. We looked into their psychological background, and came up with some interesting data. Five of the seven were girls. One of these had a clear history of hypochondria. Another was always sick or sickly. Another had a habit of hyperventilating in moments of stress. Another had come to school that day feeling vaguely ill. Another was one of Sandy's closest friends. One of the boys was a chronic discipline problem. The other boy was described by the school as highly excitable.

"The questionnaire provided some very interesting information. The comments of some of the children who reported feeling sick that morning were particularly revealing. These were mostly in response to the question 'When you got sick, did you know that other children were sick, too?' I'll read you some of the comments. One girl answered, 'Yes, because Sandy fainted.' Another wrote, 'Yes—a lot of kids. I started to feel sick between Music and Language Arts, and then they carried me outside.' Another girl answered, 'I just knew that a boy vomited.' Her only symptom was nausea. Another girl wrote, 'Yes—Sandy was sick.' And a boy— one of the few boys—wrote, 'Yes, and after Sandy got sick and there was a fire drill, and when everybody was walking out of the building, I felt like a small headache.' Well, you get the drift. We also talked to Sandy. She turned out to be pretty much as expected. I mean, she was the right type. She was attractive. She was a good student. She was precocious. And she was very popular. She was looked up to."

"She was a kind of leader?" I asked. "She set the pattern?"

"Yes," he said. "I think you could say that."

"But what about her?" I asked. "What made her get sick?"

Dr. Nitzkin looked at me. "Oh," he said. "Sandy was *really* sick. She had some sort of virus. All that standing and singing in place was too much for her. She just passed out."

[*1978*]

CHAPTER 21

A Rainy Day
on the Vineyard

A LITTLE AFTER ONE O'CLOCK on the afternoon of Wednesday, August 16, 1978, a switchboard operator at the Bureau of Epidemiology of the national Center for Disease Control, in Atlanta, received a long-distance call from a man who asked to speak to someone in authority in the parasitic-disease division. He had, he said, some information to report, and he wanted some advice. The operator connected him with a young Epidemic Intelligence Service officer assigned to that division named Steven M. Teutsch.

"I don't know about authority," Dr. Teutsch says. "But I guess I was in charge." A white smile stirs in the depths of a curly black beard. "I happened to be the only doctor in our office that afternoon. So I took the call and gave my name, and the caller identified himself. He was a physician in practice in Grand Junction, Colorado, and he needed help. He then went into a rather complicated story. It took me a minute or two to get it straight. It seemed he had a patient in a Grand Junction hospital seriously ill with pneumonia—a man I'll call Daniel Stafford. He had become ill on August 11th, with a fever of a hundred and four degrees, headache, muscular aches and pains, loss of appetite, nausea, and a cough. He had been treated with penicillin but had shown no improvement. It also seemed that Stafford's wife—I'll call her Anne Lord Stafford—had been ill for several days with some similar pulmonary infection, but was now recovered. They were both in their early thirties, and they had recently returned from a family houseparty at her father's summer cottage on Martha's Vineyard. It was the doctor's understanding that two other guests at the family gathering had become sick and were hospitalized— one at the Martha's Vineyard hospital and the other at Beth Israel Hospital, in Boston. The patient at the Vineyard hospital was Mrs. Stafford's younger sister. I'll call her Patsy Lord Hooper. Mrs. Hooper had become sick on August 7 with much the same symptoms as her relatives out in Colorado. The Beth Israel patient was Patsy Lord Hooper's brother-in-law—John Hooper. And he was the particular reason for the Colorado doctor's call. The advice of a parasitologist was needed. John Hooper's illness was thought to be babesiosis.

"Well! That was interesting—very interesting. And the doctor's concern was very easy to understand. Babesiosis is one of our newer diseases. I should say newly identified and described—it is probably as old as any of them. But it's new to medicine. It isn't even mentioned in the standard medical texts. It is primarily a tick-borne disease of wild rodents, but it is readily transmitted by infected ticks to man. The causative organism is a protozoa—a blood parasite named *Babesia microti.* Babesiosis came into prominence only four or five years ago, and most, if not all, known cases have been in and around coastal southern New England—the Cape Cod area, including Martha's Vineyard and the east end of Long Island. It's a serious disease. The Colorado doctor went on to say that he knew next to nothing about babesiosis, but if that was what John Hooper had, might it not also be the cause of Daniel Stafford's illness? If so, what should he do? That was a good question. There is a treatment for babesiosis—an agent called penamidine. But I wasn't about to recommend it in the case of Daniel Stafford. I didn't know about John Hooper up in Boston, but it didn't sound at all to me as if Daniel Stafford out in Grand Junction had babesiosis. Babesiosis is not a pneumonic disease. It tends to come on like malaria. Which is also, of course, a protozoan disease. I told the doctor what I thought. I thanked him for his call, his report. I said that his Martha's Vineyard houseparty sounded very much like something we would want to look into. I would keep him informed. And I *was* interested. I mean, all that pneumonia, all that pulmonary disease in one small group! I thought it sounded like a very unusual epidemic.

"I had a little problem there, of course. I wanted to go ahead, but I was working in the Parasitic Diseases Division and pulmonary disease was out of my official capacity. I was tempted to wonder if maybe I was mistaken about babesiosis. I didn't think so . . . and yet. Well, I decided to take at least the first step. I put in a call to Boston—to Beth Israel—and talked to one of the doctors there about John Hooper. Babesiosis was still very much on their minds, and Hooper was being treated for that. Their thinking was based on a blood smear that seemed to show a protozoan organism. He was not doing too well. He was still running a fever—up to a hundred and three. On the other hand, of course, they were keeping an open mind. They had taken blood samples for a series of evaluations. Then I called Martha's Vineyard and talked to the doctor there who was treating Patsy Lord Hooper. It was his impression, based on a chest X-ray, that she was suffering from an atypical viral pneumonia. She had run a fever up to a hundred and two, but she was now much improved. We talked about other possibilities—more specific kinds of infection—and he told me he was arranging for the indicated serum studies. I later learned that he had also treated Patsy's husband for a febrile illness—a mild case, from which he was now recovered. Patsy and her husband had spent the summer on the Vineyard at the family cottage. They were joined there on Wednesday, July 30, by the Staffords and John Hooper and by her father and a woman friend of his. The party broke up on Tuesday, August 5. Everybody had arrived well and remained well during the week of the gathering. But every one of them had become ill with pulmonary symptoms of some sort within a few days after the party. It was, indeed, an interesting epidemic.

"Well, I thought I'd made a start. In any event, I'd gone about as far as I could go on my own authority. And my superior was out of town. In an organization like C.D.C., everything is structured. Everything moves through channels. I put my notes together and got up and went down along the hall to what seemed to be the appropriate starting channel."

The immediate course and consequence of Dr. Teutsch's alerting report is noted in a subsequent interoffice memorandum channelled to Dr. William H. Foege, the director of the Center. It begins:

On August 16, 1978, Steven M. Teutsch, M.D., EIS Officer, Parasitic Diseases Division, Bureau of Epidemiology (BE), advised Arnold F. Kaufmann, D.V.M., Chief, Bacterial Zoonoses Branch, Bacterial Diseases Division; John Bennett, M.D., Director, Bacterial Diseases Division; William G. Winkler, D.V.M., Chief, Respiratory and Special Pathogens Branch (RSPB), Viral Diseases Division (VDD), BE; and Robert B. Craven, M.D., Chief, Respiratory Activity, RSPB, BE, of a possible outbreak of acute respiratory disease in humans on Martha's Vineyard, Massachusetts. Dr. Craven and Stanley I. Music, M.D., Deputy Director, Field Services Division, BE, contacted Joseph P. Rearden, M.D., Assistant State Epidemiologist, Massachusetts State Department of Public Health, and discussed this possible outbreak. . . .

Additional discussions were held with Edward W. Brink, M.D., EIS Officer, located at the Connecticut Department of Health, and Timm A. Edell, M.D., EIS Officer, Colorado Department of Health. Frequent discussions were held with the Massachusetts Department of Public Health staff over the next 12 days. . . .

On August 28, David R. Kimloch, M.D., Deputy Health Officer, Massachusetts Department of Public Health, spoke with Dr. Winkler and invited CDC to assist in the investigation. On August 29, Dr. Teutsch departed for Martha's Vineyard. . . .

"Actually," Dr. Teutsch says, "it wasn't quite as simple as that. The invitation from the Massachusetts Department of Public

Health to C.D.C. to participate in the investigation was not entirely spontaneous. There was an *éminence grise*. He was the
famous Dr. Alexander Langauir. Alex—everybody calls him Alex
—was the founder of the Bureau of Epidemiology at C.D.C. and
its first director. He's now retired, living in Boston, but it so
happens that he has a summer cottage on Martha's Vineyard, and
he was in it. He knew as well as anybody that state health depart-
ments do not automatically invite the participation of C.D.C. in
local public-health investigations. They sometimes hesitate to appear to be in need of outside help. Which, of course, is only
natural. Well, Alex had heard about the outbreak, and he knew
that I had talked with the doctor who had treated Patsy Lord
Hooper and her husband. He pulled some invisible strings and got
himself appointed a temporary consultant to the local health department. Then he set about advising the local authorities to ask
the state to ask C.D.C. to come in. And, as usual, Alex succeeded.
And he was on hand to welcome me when I arrived on the Vineyard. I even stayed at his house.

"On the drive in from the airport, Alex and I brought each
other up to date. There really wasn't much to report. Patsy
Hooper had been discharged from the Vineyard hospital, and her
brother-in-law John Hooper had been discharged from Beth Israel. But Dan Stafford, out in Colorado, had been flown from
Grand Junction to a Denver hospital. He was in intensive care, on
a respirator, and his condition was considered critical. That was
an extra incentive to us to try to identify the problem. Dan Stafford couldn't be properly treated until his doctors knew the cause
of his illness. And that was still a mystery. The first round of
laboratory evaluations were inconclusive—more or less as expected. Pneumonia of the sort we were dealing with here is diagnosed by serum antibody tests, and antibodies don't appear in the
blood during the acute stage of the illness. Antibodies rise slowly,
as the body mobilizes its defenses, and an accurate positive result
can't be expected in much less than three weeks. Arrangements
were being made now for a new collection of blood samples—
convalescent samples. The tests would be run at the C.D.C.

laboratories in Atlanta. Another complication was the sheer number of disease possibilities to be evaluated. There are a good many atypical pneumonias. Legionnaire's disease was one possibility, and one that was very much on our minds in C.D.C. It is also very much an epidemic disease. Another possibility was mycoplasmal pneumonia. The other possibilities included psittacosis, influenza A-type and B-type, adenovirus infection, respiratory syncytial virus disease, tularemia, parainfluenza types 1, 2, and 3, and the three major fungal infections—histoplasmosis, blastomycosis, and coccidioidomycosis. Alex's first choice was Legionnaire's disease, and his second choice was mycoplasmal pneumonia. His third choice was a little scary. He thought it might be something brand-new—something we knew nothing about.

"I spent three days on the Vineyard. Ed Brink, a fellow E.I.S. officer assigned to the Connecticut Health Department, came over from Hartford, and we divided up the work. Our job was to get a standard epidemiological study under way. That involved two approaches. One was a study of the Lord cottage and its environment. The other was to determine the dimensions of the outbreak. Was it confined to the seven members of the houseparty, or were they part of a larger epidemic. Alex drove me over to the cottage. I don't know if you know Martha's Vineyard. It's roughly triangular in shape, about twenty miles long from east to west, and from two to ten miles wide. The chief town is Edgartown, at the east end, and the hospital is there. The Lord cottage is at the west end, facing the ocean across an inlet called Chilmark Pond. The Chilmark area is largely rural—hilly, with scrub woods and brush. Patsy and Tom Hooper were still on the Vineyard, and we talked to them at the cottage. Here's what we learned: The nearest neighbor was another summer cottage, about two hundred yards distant, and the two houses shared a well. There had been no illness of any kind at the neighboring house. The Lord party stayed pretty close to its cottage. The only outside activity in which all seven people participated was an annual Vineyard affair called 'The Illumination,' at Oak Bluffs, at the northeast end of the island. We went through the house—large living room with a fireplace, kitchen, bedrooms, and bath. The house is heated when necessary by the fireplace, and the fireplace is equipped with a device called a Heat-o-Lator, which distributes the hot air. The fireplace was used only once during the houseparty week—on the night of August 2. There was nothing significant in the food eaten during the week. The party often went swimming, and waded across the shallow, brackish Chilmark Pond to get to the ocean beach. The Hoopers had two dogs—a Labrador and a Chesapeake Bay retriever—both young and active. Mice had been seen in the cottage early in the season. Rabbits were abundant in the area, and so were ticks, mosquitoes, and flies. During the week of the gathering, the septic tank backed up, but did not overflow. A moldy wooden post supporting the front porch had collapsed early in the week—and moldy wood suggests fungus. We took samples of the pond water, samples of the rotten log, samples of earth at the septic tank, samples of household dust, samples of debris from the Heat-o-Lator, and a sample of drinking water from the kitchen tap. We hadn't picked up anything of much importance. But maybe the laboratory in Atlanta would.

"The other phase of the investigation was hardly more productive. We talked to all the Vineyard doctors, and we burned the midnight oil at the hospital checking out their records. The records in the emergency room showed that four thousand three hundred and ninety-four persons were seen there during July and August of 1977. In the same period of 1978, the number was four thousand three hundred seventy-three. No leads there. Then we checked admissions. We came up with three patients in 1977 who had been admitted and subsequently discharged with a diagnosis of pneumonia. The total for the period in 1978 was six, of which one was Patsy Lord Hooper. We ran down the five other patients and arranged for blood samples. One of the few positive findings was that there seemed to be no possible connection between the activities of the Lord group and the five other pneumonia patients. And those five were also unconnected. So the indications were reasonably clear that we were dealing with an outbreak concentrated at the Lord cottage. The only really encouraging news came from Colorado. Dan Stafford was off the critical list. He was still very sick, still in intensive care, but they had taken him off the respirator. We couldn't take much credit for that."

Dr. Teutsch's arrival on Martha's Vineyard had been noted in a front-page story in the biweekly *Vineyard Gazette* on Tuesday, August 29. The story, written by an industrious reporter named Mary Breslauer, was headed:

EPIDEMIOLOGISTS WILL PROBE
CHILMARK FAMILY'S ILLNESS

Three days later, on September 1, on the eve of Dr. Teutsch's departure, the results of the investigation were also given front-page prominence. The story, also the work of Mary Breslauer, was headed:

EPIDEMIOLOGISTS STILL BAFFLED
BY CHILMARK FAMILY'S ILLNESS

It began: "To the investigating federal epidemiologists, the attending Island physician, and the affected family, the respiratory illness which blighted a Chilmark summer vacation is a medical mystery. . . ."

And ended: " 'We don't know what is going on here,' Dr. Teutsch added. 'It's not a classical presentation of anything at this point, including Legionnaire's disease. We don't know much about Legionnaire's, and it probably has a very broad spectrum, like most diseases. What's happened here is consistent with Legionnaire's, but it's not something that jumps out at us. Nothing is obvious to us yet.' "

"Oh, yes," Dr. Teutsch says. "That's the way it was. We were very definitely thinking of Legionnaire's disease. It had always been Alex's first choice, and Ed and I were quite willing to go along. So was everybody in Atlanta. It was at the top of everybody's list. An attack rate of one hundred per cent is most unusual—seven people, seven illnesses. It means a highly infectious organism, and *Legionella pneumophila* is that kind of organism. But then the first serologic results came through from our laboratory. We had ordered the tests in order of probability, and the first results were for Legionnaire's. And they were negative. Well! The next best possibility on our list was mycoplasmal pneumonia. Those results were also negative. Well, that was interesting, too. I think the next results were on psittacosis. Negative. So were the tests for influenza A and B. And adenovirus infection. And respiratory syncytial virus disease. And parainfluenzas 1, 2, and 3. And the fungus infections. That left only tularemia. And tularemia seemed to be it. The first four convalescent blood samples—from the three Hoopers and from Anne Lord Stafford—had very significant tularemia-agglutinating antibodies. The levels ranged from 1:40 up to 1:640, and three of the patients had readings in the upper range. When the results of the tests on the other three came through— on Lord and Betty Smith and Dan Stafford—it was very much the same. They all had significantly high readings on the tularemia organism. I have to say that those reports left us all a little bit astonished. Tularemia was the very last thing we expected. But that was it. Our epidemic was an epidemic of tularemia."

Tularemia is one of an unpleasant handful of infectious diseases that made their first scientifically recorded appearance in the United States. Its companions in this morbid union include rickettsialpox, babesiosis, Lyme arthritis, St. Louis encephalitis, murine typhus, the several Coxsackie-virus infections, and, of course, Rocky Mountain spotted fever and Legionnaire's disease. Like many of these, its name declares its place of origin. The Coxsackie viruses are native to Coxsackie, New York; Lyme arthritis is a native of Lyme, Connecticut; St. Louis encephalitis was first identified in St. Louis and environs; Rocky Mountain spotted fever is the now widely travelled native of the Continental Divide; and Legionnaire's disease made its disastrous début at the American Legion's 1972 convention in Philadelphia. Tularemia was first recognized as a specific disease in Tulare County, in south-central California, in 1911. ("Tulare" derives from the Spanish *tule,* a kind of reed that grows abundantly in California—so abundantly that "out in the tulies" has the same meaning in California that "out in the boondocks" does elsewhere.) The discoverer of tularemia was a United States Public Health Service investigator named George W. McCoy. "During the routine examination of ground squirrels," McCoy noted in a report to the *Public Health Bulletin* in April, 1911, "we have encountered an infection the lesions of which are readily mistaken for those of plague." This slippery new infection, he further noted, had been transferred to healthy squirrels, guinea pigs, domestic rabbits, rats, mice, and a monkey, and "attempts to grow the causative organism are still being made." His conclusion was a muted warning: "We do not know whether the organism causing this disease is pathogenic for man, but judging from the large number of species that are susceptible, we are inclined to suspect that man might contract the infection." The following year, his attempts to grow the organism were finally successful, and he and an associate, C. W. Chapin, announced its laboratory generation under the name of *Bacterium tularense,* from which (at the suggestion of Edward Francis, a celebrated bacteriologist and 1928 Nobel Prize nominee, who largely completed the elucidation of the disease) tularemia in turn derived its name. A year later, in 1913, *Bacterium tularense* confirmed McCoy's natural suspicions and felled its first recorded human victim—a meat-cutter in a Cincinnati restaurant specializing in wild-rabbit dishes; he is known to medical history only as "E. E." The historic diagnosis of E. E.'s illness (fever, prostration, a plaguelike lesion), and its subsequent substantiation by a perspicacious professor of bacteriology at the University of Cincinnati named William H. Wherry, was somewhat more than the mere première of human tularemia. It led to the identification, as tularemia, of a long-puzzling occupational complaint called variously "market-men's disease" and "meat-cutter's disease" and, most commonly, "rabbit fever," and it also established the wild cottontail rabbit as the chief source of the infection in human beings. It was, in fact, the threat of tularemia that eventually led most American municipalities to restrict the sale of wild rabbits as food.

Bacterium tularense—or *Francisella tularensis,* as the organism is now called for keener definition and to honor the illuminations of Edward Francis—stands well apart from most other pathogens in two rapacious respects. It has, for one thing, an extraordinary pathological vigor, a seething virulence, an almost matchless capacity to communicate disease. Most infectious diseases require an assault force of thousands, even millions, of organisms to successfully (depending upon a complex of physiological variables) invade the human body. Tularemia is one of a ferocious few (influenza, measles, Lassa fever) that can overwhelm their victims with a force of no more than a dozen. There are four general portals through which a disease organism may stage an invasion. Many infections (typhoid fever, cholera, trichinosis) are transmitted by the ingestion of contaminated food or water. Many (pneumonia, whooping cough, the common cold) are transmitted by way of the respiratory system, by inhalation. Others (malaria, yellow fever, Rocky Mountain spotted fever) are transmitted by injection—by the bite of an insect carrier or (as in serum hepatitis) the thrust of a contaminated hypodermic syringe. A few (boils and other staphyloccus infections, tetanus, conjunctivitis) are transmitted by contact—by entry into the body through a fortuitous cut or scratch. Numerous pathogens are capable of entering the body by more than one of these routes, some by as many as three. *F. tularensis* can avail itself of any available route.

The route by which *F. tularensis* enters the body largely determines the clinical nature of the disease, the organs most particularly affected, and, to a considerable extent, its severity. Tularemia is most commonly acquired (as in the classic case of E. E.) by contact—by handling the carcass of an infected animal, most commonly a wild rabbit. This can be, and often is, wholly inadvertent. In New Orleans, in 1949, a woman became infected (an investigation finally determined) by merely resting her wrist on a marble counter where a butcher had recently cut up a wild cottontail rabbit; entry was made through a little abrasion, the pinprick of a rose thorn. Contact with an infected carcass, the bite of a carrier tick or deer fly, and (though much less commonly) the ingestion of water contaminated by the carcass of an infected animal or of undercooked infected meat account for practically all of the some two hundred cases of tularemia that occur in this country each year. The tiny remainder—less than one per cent—are cases of pulmonary, or inhalation, tularemia. Practically all these are isolated cases, and practically all of them involve laboratory technicians working with cultures of *F. tularensis.* Epidemics of pulmonary tularemia—a cluster of related cases—are very rare, almost unheard of. Indeed, as it happens, an epidemic involving as many as seven cases had never been heard of until the outbreak on Martha's Vineyard in the summer of 1978.

The second phase of the C.D.C. investigation of the ill-fated Lord family reunion was undertaken by a differently constituted, more sharply focussed team. Its members all were epidemiologists with specific training and experience in the natural history of tularemia. The team was headed by Arnold F. Kaufmann, a veterinarian and the chief of the Bacterial Zoonoses Branch of the Bacterial Diseases Division (the zoonoses are diseases of animals that may be transmitted to man), and included Morris Potter, a veterinarian on Dr. Kaufmann's staff, and William J. Martone, an E.I.S. physician assigned to the Bacterial Zoonoses Branch. Dr. Kaufmann and Dr. Martone are co-authors of an authoritative monograph entitled "Tularemia."

"I guess you could say that we were picking up on Steve Teutsch and Ed Brink and the others," Dr. Kaufmann says. "They laid the groundwork for our study. But in a sense we were still at the beginning. In a lot of epidemiological investigations, maybe in most, the basic problem—the cause of the outbreak, the disease responsible—is known at the outset. The purpose of the investigation is to identify its source and control any possible wider spread. So, in a way, we were only just starting. We were investigating an outbreak of pneumonic tularemia. The seven clustered cases were, of course, the center of our study. But that wasn't our only interest. We wanted to know if the case cluster was an isolated event. Or did it represent a larger problem. Was tularemia —animal tularemia—endemic on Martha's Vineyard? We got under way on September 19. We couldn't have moved much faster. It didn't much matter, though. There was no great pressure. The clinical problem had been taken care of. As soon as we knew we were dealing with tularemia, we discussed the outbreak and appropriate treatment with Dan Stafford's doctor in Denver. Streptomycin is a specific for tularemia, and it makes a big difference in the pneumonic form of the disease. Pneumonic tularemia can be fatal. But a proper treatment is desirable in any variety. Untreated or improperly treated tularemia can drag on for weeks, even months. Tetracycline is effective, at least in treating the first rush of symptoms, but relapses often occur.

"Bill Martone led the way to the Vineyard. He left Atlanta on the nineteenth, and Morris Potter followed him up on the following day. Potter's particular job was to get source samples for testing for the tularemia organism—animal material, insects, soil, rotted wood, water, house dust from the cottage. That had been done before, of course, but now we knew what we were looking for. The sample material would be injected into laboratory mice, and a positive reaction would give us a pretty good idea of where the Lord exposure came from. And it might lead to other cases —sporadic, unrelated cases. That was Martone's job. He was working at the hospital, with the local doctors, rechecking the pneumonia-like admissions. But he was also interested in any suspicious skin diseases, anything that might resemble the classic lesion of what we call ulceroglandular tularemia—the mark of handling an infected carcass or suffering an infected bite. Anybody who seemed to qualify would be tested for antibody evidence of tularemia infection. Potter finished up on September 23, and he and I had a conference here in Atlanta. Then I went up to the Vineyard, on the twenty-sixth. Martone was still there. I can't say it was a very exciting visit. There are few things duller than checking hospital records. But Martone hadn't wasted his time. He had learned something, and when I joined him we learned a little more. We found a total of twenty cases of pneumonia-like illness and ten cases involving suspicious skin lesions. They were run down, one by one; some were locals on the Vineyard, some were summer visitors. We ended up with serologic evidence of tularemia in eight of the twenty, six of them local residents. Two of these were gardeners. One of the visitors, an itinerent sheep-shearer, was still sick, up in New Hampshire, with pleuropulmonary disease. We talked to his doctor and learned that he was being treated for tuberculosis. His doctor then put him on streptomycin, with the usual good results. The way he had been going, he might never have got out of bed. Well, all those findings told us a couple of things. One was that there is tularemia on the Vineyard, as there is in most places. And now the doctors there are well aware of it. They have it in mind as a possibility. I don't think many future cases of tularemia on Martha's Vineyard will be misdiagnosed. The other thing we learned was that the Lord outbreak was self-contained. The other cases had no relation to the cottage or to each other.

"I don't know what to say about the central problem—the outbreak at the cottage. It is officially still unsolved. We established certain things. There was, of course, no question about its nature. It was a case of airborne transmission. The intensity of the epidemic—an attack rate of one hundred per cent—suggested very strongly that the exposure occurred inside the cottage when the family was all together. We were able to eliminate certain possibilities—bare possibilities, I would say. An analysis of house dust was negative. So was an analysis of dust and what could have been mouse material found in the Heat-o-Lator system. In any event, the Heat-o-Lator didn't have much propulsion: it couldn't throw out much of an aerosol. On the positive side, we established the infection in several rabbit carcasses. We also learned that the Hooper dogs chased rabbits, and caught them more than once. Dogs seem to have some resistance to tularemia. At least, they don't usually seem to show any clinical evidence of disease. But we can detect exposure, and both of the Hooper dogs showed serologic evidence of tularemia infection.

"Well, those are the facts. I have my own ideas, and they are generally supported by the facts. I think what happened was this, and I think it must have happened on August 2: August 2 was a rainy day, and the houseparty stayed indoors. The dogs were out, as usual—in and out. It seems quite likely to me that one or both of them killed a rabbit that day. Killed and ate or mangled it. Then they came into the house. You know how a wet dog shakes itself. It doesn't just shake its body. It also shakes its head, its muzzle, its jowls. And there is always a certain spray of saliva."

[*1980*]

CHAPTER **22**

Live and Let Live

I TALKED WITH CAROL TERRY one October afternoon in 1978 in her office at the Department of Energy, in Washington, D.C., where she had been employed for almost a year as an auditor. She said she felt well, and she looked it. Mrs. Terry—Mrs. Michael Terry—is an attractive young woman, small and slim, with thick brown hair and large hazel eyes and a somewhat lopsided smile. That—her crooked smile—and an occasional catch or hesitation in her speech seemed to be the only remaining signs of her dismaying experience. Later, when she stood up from her desk to see me to the door, I saw that she also had a slight limp, a favoring of her right leg. But that was all.

It is hard to say precisely when Mrs. Terry's trouble began. In one sense, and a very real one, it began at the moment of her birth. That was on June 23, 1946, in Los Angeles. Its first manifestations, however, came later, much later—in the summer of 1971, when she had just turned twenty-four. "It was strange," she told me. "Looking back, I would call it insidious. But at the time I didn't know anything was even happening. I thought I was well and happy. I was an Army brat. I grew up all over the country. But now I thought I was finally settled down. I'd been married for two years to a young Italian from Italy named Pasquale Francone, and we had a nice apartment near L.A., in Brentwood. Pat and I were both working, and I was taking fertility pills because we wanted a baby. The beginning of it all must have been around the middle of July. I seemed to be getting kind of shaky, and sometimes my mouth felt a little tight. Then one day at work, right out of the blue, I fainted. My job was handling claims for an insurance company, and I suppose you could say it was a pressure job. But even so! They were very nice. They sent me home and told me to take a week off. I went to the neighborhood family doctor we knew. He looked me over and said it was overfatigue, and gave me a prescription for Valium. My parents were living in Salt Lake City then. I knew and liked Salt Lake City—I went to high school there—so it seemed like a good idea to spend my week off with them. No, I'm not a Mormon; I was raised a Catholic. Pat got some time off, and we drove up. I really felt very funny—shaky and very nervous. When I told my mother, she just laughed. She said I was simply growing up. She said she had been nervous all her life.

"We drove back to L.A. on Saturday, July 31. On the way, Pat and I got into a fight. I can't remember what about. Maybe it was me. But we ended up not speaking for the entire seven hundred miles. We went to bed not speaking. I woke up at five o'clock on Sunday morning. I had this feeling that I had to do something. I can't explain it any better than that. I went into the bathroom and got down Pat's razor and took out the blade and began to slash my wrists—first one, then the other. Look! You can still see the little scars. I didn't want to die especially. I just wanted to take some action. I don't know why. And I don't know why it took that form. Anyway, I was standing there at the sink with the water running and washing off the blood when Pat woke up and saw me and began to scream. I remember how he screamed. I must have been an awful sight. But he had the sense to get me dressed and rush me to the emergency room of the nearest hospital. They stitched me up, ten stitches—and gave Pat a sedative. I let them do what they wanted. I wasn't saying very much, but I had the feeling that a great weight had been lifted off my shoulders. I don't know why."

There was a pause. Mrs. Terry looked blankly down at her desk, then abruptly raised her head. "Two days later," she said, "I had my first psychiatric consultation." Her voice had an edge to it. "Pat had telephoned my parents, and they flew down and made the arrangements with our family doctor. I spent a month in a psychiatric hospital in the L.A. area. It wasn't until I was settled there that I began to realize what I had tried to do. I was so ashamed, I felt so guilty. I told you I was raised a Roman Catholic. And the thought of all the trouble I was causing was almost too much to bear. And I was depressed. I hardly said a word for three days. Then I began to feel a little better. I suppose my guilt began to wear off. So I was discharged. The psychiatrist thought it best that I not be alone, and it was decided that I'd go back to Salt Lake City with my parents. That meant a separation for Pat and me, but I don't think either of us really cared. He felt humiliated by what I had done. Our marriage was shakier than we had thought. And, of course, I had other things to worry about. The hospital psychiatrist referred me to a psychiatrist in Salt Lake City. He and the psychiatrist in L.A. are one and the same in my memory. I don't mean they looked alike. I only mean they had the same moves. I'd been on Thorazine in L.A. The Salt Lake City man started me on Tofranil, and when that didn't seem to work he put me back on Thorazine—six hundred milligrams a day. At our consultations, he expected me to talk. All he did was listen. Whenever I asked a question, he simply threw it back at me: Why do you ask me that? He seemed to have made up his mind about me before I ever walked into his office. I realize now that I already had a permanent label, a category. And those people, the psychiatrists, deal entirely in labels. I was an attempted suicide. That translated into a case of hysteria. Everything I said was interpreted according to my category. Categories behave in such-and-such a way for such-and-such a reason. He went away for two weeks, and at our next meeting I said—just to be polite, just making small talk—How was your vacation? He gave me a smug look. Oh, he said, you're angry because I went away and left you. Actually, I *was* angry. But not for his smug reason. I was angry with him because something frightening was happening to me, and it didn't bother him at all. I was having trouble with my hands, they felt shaky and clumsy. It was getting hard for me to do all kinds of manual things—like fixing my hair, or buttoning a blouse, or washing my face. I wanted to know why. He wouldn't, or couldn't, tell me anything. He would just nod and say I was angry with my husband. I remember asking, 'Why can't I make my hands stay still?' He said, 'Why do *you* think?' I got so sick of that kind of talk that I began to cancel my appointments.

"That was in early October. He interpreted my behavior as antagonistic and suggested that I might do better in a hospital environment. I let him admit me to the psychiatric ward of a hospital he was affiliated with. I think he was tired of me. He saw me there, but only occasionally, and then he only looked in. The treatment was more Thorazine and group therapy. They put me to work making things—working with my hands. And I could hardly get a button through a buttonhole. I've always been a great reader, and I was spending most of my time reading. I read *From Here to Eternity* while I was there, and also some psychology so I could try to understand myself. But then I found I wasn't really absorbing what I read. I couldn't concentrate anymore. Oh—and also about that time I began to drool. That was almost the worst thing of all. And there was nothing I could do about it. I just sat there and drooled. The strange thing—although I didn't know it then—was that the usual effect of Thorazine is dryness of the mouth. But nobody seemed to notice or care. I began to get the feeling that I couldn't talk. I don't know how to explain that. I knew I could talk, and yet I felt I couldn't. I was so miserable. I began to feel I wanted to die. Really die. That whole period is a little blurred in my memory—partly, I suppose, because of all that Thorazine. But I remember certain moments vividly. I remember that just before Thanksgiving I had the feeling I was actually going to die that very night. My mother came to see me that afternoon, and I made her take my wedding ring and gold necklace. I was afraid one of the hospital people would steal them that night off my dead body. When I woke up the next morning and found I was still alive, I refused to believe it. I told everybody I was dead. I insisted. That's all I remember of that episode. But about that time they gave me a series of tests, physical tests. I don't know why they did, or why they waited so long. They did various blood tests, a brain scan, and a liver-function test. They told me my liver was enlarged. The reason, they said, was all that Thorazine. So they took me off Thorazine, and put me into electroshock therapy. I suppose they talked to my parents and got their permission. I had three treatments, and there seemed to be some improvement. My psychiatrist said I was sufficiently improved to go home. But they were mistaken. I began to feel I was really off my rocker. Pat came up to see me, and we were driving somewhere, and I tried to jump out of the car. The psychiatrist decided they hadn't gone far enough with my electroshock therapy. He put me back in the hospital for a full course. I had eight treatments. There really seemed to be some improvement this time. I felt less groggy, less dopey. I was discharged on January 1, and went home to my parents again.

"My new routine was occasional psychiatric consultations, and group therapy. It was much the same as before. Everything was getting harder and harder for me—tying my shoes, opening jar lids, setting my hair, writing, even signing my name. But every time I tried to discuss it with the psychiatrist he simply shrugged and said what he had always said: This was my way of showing resentment toward my husband. Group therapy wasn't any more satisfactory. I wrote an account of what I thought was wrong with the way the group was run. I typed it out—hunt and peck; it took forever—because nobody could read my handwriting anymore. I asked the psychiatrist if I might read my report to our group. He refused. That finished it for me. I had stayed with it all spring and summer, but now I was through. I was totally fed up with him and his methods. He was really just laying his own hangups on us. So

I resigned. That was in September. I was at a loss. All of my
symptoms were worsening, and some new ones were coming on.
It was getting hard for me to chew and swallow. My right leg was
acting funny, I was beginning to limp. My balance was off. I don't
know how many times I tipped over and fell down. I was covered
with bruises. But, worst of all, I began to have trouble speaking,
I couldn't seem to articulate; everything came out slurred. Even
my family had trouble understanding me. Pat couldn't cope with
it. First suicide, then psychiatry, now this. He felt that I was
shaming him, that I reflected on him as a man. We both realized
that our marriage was over. We were too different. We started
divorce proceedings. At the same time, I had a desperate idea. The
only part of me that seemed still to be functioning normally was
my mind. I could still think. So maybe my trouble wasn't emotional, psychological, functional. Maybe it was basically physical.
I arranged to see a neurologist. We talked and he examined me.
He checked my reflexes and my walk and so on. I went to see him
four times. Finally he told me that my problem was entirely
psychological. He said my physical problems were side effects of
Thorazine. He put me on L-dopa, the drug they use to control the
spasms in Parkinson's disease, and referred me to a psychiatrist.
Not the one I had been going to. A different one.

"Of course, there really wasn't that much difference. In my
experience, there never is. Psychiatrist No. 3 took me off L-dopa
and put me on another tranquillizer—an antidepressant called
Sinequan. He told me that the cause of all my troubles was my
unhappy marriage. He said my shaking hands were an expression
of a suppressed desire to hit my husband. At home, I spent most
of my time lying down—I could speak better and swallow better
in that position. About all I could do was look at television. I had
practically stopped trying to read. I couldn't keep my hands
steady enough to hold a book. I began to lose weight. It was
difficult for me to swallow even water. My parents were about out
of their minds. My mother was a real nervous wreck. My father
got desperate enough to take me to a hypnotist. That was supposed to conjure me out of my tremor. For maybe half a minute

after I was hypnotized my hands stopped shaking. That was in December. My divorce was granted the following month. But that didn't faze my new psychiatrist. He simply shifted his ground. My problem was no longer related to my marriage. The cause of all my trouble now was simply that I didn't want to go back to work, I didn't want to accept responsibility. By the beginning of February, I had lost twenty-five pounds. You can imagine what I looked like. And I was becoming dehydrated. That was really frightening. I wanted help, real help, and all I got from my third psychiatrist was the usual questions and a few textbook answers. I finally told him that I thought I should go into the hospital again, but a different hospital this time, the biggest and best in town—the University of Utah Medical Center. He nodded his head. What I really wanted, he told me, was attention. I just wanted to be waited on. Oh, he was a real winner. But I insisted. I think it bothered him that I wanted to go into the Medical Center, because he wasn't affiliated there. He had to get a colleague to make the arrangements. I never saw him again. About a year later, I wrote to him and brought him up-to-date on my case. I thought he would be interested. He never answered.

Mrs. Terry (Carol Francone) was admitted to a psychiatric ward at the University of Utah Medical Center on February 12, 1973. She remained a patient there for eleven days, until February 28, when "due to discontinuation of her insurance in December and unavailability of welfare insurance," she was discharged. Her "Discharge Summary," which included a comprehensive review of her case, largely confirmed her own recollection of those many months of conventional counsel and inexorably failing health. It opened, "This is the 1st UUMC admission for this twenty-six-year-old divorced white woman from Salt Lake City referred by Dr. ——. Chief complaint is given by the father, who reports her 'becoming catatonic.' " It continued, "The patient has a psychiatric history going back several years. She was previously hospitalized at —— by Dr. ——, the first hospitalization occurring in August of 1971, following a 'nervous breakdown' in California. She was treated with electroshock therapy $\times$ 3 and was discharged in approximately 1 week, showing marked improvement. She continued with Dr. —— in therapy, but relapsed in December of 1971, receiving 8 electroshock treatments while hospitalized. Again she showed rapid improvement. Although the diagnosis of schizophrenia was never made, the patient did receive [Thorazine] treatment intermittently. The patient stopped seeing Dr. —— and was referred to a neurologist, Dr. ——, for evaluation of muscle tightening on her right side and drooling, which was felt at the time to be a reaction to Thorazine. She did make improvement, but then, under Dr. ——'s supervision, the patient began regressing, at which time the diagnosis of secondary to Thorazine was not felt to be accurate, and, without evidence of neurological disease, hysterical-conversion symptom was considered. At this time, the patient was referred to Dr. ——, who treated her with Valium 20 mg a day and Sinequan. The patient has had a very stormy marriage, which was recently ended by divorce, final January, 1973. Since that time, she has progressively gotten worse, with increasing signs of muscle stiffness, drooling, and difficulty in swallowing. Prior to admission, the patient was not eating or drinking fluids properly, and was becoming dehydrated. She was not able to communicate, because of her unformed words."

The review of Mrs. Terry's medical history was followed by an evaluation of her mental status. It read, "Alert, oriented. Patient speaks in a high, squeaky, almost unintelligible voice, which makes questioning impossible. Speech is slow, with short phrases. Follows commands readily. Psychomotor—rigid posturing of hands and fingers, hyperextended and stiff jaws. Patient moves in very mechanical fashion." The results of a routine physical examination (pulse, blood pressure, temperature, eyes, chest, heart, abdomen) were noted as being essentially normal, including "no pathologic reflexes." So, with one exception, were the results of certain laboratory tests: complete blood count, urinalysis. The exception was a test (SGOT) which determines the concentration in the blood of an enzyme whose excessive presence there is indicative of liver-tissue damage. Normal is forty to fifty units. Mrs. Terry's concentration was given as one hundred and ten.

The "Summary" then reviewed Mrs. Terry's stay in the hospital. "It became apparent," its authors, a staff psychiatrist and a resident, noted, "that the patient did have some control over her movements, at times with a gross tremor in doing a task such as trying to tie her shoe, seeking attention and help; however, when the behavior was not attended to, the patient was able to manage adequately, at times with total cessation of her tremor. The patient improved with her food and fluid intake. Under hypnosis, the patient was able to totally relax, with entire control of all body tremors or movements. During sessions of behavior modification and shaping, her voice became very rapid; responding to socialization, it was easy to have the patient change her voice with the words becoming rapidly more distinct."

It concluded, "Final Diagnosis: Hysterical neurosis, conversion type. Disposition: The patient is transferred to Granite Community Mental Health Center. The patient did not receive any medication during hospitalization nor on transfer. Prognosis: Fair."

Mrs. Terry entered the Community Mental Health Center, as a welfare patient, the same day. There is no knowing how long she might have remained there—limping, drooling, trembling, mumbling—if chance had not intervened. Its instrument was a doctor on the Center staff named David Reiser. Something about Mrs. Terry's appearance prompted him to take a second look at her file. The history of her case only quickened a suspicion he had formed. Since the Mental Health Center's diagnostic facilities were not up to proving his theory, he sent her to John Shields, an internist at a neighboring clinic that was equipped to do more elaborate tests —a liver function test, a ceruloplasmin test, a serum-copper evaluation, as well as a slit-lamp examination of the corneas of her eyes.

"I knew something important had happened," Mrs. Terry told me. "It was the way Dr. Shields and the ophthalmologist stood there looking at each other and nodding. I waited, and finally Dr. Shields turned to me. He wasn't exactly smiling, but he certainly looked pleased. He said they thought they knew what was wrong with me. He said they thought my trouble was basically not psychiatric. He said they thought I had a disease called Wilson's disease. He said he understood that it was treatable. And also, he said, it so happened that one of the leading authorities on Wilson's disease was here in Salt Lake City. His name was Dr. Cartwright, and he was the head of the Department of Medicine at the University of Utah College of Medicine. I heard what Dr. Shields was telling me. I drank in every wonderful word of it. But the thing that mattered most—the thing that put me up in seventh heaven —was that I had a real disease. I wasn't a psychiatric case. I wasn't crazy."

Wilson's disease, though something of a rarity, is only too real a disease. It is also a peculiar one. It has its origin in a genetically structured derangement of copper metabolism, and is inherited through the coincidental presence of a certain morbidly aberrant chromosome in both the mother and the father of the victim. The parents themselves, to the best of current medical knowledge, are unaffected by their possession of this baleful trait. Nor—even more baleful—is there any practical way to predetermine their carrier state.

Wilson's disease takes its name from an early-twentieth-century British neurologist named S. A. Kinnier Wilson. His classic monograph, "Progressive Lenticular Degeneration: A Familial Nervous Disease Associated with Cirrhosis of the Liver," published in 1912, is generally accepted as the first more or less comprehensive description of the disease, and it contains an impressively accurate postulation: "There is evidence to show that the disease is toxic in origin." Copper, as mid-twentieth-century nutrition importantly established, is one of the several elements essential in minute quantities to the normal functioning of the human body (it makes possible the assimilation of iron), but like some other members of this company (most notably iron), its impact in large quantities, as has long been known, can be toxic. Wilson's disease is essentially a chronic copper poisoning. In it, the natural balance between copper ingestion and copper excretion is disturbed and the copper thus retained is stored in certain organs. The liver is its first and chief repository. In time, as the storage capacity of the liver is exhausted, the continuing accumulation passes from the liver into the bloodstream and is carried to the other organs for which copper has a grim affinity. These are most conspicuously the brain and the cornea of the eye. The relentless retention of copper begins at birth, but so efficient is the liver in its protective role that years, many years —ten, twenty, even thirty years—may elapse before the first intimations of morbidity are felt.

The liver, understandably, is the first of the susceptible organs to suffer damage. This damage, however, may not immediately be clinically apparent. When it is, the manifestations are often mistaken for chronic active hepatitis. Many cases of Wilson's disease (at least forty per cent) are at this stage asymptomatic, and the first forceful indications of disease appear only when the accumulation of copper in the brain has reached a bruising concentration. These indications embrace a sweeping range of neurological disturbances—slurred speech, failing voice, excessive salivation, drooling, difficulty in swallowing, tremor (hands, head, trunk), incoordination, spasticity, muscular rigidity, progressing to bedridden helplessness. They may also mysteriously appear in the guise of functional disease, as an adolescent adjustment problem, as anxiety neurosis, as depression, even as schizophrenia. The gravitation of copper to the eye has an equally curious impact. It produces a phenomenon known to medicine (in celebration of two German investigators who were its pioneer observers) as the Kayser-Fleischer ring. A Kayser-Fleischer ring consists of a more or less complete ring of rusty-brown pigmentation—a literal implantation of copper—around the rim of the cornea. It is not known just how the copper is deposited there. The apparition, however, is at once a harmless peculiarity and a signal of the utmost clinical significance. The liver damage that occurs in Wilson's disease and its subsequent neurological disruptions may be perhaps misunderstood by the indifferent diagnostician, but only ignorance can see and ignore the Kayser-Fleischer ring. The Kayser-Fleischer rings are definitive signals of Wilson's disease, and failure to comprehend their meaning is a fateful diagnostic error. For unless Wilson's disease is opportunely recognized and promptly and appropriately treated it is invariably and agonizingly fatal.

I talked with Dr. Cartwright—Dr. George E. Cartwright—in his office at the hilltop Utah Medical Center, a big, sunny room with a view of Salt Lake City spread distantly out below. Dr. Cartwright, a graduate of the University of Wisconsin and of the Johns Hopkins University School of Medicine, is a sparkling, white-haired man in his late middle years, small and wiry, with three passionate enthusiasms—backpacking, skiing, and Wilson's disease. Indeed, as Dr. Shields was providentially aware, he is generally considered (along with Irmin Sternlieb and I. Herbert Scheinberg, both of Albert Einstein College of Medicine, in New York, and J. M. Walshe of Cambridge University) to be among the most eminent contemporary students of the disease.

"Of course I remember Carol Terry," he told me. "Carol is one of my best successes. But it was a very near thing. Dr. Reiser caught her just in time. He had seen a case of Wilson's disease in medical school, and remembered what it looked like. I've somehow happened to see a lot of Wilson's disease—relatively a lot, I should say. We estimate the total number of cases in the United States at only around a thousand. I've seen a dozen or so. Carol was in many ways a classic case. I appreciate that Wilson's disease is not an easy disease to diagnose. Not every doctor has it prominently in mind. And I appreciate that Carol's eyes are not blue or gray or green—she has dark eyes, hazel eyes, which made her Kayser-Fleischer rings a little hard to notice at a glance. But—my lord! I remember the first time I saw her. Dr. Shields had called me and told me about Dr. Reiser's providential memory, and I arranged to have her admitted here as a research patient. I'll never forget the look of her. She *looked* like Wilson's disease. She had the typical masklike face and the fixed and twisted smile. She had what we call the wing-flapping tremor in her arms. Her fingers were constantly moving in what we call a pill-rolling tremor. She was drooling. I thought her Kayser-Fleischer rings were grossly visible. When I said hello to her, she answered me in that distinctive slurred speech, and in that typically squeaky voice. And then she gave that laugh they have. It's the damnedest laugh—it doesn't come in the usual way, it comes on inhalation. It was all there. She was practically a textbook presentation. I think my secretaries could have made the diagnosis.

"But, of course, we ran the standard laboratory tests. They confirmed Dr. Reiser's diagnosis and the evidence of my own eyes. Her serum copper determination was fifty-eight. Normal is between eighty-one and one hundred and forty-seven. Her ceruloplasmin—a definitive test—was four point six. Normal is between twenty-five and forty-three. Her urine copper was four hundred. Normal is between five and twenty-five. The tragic thing is that her liver involvement had been noted in all of her hospital stays. Noted, but either ignored or misinterpreted. The trouble is, of course, that cases like Carol's, which present both psychological and neurological symptoms, are almost always diagnosed as psychiatric problems. At least fifty per cent of the cases I've seen or know something about were first diagnosed as functional. And a psychiatric diagnosis tends to stick. I fault psychiatry on that. The psychiatrist should not assume that every patient in the six-to-thirty-two age group who seems to be in need of psychiatric help is actually a psychiatric patient. He should stop and consider the possibility of somatic illness. And in a referral, he shouldn't simply accept the previous diagnosis. He should see for himself. I'll go further. I would suggest that all individuals admitted to psychiatric wards between the ages of six and thirty-two be screened for Wilson's disease with a ceruloplasmin test. It's a simple test—all you need is a blood sample. That would save a lot of misery, even lives.

"Wilson's disease has always been a rare disease. I'm sure it always will be. But it is even rarer in another respect. It's not only treatable, it is also preventable. That sets it apart from the great majority of diseases. By preventable, I mean this: If the disease is diagnosed early enough—if treatment is started in good time—

then its terrible progression of symptoms can be stopped before they start. That's the point of routine ceruloplasmin testing. Treatment isn't to be confused with cure. There is as yet no cure for Wilson's disease, and none in prospect. It is difficult even to conceive of one. It would require some extraordinary genetic manipulation. The treatment of Wilson's disease is basically very simple and straightforward. The idea is to remove the toxic concentration of copper in the body and prevent its reaccumulation. The first relatively effective therapy was initiated by a group of British investigators in 1948. That was the same year in which the role of copper in the disease was firmly established. The investigators used a chemical, dimercaprol, generally known as Birish antilewisite, or BAL. BAL was developed at Oxford during the Second World War as an antidote to an arsenical war gas called lewisite. It was later found to be effective against other heavy metals, of which copper, of course, is one. BAL was of use in Wilson's disease. I used it for some years. Unfortunately, it has its drawbacks. It is administered by injection, and they are very disagreeable injections—long and painful. And also BAL has frequent severe side effects. The treatment now in use was developed by J. M. Walshe, of Cambridge, in 1956. His paper 'Penicillamine, a New Oral Therapy for Wilson's Disease' is a classic. Penicillamine is a derivative of penicillin, but it's not an antibiotic. It's what we call a chelator. It has the ability to combine with—to bind—a metal. In Wilson's disease, it mobilizes copper from the tissue and excretes it in the urine.

"Penicillamine is truly a life-saving drug. The results can be dramatic. I'm not talking about overnight improvement. But unless the patient has suffered irreparable destruction, the damage can be repaired. Little by little, the liver returns to normal, the Kayser-Fleischer rings fade away, and the neurological manifestations disappear. The standard regimen combines penicillamine and potassium sulfide. Potassium sulfide acts to prevent the absorption of copper by forming an unabsorbable copper sulfide in the gut. Since penicillamine is a treatment rather than a cure, it is continued for the lifetime of the patient. Potassium sulfide is usually discontinued within six months or a year. We started Carol on therapy on April 11—twenty milligrams of potassium sulfide three times a day and one gram daily of oral D-penicillamine in three equal doses. She responded well. The first notable change was the disappearance of her so-called psychiatric symptoms. They departed as soon as she learned the true nature of her trouble. In less than a month, she was doing well enough to continue her convalescence at home, and I discharged her on May 8. I saw her as an out-patient at intervals through the rest of 1973. I readmitted her for a week of thorough evaluation on December 2. She had continued to progress satisfactorily, and I discharged her. Here's what I mean by satisfactory. I'll read you a few lines from her discharge summary. 'Patient is currently without specific complaint referable to disease, and she has improved dramatically since March, when her disease was discovered and she was placed on penicillamine therapy. Her only residual of the disease is a limp on the right and some mild dysarthria and an occasional tremor, usually brought on by anxiety states.' Dysarthria means slurred speech. Her recovery was really impressive.

"I was also impressed by her emotional recovery—her character. Carol had had two years of college in California before her first marriage. She went back to school in January of 1974. She enrolled at the University of Utah here. Not only that. She soon began doing volunteer work at the hospital, and then she got a part-time job, as a file clerk with the Bureau of Land Management of the Department of the Interior. She met Michael Terry there, and they were married in June of 1976, and in December she got her degree. And moved to Washington. I see her once a year, though. She was here just a few months ago, visiting her parents. I'm not sure that her walk and her speech will ever be entirely normal. We'll just have to wait and see. But when I remember the way she looked the first time I saw her . . ."

"Well, yes," Mrs. Terry told me the last time we talked. "I've thought about a malpractice suit. I'm told I have a good case.

Those psychiatrists—I wouldn't wish them on my worst enemy. But then—I don't know. I guess they're only human. So I think I'll live and let live."

[*1979*]

CHAPTER **23**

The Fumigation Chamber

A WOMAN I'LL CALL BETTY PAGE was awakened in the middle of the night, around three o'clock on the morning of Wednesday, May 30, 1984, by a spasm of nausea. She sat up with a lurch and a groan. Her husband, Lewis, lifted his head and asked what was the matter. She told him. She said she didn't understand it—she just felt sick and awful. She crouched there, wondering. Then she had an impulse. She had a sudden craving for milk—for a cup of warm milk. She went down to the kitchen and put some milk in a pan on the stove to heat. She drank a steaming cupful, sipping slowly, cautiously. She began to feel better. She went back to bed and fell asleep, and slept until morning.

The Pages live in a comfortable fieldstone house in Jenkintown, Pennsylvania, a pleasant suburb of Philadelphia. They also have a weekend cottage on a lake in northern Pennsylvania—a three-hour drive from their home—and a condominium apartment in Aspen, Colorado. Betty Page is a small woman, grayly blond with blue eyes and a wide, smiling mouth, and in May of 1984 she had just turned fifty-seven. Lewis Page—big, bald, serenely poised—

was sixty. The Pages are both physicians. Lewis Page (Harvard College, the University of Pennsylvania School of Medicine) is an obstetrician and gynecologist in private practice. Betty Page (Wellesley College and Temple University School of Medicine, where she was elected to Alpha Omega Alpha, the Phi Beta Kappa of medicine) is an internist, and at that time was an assistant professor of clinical medicine at the University of Pennsylvania School of Medicine. She is now, and has been for three years or more, on leave from her academic duties. And for good reason.

"Lewis and I have been married for thirty-nine years," Betty Page told me in a conversation we had not long ago in her home. "They've been happy years. We have a lot more in common than just medicine. But we have our differences. I like to ski, and he prefers scuba diving. His hobby is mineralogy. Mine is archeology. His glass is always half full. Mine is always half empty. That was the way it was on that Wednesday three years ago, after that night. Lewis was satisfied that it was just one of those things. I wasn't so sure. I was still feeling a little funny. But I had some big distractions that took my mind off myself. Not just my work. Our daughter was getting married in July, on the twenty-eighth, and, as the mother of the bride, a lot of the arrangements devolved on me—for one thing, the invitations, the calligraphy. Plus I was making my own dress. Still, I got through the day on that Wednesday. I didn't feel too bad. At night, I fell right to sleep—and then it was the night before all over again. Except worse. This time, the warm milk wasn't really very effective. I took some Tigan, an anti-nausea medicine, and that helped, but only a little. And along with the nausea I had a little crampy feeling, abdominal cramps— what used to be called green-apple cramps. I got through the night. The next day, I did what I had to do, but it wasn't easy. I had an attack of diarrhea. I tell you, I felt like hell. Thursday night was another bad night. About the only thing that kept me going was the thought of our cottage on the lake. We had opened the cottage on the Memorial Day weekend—May 25, that year— as we always do. Actually, I had done most of the opening, starting on the Friday. Lewis sees patients all day Friday, and until late on Friday evening. He does that to keep his weekends free. So now, a week later—that would have been Friday, June 1—I took off for the lake. I love the lake. I always feel wonderful there. It beckoned to me like a haven.

"I felt some better when I got there. I began to think, to hope, that I was overreacting. I did the necessary marketing and got settled for the weekend. There was a lot of work to do. It was a little discouraging. I thought we had done all that the weekend before. The cottage is set in a lovely hemlock wood, and when we went up on the Memorial Day weekend there was hemlock pollen everywhere in the cottage. And now, a week later, there was another dusting of pollen all over the place. So I went to work again. Lewis arrived sometime after midnight. I woke up and we talked a moment, and then another wave of nausea came over me. Milk helped, enough to let me get back to sleep. In the morning, it was good being at the lake, but I really wasn't feeling any better. The nausea came and went, and my mouth was always filling with saliva. I kept having to spit. My abdominal cramps continued. And the diarrhea. I didn't complain to Lewis. I tried to keep it even from myself. I took my temperature. It was normal. That was something, but I still felt just punk. Lewis drove back home on Sunday evening, as usual, and I stayed over for another day of rest and then to close up. I drove down home on Monday afternoon. I hated to leave the lake and go back to work. I was sick, and I knew it.

"Well, now we were into June. I kept thinking, hoping, that tomorrow or the day after or the day after that I'd wake up feeling better. But nothing changed. I never vomited, but I was nauseated almost all the time. I had clammy cold sweats and an awful metallic taste in my mouth. I lost my taste for food. A little milk, some Cream of Wheat—I lived on baby foods like that. My normal weight is around a hundred and ten. Now I was losing something like a pound a week. And I was getting worried—really worried. I had a breast cancer back in 1963, and a radical mastectomy. There had never been any suggestion of a recurrence. But there is something that everybody who has ever had any kind of cancer knows. There is always a sort of subliminal anxiety. An ache, a pain, a feeling not quite normal, and you think, Is it cancer? So I pulled myself together and called a gastroenterologist I knew and made an appointment. Doctors are like lawyers. Lawyers don't represent themselves in court. Doctors don't treat themselves for anything that might be serious. I kept my appointment, and the gastroenterologist heard my complaints and was very concerned. He gave me a full examination—the complete mouth-to-anus workup. The results were negative, entirely normal. That should have been good news. But actually it was frustrating. And the gastroenterologist was frustrated, too. So he did what so many doctors do in a situation like that: he went the functional route. What else? He reminded me that I was almost sixty. He said, 'Betty, your trouble is that you're not growing old gracefully.' He said, 'Wait until the wedding is over. You're simply a nervous mother. Your histrionics are getting out of hand.' He said, 'Take a little nap in the afternoon. Try to relax.' He prescribed a medication. It was a mixture of his own—a mood elevator and a tranquilizer. I took his nostrum, and I thought I was going to die. I could hardly breathe. It did something to my chest muscles. I didn't have the power to move air in and out of my lungs. I got rid of the nostrum, and at once felt better. I mean, at least I could breathe again.

"I may have been a nervous mother. I may still be. But that was hardly the cause of my illness. The day of Elizabeth's wedding came, and everything went off very smoothly, very beautifully—and I still felt sick. I had to take a leave of absence from my job. I couldn't manage anymore. I wasn't exactly idle. I had undertaken an ecological study of our lake. There is an association of families who summer there, and I was expected to read my report at the annual meeting, on Labor Day. I thought if I paced myself I could make that deadline. My long weekends at the lake gave me plenty of time to do the necessary research. And rest. I needed rest. The symptoms that sent me to the gastroenterologist were still with me, and I also had some new ones. I was feeling a very unpleasant tightness in my chest, a viselike feeling, and my heart, even at rest, seemed to be skipping beats. You know how we doctors have fragmented ourselves with our medical specialties. My reaction was typical. I was choosing a doctor on the basis of my symptoms. When my symptoms were gastrointestinal, I went to a gastroenterologist. Now my heart seemed to be acting up, so I went to a cardiologist. I don't think he prescribed a nostrum. But the results were much the same as before. He gave me a thorough examination—everything known to cardiac technology. There was apparently nothing wrong with my heart. Then some new symptoms began to develop. I had difficulty reading. After perhaps half an hour, I would begin to get double vision. Another problem was muscle weakness. Another was involuntary twitching in my legs. They twitched like a twitching eyelid. Do you remember Fourth of July sparklers? How we used to light them and then pass our hands through the sparkle and get that funny, pinprick feeling? I began to have that feeling on the bottoms of my feet. That was more than a little frightening. I couldn't help but think of amyotrophic lateral sclerosis—Lou Gehrig's disease. Well, the doctor I went to this time was able to reassure me. It wasn't that. Believe it or not, I was still playing tennis. Or trying to. I wanted to keep up my muscle tone and my muscle strength. I thought that was the way to fight the weakness in my legs. But I was a blob. And my spells of double vision didn't help my game. Pretty soon, it was all I could do to climb the stairs to bed. Even marketing was exhausting. I remember one day at the supermarket. I was pushing my cart along and I just ran out of steam. I left the cart with all my groceries right there in the aisle and crept back home.

"My weight went down to ninety-five pounds, but I made it to Labor Day. I went up to the lake on Friday, as usual, and Lewis came up that night. I gave my ecological report at the association meeting, and it seemed to be well received. I was nominated to serve on the board, and, weak as I was, I accepted and was elected. I was trying not to be sick, I was trying to be part of the living world. Labor Day weekend is always a little sad. It means the end of the season, the closing of the house until Memorial Day. We put up the boat and took down the tennis-court net and emptied the fridge and drained the pipes. I drove back home on Tuesday, and settled down for the winter, and everything went on as before. Except that by the end of September I began to feel a little better. I seemed to be in remission. I was still as weak as ever, but I seemed to be getting over that awful middle-of-the-night nausea.

"At about that time, I got a notice from the association. There was a board meeting scheduled for the second weekend in October. I was expected to attend. Good. It meant one last visit to the lake. Lewis wasn't interested. I drove up alone on Friday. The meeting was Saturday morning. I had my dinner and puttered around and went to bed. And woke up in the night as sick as a dog. I was twitching and salivating and shivering with cold sweat. I couldn't believe it. I was stunned. I crawled out of bed and went into the kitchen and fixed myself a cup of warm milk. I sat at the kitchen table, as depressed as I've ever been. There was a paper on the table at my elbow. It was the exterminators' regular monthly invoice. I'd seen it earlier, but I hadn't really noticed it. It was too ordinary. When we bought the cottage, back in 1971, the then owners told us they had had regular exterminator, insect-pest-control service, and advised us to continue it. Which we did. It seemed like a good idea. We even increased it during the summer of 1983, when we began to see a lot of carpenter ants. We had always had some, but now there were hordes, and they were huge—really enormous. So we asked the exterminators to add ants to their regular insect controls. Well, as I say, I was sitting there at the table with my warm milk and looking at that invoice. My brain sort of went into gear. I had been feeling better, but now I was sick again. And these people had been exterminating every month for ants. The light came on. It was three o'clock in the morning, but Lewis is a night person. He would probably still be awake. I called him up, and he answered right away. I said, 'I think I know what's wrong with me. There's the exterminators' invoice here, and I'm as sick as I've ever been. Go to "Goodman & Gilman" and look up insecticides.' 'Goodman & Gilman' is 'The Pharmacological Basis of Therapeutics'—the standard text on the subject. I stayed on the phone, and Lewis came back with the book open to insecticides, to the organophosphates. He read off the clinical picture. Here's what he read: 'Respiratory effects consist in tightness in the chest and wheezing respiration, due to the combination of broncho-constriction and increased bronchial secretion. Gastrointestinal symptoms occur earliest after ingestion, and include anorexia, nausea and vomiting, abdominal cramps, and diarrhea . . . localized sweating and muscular fasciculation . . . fatigability and generalized weakness, involuntary twitchings . . .' Well, that was it. That was me. I was a textbook case.

"I went to the board meeting, and left the minute it ended. Sunday, at home, was a long day. I was waiting for Monday. The first thing Monday morning, I telephoned the exterminators at the lake. I talked to the manager. I told him what I suspected, and asked what chemicals they were using to control the carpenter ants. He seemed astonished. He couldn't see any connection between the cottage spraying program and my illness. However, he called me back with the information I wanted. I thanked him, and told him to discontinue the spraying program. My hunch seemed horribly right, but I was trembling. The insecticides that his technicians—he called them technicians—had used were Ficam and Dursban. Dursban is an organophosphate. Ficam is methyl carbamate. You may remember reading about the so-called nerve gases that were developed in Germany just before the Second World War. Their active ingredients included one or more of the organophosphates. And that's what I had been cleaning up at the cottage on those summer weekends—the residue of an organophosphate spray. That was my hemlock pollen.

"There are dozens of organophosphate insecticides on the market. They are wonderfully effective insecticides. They are very quickly and completely absorbed by all routes—through the skin, through the lungs, and by mouth. The organophosphates—and the carbamates, too—are what are called cholinesterase inhibitors. Cholinesterase is an enzyme found in the nervous tissue and in the blood—in the plasma and the red cells. It acts as a neural moderator. It controls the accumulation of an ester that governs the transmission of nerve impulses. It transmits the signal to raise my arm, say, or move my fingers. When that action is done, normally the impulse is erased. Carbamates and organophosphates block that normal procedure. They defuse cholinesterase. The result is a continued stimulation of the nervous system. That produces, directly or indirectly, the wide range of symptoms and signs in organophosphate intoxication.

"I knew, or thought I knew, the nature of my illness. But, of course, I wasn't going to treat myself. I wanted expert advice. I inquired around and got the name of an expert—a Ph.D.—in pesticide operations in the Pennsylvania Department of Agriculture, in Harrisburg. I telephoned him and identified myself. And then I got a little tricky. I let him think I was asking for advice in the treatment of a patient. I thought I might get a more objective response if he didn't know that he was talking to the patient herself. As it turned out, he penetrated my little subterfuge very quickly, but I think it was a useful ploy. He was most sympathetic and helpful. He listened to my story. He agreed with my diagnosis. He pointed out a standard treatment. That was oral atropine sulfate, six-tenths of a milligram, every four hours around the clock. Atropine, or belladonna, can control many aspects of organophosphate poisoning, although not the muscle involvement. There *is* a drug that reverses the muscular weakness, but it can be used only in acute cases. My trouble, unfortunately, was chronic. The unhappy fact is that there is not much information about chronic exposure to subacute doses of organophosphates. As for the weakness and soreness in my thighs, the only therapy for that was hope and time. That and the avoidance of any further exposure. I remember he spoke of the accumulation of the insecticides in the ambient air of the closed cottage. He said that every time I walked in there on those summer Friday afternoons it was like walking into a fumigation chamber.

"Well, there was no chance of that happening again. The cottage was closed for the winter, and, on the pesticide specialist's advice, we were going to arrange for a professional team to give the cottage a thorough scrubdown before we moved in on Memorial Day. Meanwhile, I was feeling very much better. The atropine was wonderfully effective. The only problem was still my legs, my weakness. But even that was manageable. Lewis's birthday is in November, and we had a long-standing plan to celebrate it with a tour of the Loire Valley. Which we did. We ate the wonderful food and drank the wine and saw the beautiful sights, and I managed pretty well. Maybe I overdid it. Because I began to get some twinges, some pain in my left knee. Still, I felt good enough to go out to Aspen in January. Now we're in 1985. I went to Colorado alone, as usual. Lewis went to Grand Cayman, to scuba dive. I'm a good downhill skier. But I wasn't this time. My legs were weak, and there was that pain in my knee. I consulted an orthopedist there, a man who knows about all there is to know about ski medicine. He did an X-ray. Negative. He found some contracture in my left hamstrings. There was no ligamentous instability. He advised me to warm up before skiing, and suggested that an arthroscopy—a kind of diagnostic surgery—might be desirable. When I got back home, I looked up a recommended orthopedist. I won't go into the hell that man put me through. He took one look at my history, saw the note about my mastectomy, and satisfied himself that I was suffering a recurrence of cancer. He put me through two agonizing procedures—an electromyography and an arthrography. Chinese tortures aimed at my knee. To no purpose. And, what was almost worse, he was always forgetting that I, too, was a physician. 'Oh, yes—that's right,' he'd say. 'So sorry.' In other words, he treated me like a woman. I mean, like brainless. I consulted a neurologist, a woman. There were more tests. They showed a fifty per cent decrease in the strength of my left quadriceps, spontaneous muscle twitchings, and some reduced sensory response in my lower left leg and foot. Her diagnosis was peripheral neuropathy, most probably caused by organophosphate intoxication. She prescribed oral Vitamin B complex and Vitamin B_{12} by injection, and some muscle-strengthening exercises. I had the feeling I was beginning to emerge.

"I might mention that with that diagnosis established, I called one of the doctors I had consulted earlier. I thought he would be interested. He was. He said, 'Hey, my wife has a greenhouse and she uses malathion. Is that a risk?' Well, I was certainly having what they call a learning experience. About the contemporary practice of medicine."

Dr. Page made a face. "But," she said, "good medicine or bad, tests—and I had them all—or not, I seemed to be getting a bit better. Some days were better than others, but I had only one sort of reversal. I woke one morning in March with real pain in my knee and some real weakness in my legs. The reason, when I thought about it, seemed clear enough. We had gone with some friends the night before to the annual Philadelphia Flower Show, at the Civic Center. To the opening night—very social, very black-tie. And, unfortunately, a tremendous amount of walking. The show contained acres—literally, acres—of the most fascinating displays, and I think we must have walked through them all. I just very foolishly overdid it. But that was the only setback. So then it was spring, and then Memorial Day was coming up, and I had the cleaning of the cottage to arrange. The state pesticide expert had insisted on that. I got in touch with a firm of industrial cleaners, and we set a date in May. We met at the cottage, two men and myself. It was a thorough scrub-down. They did the floors and the walls and the ceiling—it's a cathedral ceiling, with exposed beams. Plenty of surfaces. It's not my nature to just stand there. I pitched in and did the furniture. We also stripped the beds and took down the curtains and took off the slipcovers, all those things, and put them through the washer. Not only that. Before the scrub-down and after, we took samples by vacuum cleaner from all the likely places—door tops, kitchen cabinets, the baseboard radiators. The samples were sent to a commercial industrial lab—a company the state man knew about—and they did a professional analysis. When the results came through, they were shocking. All the samples—the before *and* the afters—showed significant amounts of organophosphates and methyl carbamates.

The exterminator's technicians had certainly done a thorough job. They hadn't missed a nook or a cranny. And they must have simply drenched the place.

"I say the lab reports were shocking, and they were. But I wasn't taken totally by surprise. Because by the time they came through I was sick again. I suppose what I did at the cottage was foolish. But it never occurred to me that there was still a risk there. After all, the spraying had ended back in October, and the cottage had been opened and repeatedly aired. Anyway, I went back on atropine, and it worked its usual wonders. The knee pain, the weakness, the twitching, the burning, all the neurological symptoms—they were just something I had to live with. I saw another neurologist. All the likely causes of polyneuropathy were excluded by appropriate studies. I didn't have diabetes, pernicious anemia, lead intoxication, collagen vascular disease, porphyria, or a malignancy. The diagnosis of organophosphate intoxication was reconfirmed. I talked with the state pesticide man, and he felt it was safe to use the cottage again if stringent precautions were taken—windows open, plenty of circulating air, fans. Which reminds me. Very early on, there was the question about Lewis and the cottage. Why wasn't he affected? The answer we decided on was fairly simple. By the time he arrived on Friday night, I had opened and aired the cottage. It was I who walked into the fumigation chamber every Friday afternoon. He didn't. Well, the summer of 1985 passed without any serious problems. I walked, and I swam a little, and I tried to play tennis. My left leg continued to hurt, but it's wonderful how one can get more or less used to pain. We closed the cottage, as usual, over Labor Day. Later in September, my knee flared up. It was really disabling. I had a friend and colleague, a rheumatologist, and I consulted her, hoping for some relief. And found it. She gave me a steroid injection, and for about ten days I was feeling great again. She referred me to an orthopedic surgeon for the procedure called arthroscopy—an examination of the interior of my knee. The surgeon did some repair work, but found no evidence of degenerative joint disease. I was put on an exercise program. Leg raising with weights. Hamstring lifts. Bicycling. Various things. I began, at long last, to improve.

"I think I'll move ahead for a minute. I'm thinking of the problem of the cottage. It was the beginning of everything, but it wasn't the end. It *was* the scene of my last naïveté. That was the Memorial Day weekend in 1986. I went up to the lake on Friday afternoon as usual. I aired everything out—that was second nature now. The weather was cool for May, chilly, even. Lewis arrived late Friday night. We spent Saturday working around the place, preparing for another summer. Saturday night was almost cold. The heat came on—our baseboard electric heating system. I woke up in the middle of the night. Nausea. Abdominal cramps. Twitching. All the old familiar symptoms. I realized at once what had happened. In spite of everything, there must still be some organophosphate residue around. On the radiators, perhaps. And the heat had volatilized it and blown it around the cottage. Well, we decided to volatilize it right out of the cottage, so we turned all the radiators on high, opened all the doors and windows, and started up a couple of fans. I heated some milk. I had my atropine. Then we drove back home. The whole thing was unnerving. We were in a real quandary. What were we going to do about the cottage? Abandon it? We couldn't sell it. That would be practically criminal. What we finally did—what we have only just finished doing—was another, and really drastic, cleanup. More than a cleanup, actually. We brought in another toxic-waste-disposal team. We discarded all upholstered furniture and bedding. We removed and replaced the baseboard radiators. We removed and replaced all linoleum flooring. We washed in a detergent solution everything that could be washed—walls, floors, ceilings. Then we sealed the entire inside of the house with two coats of polyurethane sealant. I had to be there to supervise, but I wasn't taking any chances. I wore a toxic-cleanup suit with a double-filtered mask, gloves, boots, goggles. I looked like a monster from space. But I survived. We left the house open all summer. Doors and windows—everything. I don't know exactly what we're going to do. I don't want to sell the cottage. I've always loved it. We'll probably keep it. It is almost literally a brand-new building."

"That last experience at the cottage taught me something. The contamination that night from the radiators couldn't have amounted to much. It certainly had no effect on Lewis. But I was different. I was susceptible. I was exquisitely sensitized. I said the cottage was the beginning but that it wasn't the end. I don't know if there ever will be an end. I recovered from that last cottage exposure. I heard about a doctor who specialized in sports medicine, an expert on knees. He turned out to be wonderful. I began to get going again. He kept me on my exercise program, and encouraged me to branch out. Toward the end of the summer— the summer of 1986—I started seriously getting back into tennis. I mean, what's life without tennis! I signed up at a tennis school not far from home, a year-round school, with indoor courts. Actually, it was a general health facility. I worked with a coach a couple of days a week all through September and October and into November. Then, little by little, I began to get sick again. I remember one day I dragged myself over for a lesson and had to stop in the middle of play, I was so nauseated. I remember standing there leaning against the backstop and looking down at the court. It was a concrete slab on grade. And along the slab at the backstop were dozens, hundreds, of dead water bugs. I looked at all those dead bugs, and I had a horrible thought, a horrible suspicion. I called the therapist—the coach—over, and asked him if they by any chance had had any exterminating work done. He said, well, yes, some of the members had been complaining about an infestation of water bugs. The exterminators had made four visits in the course of the past twenty days. That was about the length of time I had been relapsing. I asked him if he knew what insecticides had been used. He didn't know, but there must have been something about the way I looked, and he said he could probably find out. So he made a phone call, and talked for several minutes, and came back with the answer. They used something, he said, called diazinon. Diazinon! Another organophosphate! For some harmless little water bugs.

"I tell you, I was shaken. I was frightened. And when I got home I got to thinking. I don't know how such things happen. But I got to thinking about that flower show, about how I'd felt the next day. And the more I thought about it, the more I wondered. I went to the phone and called the Pennsylvania Horticultural Society and talked to a woman who had something to do with the annual show. I asked her about the use of pesticides and insecticides. She said the society had nothing to do with that. That was up to the individual exhibitors—and there were hundreds of them. She thought it was very probable that some of them, maybe all of them, went in for some sort of chemical control. I thought so, too."

Dr. Page smiled her wide smile and shook her head. The smile faded away. "I'm just beginning to realize that the world is a very dangerous place. It's something nobody really wants to think about. I mean the thousands and thousands of toxic chemicals that have become so much a part of modern living. I mean the people who use them without really knowing what they can do. I mean the where and how and why they use them. It's frightening. I think I'm pretty much recovered now. I haven't had any trouble for over a year. But you never know. The only thing I'm sure of is that I'm going to have to be very careful for the rest of my life."

[1988]

CHAPTER 24

A Lean Cuisine

ONE COLD MARCH MORNING in 1985, Dr. J. Michael McMillin, professor of medicine and head of the Division of Endocrinology at the University of South Dakota School of Medicine and associate chief of staff for research and development at the Veterans Administration Medical Center in Sioux Falls, was walking along a corridor in Sioux Valley Hospital there when he was hailed and stopped by a colleague named James Felker. Dr. Felker was a Sioux Falls internist with a practice that extended throughout the surrounding countryside, and Dr. McMillin knew him to be something of a diagnostic perfectionist.

"Mike," Dr. Felker said. "I've got a little diagnostic problem. What do you know about thyroiditis?"

"What do you want to know?" Dr. McMillin said.

That was a reasonable response. Thyroiditis, as its name suggests, is an inflammation of the thyroid gland. The thyroid, which is common to all mammals, shares membership with the parathyroid, the pituitary, the adrenals, the pancreas, and the gonads in the masterly hormonal manufactory known as the endocrine system, and is situated astride the throat, just below the Adam's apple. Thyroid disease, in general, reflects either an inadequate production of the thyroid hormones (two amino acids that contain iodine) or an excessive supply of them. The former condition is called hypothyroidism. The latter is called hyperthyroidism or, more descriptively, thyrotoxicosis. The usual cause of thyrotoxicosis is a derangement of the body's immune system, producing antibodies that stimulate the thyroid to an excessive hormonal output. This is the condition more familiarly known—in commemoration of the Irish clinician Robert James Graves (1796–1853), who drew a full-length portrait of it—as Graves' disease. There are, however, other avenues by which the body can be oversupplied with the thyroid hormones. One of these is by direct ingestion. Many health stores, for example, offer their credulous clientele a thyroid preparation as a bracing dietary supplement. Another, and more common, cause of thyrotoxicosis is an inflammation of the thyroid gland, which forces a leaking into the bloodstream of a quantity of hormone normally held in storage. Thyrotoxicosis, whatever its origin, is marked by a panoply of discordant signs and symptoms that include restlessness, irritability, weight loss, increased appetite, rapid heartbeat, a pronounced sensitivity to heat, shortness of breath, and, occasionally, pain or tenderness in the area of the thyroid gland.

Dr. Felker's response to Dr. McMillin's question was also reasonable enough. He said he had a patient over in Valley Springs who clearly was suffering from thyrotoxicosis. Valley Springs is a farming hamlet (pop. 801) ten miles east of Sioux Falls and less than a mile from the Minnesota border. As a matter of fact, the patient was the Valley Springs postmaster, Richard Jacobson. The diagnosis of thyrotoxicosis derived from the presence in his blood of high levels of thyroid hormone. But it was a thyrotoxicosis that confounded Dr. Felker's understanding of the disease. For one thing, the standard evaluation of thyroid function (a test involving a small dose of radioactive iodine and surveillance by Geiger counter) showed that Jacobson's thyroid wasn't overexerting itself—it was, in fact, underactive. The other thing that puzzled him was that Jacobson insisted that he had no pain in the area of his thyroid. It wasn't even tender.

"I could understand Jim's confusion," Dr. McMillin told me during a talk we had not long ago in his big, cluttered, librarylike office at the Center. He is a tall man, sandy-haired and nearing fifty, with a narrow, searing blue-eyed gaze, a quick and easy smile, and a way of accompanying every remark with an illustrative gesture or grimace. He ducked his head now in an expression of concern. "The thyroid is a very complex organ. But I'm an endocrinologist and I know some of its secrets. The thyroiditis that Jim seemed to have in mind is called subacute. The inflammation is thought to be the result of a viral infection, and the disturbance to the gland is temporary. As the infection clears, the preformed hormone is gradually excreted in the urine, and the condition reverses itself. Then there is an insufficiency of hormone in the blood. This alerts the pituitary, which governs thyroid activity, and the production of hormone is resumed on a normal basis. It was my feeling that Jacobson's lack of pain or tenderness didn't mean a whole lot. Some people have a high threshold of pain. And many men tend to be stoic. They won't admit to pain. But, as I told Jim, there is another type of thyroiditis that we are only beginning to recognize, called silent thyroiditis, in which there really is no pain. And it, too, is self-limiting. I advised Jim to treat Jacobson with an anti-inflammatory, like aspirin, and perhaps a beta blocker to quiet his heart, and let time and nature do the rest. Then I went about my business. Jacobson dropped out of my mind.

"A couple of months went by. Then, in late May, I happened to run into Jim Felker again. He said something like 'Remember that guy with thyroiditis I told you about? Well, he isn't any better. In fact, he's worse.' He went on to say that he was admitting him to McKennan Hospital, another hospital in our system, and would I go over and see him? Which I did. I went over the next day, May 30. Jacobson looked sick, all right. His face was flushed, he had a very rapid heartbeat and very rapid reflexes, and he had a fine tremor. He said he had lost some weight, and he was diarrheic and just generally felt weak. I felt his thyroid. It wasn't enlarged—if anything, it was rather small—and he still insisted that it wasn't tender. He told me that he had been more or less sick ever since February. His symptoms seemed to come and go. I began to share Jim's puzzlement. I began to think this was one of those health-food cases. But he didn't really seem to be that sort of psychoneurotic type. I couldn't see him eating kelp or any other iodine-rich substance. And he denied that he had—although that meant nothing. I made arrangements for some further laboratory tests, and suggested to Jim that he discharge him pending the test results. We might know better then.''

Dr. McMillin is an active member of the American Diabetes Association, and a few days after his visit with Jacobson he traveled to Baltimore for the Association's annual meeting. It was a ten-day conference, but he cut his stay short to cover for one of his fellow endocrinologists, who wanted to attend the second half of the meeting. He returned home at the end of the first week in June. In the accumulation of messages awaiting him in his office was one from Richard Jacobson, in Valley Springs. Dr. McMillin returned his call. Jacobson thanked him, and reported that he wasn't feeling much better or much worse, but the reason for his call was something else. He had a question. He said, "Doc, can you tell me why there are four other people in this little town who have the same trouble I have?"

"That was interesting," Dr. McMillin told me, "but I wasn't too impressed. People are always calling to tell about a cluster of some disease or other. It usually turns out that there is a confusion of diagnoses. Besides, I told Jacobson, thyrotoxicosis isn't a disease that occurs in epidemic form. It isn't a communicable disease, like measles or influenza. The only outbreak of thyrotoxicosis on record in this country that might be called an epidemic occurred back in the twenties, when iodinated salt was first introduced. It brought on a lot of Graves' disease in susceptible people. Jacobson said something like 'Well, maybe so. But one of those four people is my own mother.' I still wasn't much impressed. And then, that night, something strange happened. I had brought back from Baltimore a number of abstracts of papers to be given later in the meeting, and I was leafing through them at home. I had a grand rounds to present the next day, and I thought I might pick up something new and interesting to add to my presentation. And lo and behold! My eye caught a title—'Painless Thyroiditis: A Community Outbreak in Nebraska.' The outbreak occurred between January and March of 1984 in a seven-county area of southeastern Nebraska, and it numbered fifty-four cases, most of them in the county of York. There was no thyroid enlargement or tenderness. There were no deaths, but six patients were hospitalized. The ages ranged from six to eighty-two. The outbreak ended as mysteriously as it had begun, and the cause was never determined. Well, I could hardly believe it. Now I really was impressed. If it could happen in Nebraska, it could happen here. The minute I finished my grand rounds the next morning, I went over to the Nuclear Medicine Department at McKennan Hospital and looked through the logbook of procedures. Sure enough, there had been a recent increase in thyroid studies, with results pointing to thyroiditis. Some of them were even my own patients. They came from various places: Valley Springs, of course, and several towns or villages just across the line in Minnesota—Beaver Creek, Hills, and Luverne, a town of around five thousand. I checked with Sioux Valley Hospital, and the logbook there showed much the same picture. Oddly, none of the cases were here in Sioux Falls, and Sioux Falls is the largest city by far in a rather large surrounding area in South Dakota, Minnesota, and Iowa. I arranged for a list of all the names and addresses, and drove back to my office and got out that Nebraska abstract. It gave a list of the participating investigators. Most of them were from the Centers for Disease Control, in Atlanta. I called them, one by one, and they were all out of the office. I left my name and telephone number, and waited. I had to think that I was really onto something.

"You know how it is when you're waiting for a particular call. The phone keeps ringing, but it's never the call you're waiting for. Then—at last. It was a young doctor named Daniel B. Fishbein.

He had been at C.D.C. for a number of years, and his major work there was in rabies, but, yes, he had been to York and was still very much interested in that unresolved outbreak. I told him what seemed to be happening here. He said the feeling of the team at York was that the cause was probably a viral illness. A good many of the thyroiditis patients reported an earlier upper-respiratory infection. I told him that that was what I had in mind in our problem here. I asked him if he would be interested in giving me a hand. He was—very much so. But C.D.C. is a federal agency, and there is a very strict protocol governing its investigations. It can come into a state only upon invitation from the state. Fishbein said he would sound out C.D.C. and the South Dakota Health Department. We arranged for the invitation, and Fishbein—or Dan, as I came to know him—got permission from his people. I gave him everything I had—names, addresses, and study results. We set a date: June 15.

"Meanwhile, I got a call from Jim Felker, asking me for a consultation on Jacobson's mother. She was hospitalized at McKennan. I went to see her. She was very definitely thyrotoxic. She had all the symptoms, and very high blood levels of circulating thyroid hormones. But she also confused things. Her thyroid gland was enlarged. That was atypical of this outbreak. And she had a previous history of thyroid disease—she had a goiter and had been on thyroid-hormone therapy for some years. Her thyroid gland should have been anything but enlarged. That enlargement suggested that something was stimulating it. I began to wonder if I was dealing with a genetic problem—a genetic susceptibility. Mrs. Jacobson had her own house. She didn't live with her son. But they both had what looked like the same illness. On the other hand, that hardly explained the other cases.

"Well, Dan Fishbein arrived. He had with him an enthusiastic assistant—a young woman medical student named Janet Farhie. She was working at C.D.C. as an interne. Dan was originally from California—from Hollywood, no less! She was from upstate New York. Dan and Jan—as they came to be—had with them another helper: a copy of 'The Thyroid,' by Drs. Sidney Ingbar and Lewis

Braverman, the standard text in the field. They had been giving themselves a crash course in endocrinology. But they had something valuable that I didn't have. Dan was trained in epidemiology, with years of experience, and she was learning. They moved into the Holiday Inn here, and we got down to work. We agreed, at least for the moment, that what we seemed to be seeing here was the same as the York outbreak. The fact that the York County outbreak ended so abruptly and inconclusively gave us a sense of urgency. We didn't want to be left dangling here. And our outbreak was growing pretty fast. By the time Dan and Jan arrived, the reports I was getting indicated that we had at least a dozen cases. The first step in an investigation of this sort is to set up what is called the case definition. We had to establish the criteria. We set two grounds for inclusion. Cases were defined as patients with two or more symptoms of thyrotoxicosis and with concentrations of thyroid hormone in the blood at least twenty-five per cent greater than the upper limit of normal established at the testing laboratory. Patients with a previous history of thyroid disease, of course, were excluded. Then, because of the cases that were appearing across the line in Minnesota, there was another matter of protocol. That, too, was arranged, and as it soon became clear that most of the new cases were occurring in Minnesota, the administrative leadership of the investigation was taken over by the Minnesota Department of Health, in Minneapolis. Craig W. Hedberg and Michael T. Osterholm were the department people in charge, and some others came in later. But at this stage, at least, much of the real work, the legwork, devolved on Dan and Jan.

"At about that time, I had to leave town again. I'm on the national board of directors of the American Cancer Society, and there was a meeting scheduled for the third week of June, out in California, in Beverly Hills, and I was committed to attend. I was sorry about that, but Dan and Jan were well set up to manage on their own. Everything was falling into place. We had our case definition. The next step was to visit the identified cases and interview them about the nature and onset of symptoms and take blood and other samples. Dan and Jan were both, in their jargon, experienced vampires. The samples were to be divided—half to be frozen for highly sophisticated viral studies, if necessary, at the C.D.C. laboratories. The lab work on the other half was to be done here in Sioux Falls by Dr. Mary Jo Jaqua, a microbiologist at the medical school and the V.A. Medical Center. Our feeling was that we were dealing with a viral outbreak, and Dr. Jaqua was to culture the specimens for that. Another aspect of the epidemiological study was, of course, to establish the geography of the outbreak.

"I was gone a full week. But half of my mind never left Sioux Falls. Thyrotoxicosis is only rarely fatal, and there had been no deaths in York, but even so . . . I kept in touch with Dan by phone, and he was always reassuring. There were no ugly surprises. As soon as I got back home, I called Dan, and he and Jan and I sat down together in my office. They'd had a busy week. The case total was now twenty-three. Jan reported on the demographic data. The ages of the confirmed patients were much like those in the York outbreak, ranging from the pediatric to the geriatric. There seemed to be an equal number of men and women involved. I found that strange. Endocrine diseases tend to afflict women four times as often as men. We were definitely dealing with something unusual. As might be expected in an agricultural area like this, about half the patients were farm people. The geography was also interesting. There were still no cases in Sioux Falls, and Sioux Falls is the major trade and population center. The outbreak was confined between Valley Springs, on the west, and Worthington, Minnesota, on the east. The major case center seemed to be Luverne. About a third of the patients reported symptoms that included muscle aches and pains. That certainly sounded like a viral illness. Unlike a viral illness, however, there was no evidence of person-to-person transmission. There was no evidence of disease among people who worked together, or among friends who saw a lot of each other. The earliest onsets recorded by Dan and Jan were back in January. So this had been going on for quite a while. That was about where the investigation stood at the moment. Except that, because of the concentration of cases in and around

Luverne, Dan and Jan had moved out of their rooms at the Sioux Falls Holiday Inn and were at rest in Luverne—at the Cozy Rest Motel."

The somewhat puzzling geographical picture of the outbreak drawn by Janet Farhie served as an accurate likeness through the last week of June and into the first week of July. Then it was abruptly blurred. On the afternoon of July 3, Dr. McMillin received a telephone call from a woman who identified herself as Rhonda Peskey. She and her husband and their young son lived in Sioux Falls. She had heard that there was an epidemic of thyroid disease going around, and she thought she and her family might have it. She hadn't seen a doctor. She had heard about the epidemic from her parents, who lived in Valley Springs. Dr. McMillin probably knew them. They were Larry and Margaret Long, and they owned the L & M Clover Farm Store there. But what she was really calling about was this: she was working as a waitress here in Sioux Falls, and she wanted to know if her illness was contagious. Should she quit her job, or what?

"I wasn't exactly alarmed," Dr. McMillin told me, with a shrug and a tight little smile. "But it was a disturbing piece of news. If the outbreak was spreading into Sioux Falls, we could be in for big trouble. Our investigation so far showed that in the area Jan had mapped the disease was occurring at a rate of at least one in every hundred persons. If it spread to Sioux Falls at that rate, we could be seeing patients in the thousands. So I had some serious questions for Mrs. Peskey. I asked if she and her family spent much time in Valley Springs. No—not really. Did they ever eat or drink the water there? Or stay overnight? Well, they never stayed overnight. They hardly ever ate anything there. Her parents had a little coffee shop attached to the main store, and she and her husband sometimes had a cup of coffee. She thought she knew most of the Valley Springs patients, but not as friends, and hardly ever saw them. She did sometimes bring some groceries home from the family store. I advised her to see her doctor—her and her husband and the little boy—and have the standard thyroid studies done, and to ask the doctor to report the results to me as soon as possible. I said I thought she could stay on her job. But I thought she should be very careful in handling food, and be careful about washing her hands. I hung up and did some thinking. The Longs, as far as I knew, were not among our Valley Springs cases. We knew there was little, if any, evidence of person-to-person transmission of this disease. Dan and Jan and I had talked that over and over. It was beginning to look less and less likely that we were dealing with an ordinary viral disease. Or any kind of viral disease. In that case, the possibility of some common source of infection was strongly indicated. There were several possible vectors. But the most likely—the most common—would be some food or drink that was shared by all."

"We're now well into July. We had made a point of keeping as quiet as possible about the outbreak. We didn't want a lot of community excitement until we knew a lot more than we did. That could only muddy the waters. But the news leaked out, as it always does, to the media. And nothing much happened. It was only a one-day story. We had been in touch with the Minnesota Health Department all along, of course, but now all hands decided it was time to meet. We set up a conference for July 8 in Worthington, which is between here and Minneapolis. Hedberg and Osterholm were there, along with various other health workers in the community. Hedberg presided. We brought each other up to date. We heard reports that a lot of the patients were worried that the water supply might somehow be to blame. Or insecticides. There was even a group that thought the trouble might have been brought on by some recent tornadoes. We discussed the question of genetic susceptibility. As it turned out, many people in the outbreak area had evidence of thyroid hormone in their blood but were not clinically ill. Even fairly high levels of hormone can be tolerated by some young people and adults in good health. We decided to send a letter to physicians in a wide area of the two states soliciting information about cases of painless hyperthyroidism. The South Dakota letter went out over the signatures of Kenneth A. Senger, director of the Division of Public Health, and myself. The results, I might as well say now, were positive in a negative sense. The response left the geography of the outbreak much as it had been earlier defined. The number of cases rose from week to week, but there were none really outside the area.

"I received a phone call from Rhonda Peskey's doctor. He had confirmed that she and her son were both thyrotoxic. I decided it was time to treat myself to a little shoe-leather epidemiology. The next morning, I drove over to Valley Springs. The Clover Farm store wasn't hard to find. It's practically the only store in town. With Sioux Falls only ten miles away, a convenience store is about all a little place like Valley Springs can support. I went in and introduced myself to Larry Long. We had a cup of coffee together and talked about where he bought his stock. We talked about bread and eggs and meat and milk and soft drinks and staples. Practically everything came from distributors in Sioux Falls or one of several other large distributors in the region. There was nothing surprising in that. And, of course, the same distributors supplied the various stores in Sioux Falls. Practically all the Clover Farm meats came from Sioux Falls. Some came from a plant in Luverne. That was a special kind of beef that Larry Long ground himself, for customers who liked their hamburger lean. He called it his lean cuisine. It was extra lean, almost no fat at all. We talked for about an hour, and Margaret Long joined us for another cup of coffee. We talked about the outbreak. Some of their customers were sick. Some weren't. They themselves were well. I drove back to Sioux Falls, thinking about what I had learned. There was only one thing that stood out. That was that extra-lean hamburger. Was that a clue? But how? When I got back to my office, I found a message from Dan. The number of confirmed cases was climbing. There were more than fifty now in Luverne alone.

"We were all more or less convinced that we were dealing with a food-borne agent centering on Luverne. What was needed now was a case-and-control study and a review of food sources. As it turned out, we did two such studies. Fifty confirmed cases were matched with fifty well individuals for age, sex, neighborhood, and so on. The first study narrowed the suspected food sources to commercially processed chicken and lean ground beef. The second eliminated the chicken and identified the ground beef as coming from the plant in Luverne that Larry Long had told me about. The store in Valley Springs and similar stores in Beaver Creek, Hills, and Luverne all sold that lean-cuisine beef. We had these findings by the end of July or early August. Our case total by then included two families in Sioux Falls. One of these was the Peskey family, and that presented no epidemiological problem. We knew where Mrs. Peskey got her ground lean beef. The other family could have been alarming—or, at least, confusing. Dan went to see them. He asked if they ate ground beef. They said yes. And where did they buy it? They said there was a little store over in Valley Springs that carried the best hamburger meat they'd ever had. It was *so* lean. We laid another worry to rest. At about the same time, we got an even more interesting report on a family living some little distance east of Luverne, outside the charted outbreak area. It was an extended family—three generations, numbering about fifteen people. All ages. There was no question about their illness. They were all confirmed cases of thyrotoxicosis—all but one, a boy of twelve. They got their ground beef from the plant in Luverne. One of the family worked there, and he brought the meat home. What about the boy? Oh, he was a vegetarian. Well, that was the 'My God— that's got to be it!'

"We had the vehicle, we were sure of it. There was no doubt in my mind or in the minds of Dan and Jan that the source of the outbreak was something in the extra-lean beef from the Luverne plant. Some of the others thought otherwise. Very well. But just what was that something? We didn't see how it could be a viral or bacterial contamination. The control studies showed us that practically all the cases liked their hamburgers well done. This isn't steak-tartare country. And, of course, cooking would have destroyed any microorganisms. That left the possibility of some chemical agent. But that couldn't be. I mean, there was no chemical agent known to us that could cause thyrotoxicosis. We came to a halt. The next move would be an investigation at the plant. The public-health officials in Minnesota were quietly making the necessary arrangements. That sort of thing is a delicate business. If the plant was deliberately doing something wrong, an impetuous move could alert it. But once the Luverne lean beef was clearly indicated I decided to take a look at the meat myself. I made another visit to the Clover Farm store in Valley Springs. Larry Long showed me into the walk-in cooler and pointed out the extra-lean beef. It was in chunks—dark-red meat. There was nothing visibly wrong with it. It's been a long time since I studied anatomy, but I could remember enough to recognize that what I was looking at was neck muscle. Well, the thyroid gland and the neck-strap muscle are very close neighbors. I had to think there was a good possibility that some of that meat was thyroid. In fact, there were traces of white tissue along the edges of some of the chunks. I bought a pound of that. Then, as I was leaving, I asked Larry Long something I probably should have asked long before. He and his wife were not on our list of cases, and he said they had never been sick. I asked him if he often ate that extra-lean beef. Often? He gave a little laugh. He said he and his wife owned the store. They didn't have to eat hamburger. They ate sirloin.

"I took my hamburger over to Jerry Simmons. Dr. Simmons is the pathologist at the medical school and the Laboratory of Clinical Medicine in Sioux Falls. I showed him my meat and told him what I had in mind. I pointed out those areas of white. Jerry gave me a look. He said, 'I could put this under the microscope, but it would be a waste of time. That white isn't thyroid. Thyroid gland is red—very dark red. That white is salivary gland.' I suggested we go back to Valley Springs together, and he would pick out some better samples. So we did, and he got some likely chunks, and we went back to the lab, and he prepared some samples for freezing. When the meat was frozen, he would slice off some razor-thin sections and stain them and put them under the microscope. He'd let me know the result. Which he did, about an hour later. He said, 'Mike, I think I can help you with your problem.' "

A Public-Health inspection of the Luverne plant was undertaken in the third week of August, under the general direction of Dr. Robert B. Janssen, from the Centers for Disease Control. A few days earlier, the plant had agreed to halt production of neck trim and to recall all known supplies from the stores. The findings of the investigating team were definitive. The plant was not a packinghouse. It was primarily a slaughterhouse. Dressed carcasses were shipped to another plant for processing. Only the heads were retained at Luverne. Neck muscle was trimmed from these heads and distributed to local retailers for sale as ninety-per-cent-lean beef. As was usual in the trade, every effort was made to recover as much trim as possible. The strap muscles were trimmed from their attachment to the trachea. It was apparent to the investigators that thyroid tissue was often inadvertently included in the beef trim. The plant produced an average of nine hundred pounds of trim each day. Certain changes had recently occurred at the plant. Until April of 1983, the plant had selectively removed the thyroid glands and sold the tissue to manufacturers of thyroid extract. That market vanished with the development of a synthetic thyroid extract. It was also learned that the plant had operated for many years as a kosher killing plant. That ended in November of 1984. Kosher killing made the thyroid gland readily recognizable. The neck meat remained red after the blood was drained off, but the thyroid turned pale. It was presumed that the end of this practice made it difficult for the trimmer to avoid including thyroid-gland tissue in the trim. Samples of trim were taken for extensive examination by three investigators in Massachusetts— Drs. Sidney Ingbar and Lewis Braverman (of "The Thyroid" fame) and Dr. Bruce Meyers. Their findings confirmed that the suspected beef trim produced and sold by the plant contained high concentrations of thyroid hormones. This confirmation was further confirmed by a test involving human volunteers. That test was performed, with the approval of the Human Subjects Institutional Review Committee, at the University of Massachusetts Medical School. Four volunteers ate samples of the implicated beef trim that had been ground and cooked into hamburgers. The samples had previously been cultured to exclude such disease organisms as salmonella; shigella, and campylobacter. Blood samples for thyroid-related hormone analyses were obtained before and after eating. All the volunteers remained asymptomatic, but the serum tests showed the presence of thyroid hormone. The recall of the meat from the Minnesota and South Dakota stores brought an end to the outbreak. The known cases of thyrotoxicosis totaled a hundred and twenty-one, but it is probable that there were many cases in the affected area that were unknown to the investigators. There were no deaths, but a number of sufferers, mostly elderly people, were hospitalized. The investigation also led to the resolution of the uncertainties of the York County, Nebraska, outbreak of the year before. A significant majority of its victims were found to have eaten ground lean beef purchased from a single supermarket.

The 1985 thyrotoxicosis outbreak and its successful investigation had consequences of more than local interest. One of these was that on August 29 the United States Department of Agriculture issued an advisory to all U.S.D.A.-inspected plants which temporarily prohibited trimming near the gullet in beef (and pork) used for human consumption. That prohibition was later made permanent.

[*1988*]

CHAPTER 25

The Foulest and Nastiest Creatures That Be

———◆———

I SAW MY FIRST TICK in the back yard of a rented house in East Hampton, on the oceanic East End of Long Island, one afternoon in July of 1949. It was on my wife's back. We had just come home from the beach, and the tick was conspicuously poised on the bare skin between the top and the bottom of her two-piece bathing suit. We had only recently moved east from Missouri, where the insect pest of summer is not the tick but the chigger. Still, I knew what a tick looked like, and I knew—or had heard—that there was an approved way of removing one from the flesh of its victim: apply the lighted end of a cigarette. I had a cigarette in my hand (as I often did in those innocent days), and I carefully applied the lighted end to my wife's nicely tanned back. She gave a scream. The tick did not, as the approved method promised, drop to the ground. It merely crawled a millimeter or two to the right. Instinct guided me next. I reached out and plucked it off with my thumb and forefinger. Before I killed it, before I crushed it with a pebble, I took a closer look. It was almost the size of a ladybug, only flat, with a shiny brown carapace and a yellowish capelike collar. It had eight spindly legs and a tiny snout of a head. It was ugly, but it looked—though I knew it wasn't—harmless. That, as I say, was my first tick, but hardly my last. I have since seen hundreds, perhaps thousands, in the wild and—crawling, sitting, or embedded—on me, on my wife, on my son, and on several generations of dogs.

The tick is not, to be puristic, an insect. It belongs, in the nomenclature of science, to the phylum Arthropoda and is a member (along with spiders, scorpions, mites, and the horseshoe crab) of the class Arachnida. Ticks are found in incomprehensible numbers throughout the world, and man has probably been unpleasantly aware of them since his beginnings. Human detestation of the tick easily surpasses that aroused by snakes and spiders. "Ill-favored ticks," Pliny the Elder (23–79 A.D.) cried out in his "Natural History." "The foulest and nastiest creatures that be." I know of no better confirmation of the theory of evolution than the tick. That is to say, it is hard to think of the tick, which lives on the blood of other creatures but is itself food for none, as a deliberate creation, as one of the creatures in Genesis "that creepeth upon the earth," and to believe that it had a Creator who "saw that it was good." My first tick—and, with one considerable exception, all of the many ticks I have seen—was of the species *Dermacentor variabilis,* the dog tick. It is, like all ticks, parasitic, but is itself often host to the bacterium *Rickettsia rickettsii,* the agent of Rocky Mountain spotted fever. Rocky Mountain spotted fever takes its name from the region (the mountains of Idaho and Montana) in which, in the eighteen-nineties, it was first identified, but it is no longer seen as merely a provincial menace. Its geography is now known to embrace most of the Western Hemisphere. It has, however, its favored haunts. These include, along with the mountain West, much of the states of Georgia, North Carolina, Virginia, and Maryland, coastal Connecticut and Massachusetts, and eastern Long Island.

Rocky Mountain spotted fever is a dangerous disease. The average incubation period—the interval between the bite and leisurely blood meal of an infected tick and the appearance of illness—is about seven days. The onset is often abrupt, with high fever, chills, headache, prostration, and other more or less equivocal manifestations. The characteristic spotted rash seldom appears before about the fourth day. Immediate diagnosis is consequently difficult unless the disease is suspected—unless the diagnostician is morbidly alert to its possible presence or unless the victim can inform him of a recent tick bite. This last is only sometimes feasible. Ticks seem to seek out secluded feeding sites—the armpit, the scalp, the navel. An effective treatment for Rocky Mountain spotted fever, involving a tetracycline antibiotic, has been available since the middle nineteen-fifties. Untreated or improperly treated cases have a high mortality—around twenty per cent. I was aware of Rocky Mountain spotted fever at the time of my introduction to *D. variabilis,* and I made it my business to learn more. I learned from local lore that the vector tick is ubiquitous in range, finding a comfortable habitat in fields and pastures, in brush and woods, in the grassy dunes that border the most pleasant beaches, and even in the best-kept lawns, and that its preferred hosts are rabbits, mice, squirrels, and dogs. Man is merely *faute de mieux.* I learned from a local doctor that there was no effective treatment for Rocky Mountain spotted fever (the tetracyclines were yet to come), but that there was a vaccine. My wife and son and I presented ourselves at the local clinic and were immunized. I have a vivid recollection of that vaccination: a moment after the injection, my arm received what felt like a blow from a baseball bat, and an ache that persisted for some hours. But it seemed well worth that little discomfort. I lived the next thirty-odd years—sunning myself on beachy dunes, strolling in brushy pastures, cutting firewood—in the carefree knowledge that my wife and son and I were immune to Rocky Mountain fever, and I removed any ticks I attracted as nothing more than nuisances. It was only recently that I learned that all *R. rickettsii* vaccines had long since been found to be unreliable and had been withdrawn from the market.

I understand now, having made some inquiries, how I happened to escape contracting Rocky Mountain spotted fever in spite of my unconcerned wanderings for so many years in an area in which the disease is historically endemic. It was, of course, in large part luck. But the odds were very much in my favor. It is estimated that even in heavily infested areas only about five per cent of the vector ticks are carriers, and it is a comforting fact that the dog tick (and its carrier cousins) is an indecisive feeder. It usually crawls around the skin of a human host for some little time before it chooses a place to eat. During those preliminary perambulations, it presents no kind of threat, and even when it settles down and thrusts its proboscis into the skin the migration of the *R. rickettsii* bacterium to its new host seems to take time—perhaps several hours. The moment for concern is at hand when the tick is not only attached but visibly enlarged, engorged. A well-fed dog tick is only too repellent a warning. I have never found an engorged tick on myself or on my wife or son, but I have removed any number from our dogs. They have varied in size, but the most gluttonous feeders have been about as big as a grape, and of a sickly leaden color.

I was, as it turned out, far luckier than I knew in those years of imagined immunity. I may have been to some extent protected against Rocky Mountain spotted fever, at least in the earlier years, but its carrier is not the only toxic tick that finds a congenial home on eastern Long Island, and Rocky Mountain spotted fever is not the only tick-borne disease that is endemic in my neighborhood. This other tick is called *Ixodes dammini,* and it is host to the agents of two newly noted but otherwise entirely distinct diseases. A malaria-like disease called babesiosis is one of these. Babesiosis takes its name from that of the causative organism, a protozoan named *Babesia microti,* which, in turn, takes its name from that of its discoverer, the Rumanian bacteriologist Victor Babes. It is, in general, a relatively mild disease, and its victims usually recover without chemotherapeutic assistance, but it has a grave potential. Victims whose spleens have been removed or damaged, the elderly, and cirrhotic alcoholics are always at risk, and sometimes die. *B. microti* turns up as a tenant of *I. dammini* a little more often than *R. rickettsii* does of *D. variabilis.*

The other disease that *I. dammini* may harbor is Lyme disease. Lyme disease is rarely fatal, but unless it is promptly diagnosed and appropriately treated its victims may recover from its acute assault only to give way later to a variety of debilitating chronic ills. I have encountered *I. dammini* only once, a couple of years ago, and, like my first tick, it was attracted to my wife. She found it settled on, but fortunately not attached to, her thigh after a walk in a weedy pasture. She lifted it off and showed it to me in the palm of her hand. It was a tiny thing. *I. dammini* is commonly described as about the size and shape and color of a poppy seed. That, to me, is an exaggeration. My—or my wife's—tick was about the size and shape and color of the period that ends this sentence. My wife supposed at first that it was a little scab or freckle. But it moved, and under a magnifying glass its eight arachnid legs were just visible. We twisted it up in a Kleenex and, on a hunch, took it the next day to a dermatologist in the neighboring village of Southampton—Bernard W. Berger, whom we knew to be an authority on Lyme disease and an active investigator into its nature. Dr. Berger gave it a glance, and identified it not only as *I. dammini* but also as an *I. dammini* nymph. It is primarily in the nymph, or middle, phase of its life cycle that *I. dammini* transmits Lyme disease to humans. It was well, as I say, that our tick had not yet attached itself: at least sixty per cent of *I. dammini* on eastern Long Island, Dr. Berger told us, are carriers of Lyme disease. That was bad enough. But I learned only the other day that ticks infested with the Lyme-disease organism can also be infested with the organism of babesiosis.

The coincidence of the organisms of Lyme disease and babesiosis in a single obliging tick was first reported in 1983, in the *New England Journal of Medicine,* by a pediatrician named Edgar Grunwaldt, in general practice on Shelter Island. Shelter Island is a largely wooded island, some thirty square miles in area, that rises between the north and south forks of eastern Long Island. Dr. Grunwaldt was ideally situated to observe the comings and goings and the morbid proclivities of *I. dammini.* Although Lyme disease first caught the eye of science in 1976, in the little Connecticut community from which it takes its name, it has, like Rocky Mountain spotted fever, long since outgrown its regional origins. Lyme disease has now been reported in much of the continental United States, with deep and probably ineradicable roots along the upper Eastern Seaboard, in Minnesota and Wisconsin, and in California, Oregon, Utah, and Nevada. There is, however, no place yet known where its roots go deeper than in Dr. Grunwaldt's Shelter Island. In parts of the island, especially in some of its most idyllically pastoral areas, the incidence of infested *I. dammini* ticks approaches ninety per cent.

"Shelter Island has been a laboratory for much of the investigation into Lyme disease," Dr. Grunwaldt told me in a talk we had at his home one Saturday afternoon—that being a time when he doesn't see patients. "We have the tick in abundance, and we have the disease. We also have babesiosis. As a matter of fact, it was babesiosis that brought me into the Lyme-disease investigation. A paper I wrote that was published back in 1977 in the *New York State Journal of Medicine* described three cases of babesiosis diagnosed here—the first cases reported in the state—and it came to the attention of Jorge L. Benach, of the state Department of Health and the department of pathology at Stony Brook. He was looking for a good source of ticks for research, and I told him he couldn't do better than here. It was through my interest in Lyme disease that I came to know Allen Steere, of the Yale Medical School. It was Steere and his associates at Yale who pioneered in Lyme disease and published the first report on it. And gave the disease its name. That was in 1976. You may remember the story. It's interesting. A woman in the Lyme area was the real pioneer. Back in the summer of 1972, her child developed a painful arthritis in the knee, and in talking with her friends and neighbors she discovered that a number of other children around there were suffering from the same thing. She apparently knew enough about arthritis to realize that a cluster of cases of a disease like that was unusual. She got in touch with the state health authorities, and Steere heard about the outbreak from them. Arthritis, of course, is only one of the forms that Lyme disease can take, but it was what Steere and his associates first observed. And at first they called it Lyme arthritis. I understand that the old-timers over in East Hampton had a disease they called Montauk knee. Lyme disease has probably been around for a long time. So has babesiosis. There was a disease along the New England coast that was known for many years as Nantucket fever.

"It was chance that brought me here to Shelter Island. I was born and raised in Argentina. I studied medicine in St. Louis, at the Washington University School of Medicine, and trained and practiced in California for about ten years. I married a Long Islander, and we decided to settle in the East. It so happened that Shelter Island needed a doctor. I started practice here in the summer of 1975, and one of the first diseases I saw was what I'm now sure was Lyme disease. I had several patients with a specific rash called erythema chronicum migrans, which has been known and described, particularly in Europe, for a long time. Erythema chronicum migrans is a distinctive marker of Lyme disease. The rash begins with a lesion at the site of the tick bite and slowly spreads outward, in a circular pattern with a red rim. It can be quite large—fifteen or more inches in diameter. I saw the same rash again in the summer of 1976, and I took the trouble to search the literature. Most of the reports I found were Scandinavian, some of them going back to the nineteen-twenties. I found a 1962 paper by three researchers at the University of Helsinki which discussed a possible relationship between *Ixodes* ticks and the erythema chronicum migrans that was then associated with meningitis. But you can't count on the rash as an infallible clue. For one thing, as we know now, it isn't always present. For another, it only appears several days after the bite and the initial symptoms. In any event, I started treating my cases with antibiotics, and that seemed to do some good. I followed Steere's work as it was published, and that's when I realized that what I was seeing was his Lyme disease. I remember a telephone conversation I had with Steere in 1978. He had decided that in certain cases Lyme disease was self-limiting—that it cured itself and after a while just van-

ished. That's true, of course, in a way. It seems to go away, but
it really just goes underground, and then emerges in a much more
serious way. These later complications usually take one of three
forms. One, of course, is arthritis of the large joints—most often
the knee. Another manifestation is neurological. It can resemble
a form of meningitis or the facial paralysis called Bell's palsy.
Those early Scandinavian investigators may very well have been
seeing our Lyme disease. The third form affects the heart. When
Steere assured me that the disease was self-limiting, I stopped
using antibiotics. But then he dug deeper and changed his mind.
We now know that prompt treatment with a penicillin can generally prevent the later manifestations.

"I was fortunate enough to have a role in the investigation of
the cause of Lyme disease. Benach was involved, and Jeffrey
Davis, of the Wisconsin Department of Health and Social Services. The laboratory work—the most significant work—was done
at the Rocky Mountain Laboratories, in Hamilton, Montana, by
a team headed by Willy Burgdorfer, of the National Institute of
Allergy and Infectious Diseases. They collected their ticks here on
Shelter Island and isolated a spirochetal bacterium that they were
later able to demonstrate was the causative organism. It was
named for Willy Burgdorfer—*Borrelia burgdorferi.* As you know,
the cause of syphilis also is a spirochete. Not the same one, of
course. And untreated syphilis, like Lyme disease, can later reappear, with very serious consequences.

"The laboratory findings were published in *Science* and in the
Journal of Clinical Investigation in 1982 and 1983. And that's
about where we stand right now. We have the disease as an entity,
we have the causative organism, and we have the vector—the tick.
And we have an effective treatment. There are still some loose
ends. The most important need is for an effective means of control
or prevention. There is a search going on for an immunizing
vaccine, but so far without much success. There is a growing
demand in the endemic areas for a program to eradicate the tick,
for a sanitizing spray. Many of the endemic areas are, of course,
resort areas. There are good environmental reasons for opposing

398 § THE MEDICAL DETECTIVES

that approach. But there is another good, hardheaded reason. A safely selective spray is hard to imagine. And even if there were one it would hardly be worth the trouble. Advocates of a chemical attack on the tick don't seem to fully understand the nature of the tick and its life cycle. *I. dammini* is often called a deer tick. Its principal host is the white-tailed deer. That's where the ticks mate. Mating occurs in the fall. The males die after mating, but the females live on through the winter, and in the early spring they lay their eggs in the wild. Then they, too, die. The eggs hatch into larvae, and at some point in the summer the larvae, if they can, attach themselves to a host, usually the white-footed mouse—the field mouse—and help themselves to a big blood meal. Then they rest through the fall and winter. That one meal is all they need. In the spring—the second spring of the cycle—they develop into nymphs. It is in the nymph phase that the tick usually brings the disease to us—if it carries the spirochete, and if it chooses one of us for its meal. The nymph feeds like the larvae—once in a lifetime. But that meal is a big one—a long one, anyway. It gorges for several days. And it seems that only at the end of the meal is the spirochete transmitted to the host. Feeding time can be any time during the summer and early fall. Then the nymph matures into an adult, and mates. It's true that the nymphs can be found on your lawn. But the reservoir is the wild—the field mouse. The field mouse is a burrowing mouse, and you don't often find its burrows in your front yard. You find them out in the woods and scrub. A spray would have to be a powerful spray to penetrate the scrub and soak down into the burrow. It has been generally supposed that the tick finds a suitable host by sensing its animal warmth. It waits on a blade of grass or a shrub, feels the passing warmth, and drops. There is a feeling now, though, that more than heat is involved. It has been suggested that a preferred host exudes a scent, a chemical force of some sort, that incites the tick to drop. Maybe some of us are more attractive to ticks than others. My old dog has had Lyme disease three times. I've treated her just the way I treat my other patients. Of course, dogs range. Proximity to the tick is everything. I've had several cases of Lyme disease in elderly women—old ladies who never got any closer to nature than the front porch. They puzzled me for quite some time. Then I finally figured it out. They all had cats. The cats, being cats, ranged. And cats are mousers. My feeling is that those cats did their mousing at the source—at the burrow. Then they came back home and jumped up on an unsuspecting lap."

I had been thinking about *I. dammini's* principal host and mating place—the deer. I wondered if the deer might be a crucial factor in any attempt at control. I wondered if the elimination of the deer here on Long Island, or even just on Shelter Island, would break the cycle and abort the disease. Dr. Grunwaldt thought for a moment, and shook his head. "I doubt it," he said. "I think the tick would probably find another host. And, besides, I can't see much public support for a deer-eradication program."

I knew what he meant: *What? Bambi?*

I know of no one—friend or acquaintance or neighbor—in my part of Long Island who has had Rocky Mountain spotted fever. Nor do I know anyone who has had babesiosis. Lyme disease is a different matter, and this is not surprising. Dr. Grunwaldt told me that he had seen only one case of Rocky Mountain spotted fever in his thirteen years of practice on Long Island, and only twenty cases of babesiosis. But he has seen and treated at least four hundred cases of Lyme disease. I have a number of friends who have suffered its protean rigors. One of them, and one of the local pioneers in this morbid respect, is a woman named Priscilla Bowden (Mrs. Jeffrey Potter), an artist, an amateur flutist, and a knowledgeable gardener. Miss Bowden, a slim, dark-haired woman in her forties, lives with her husband on a verdant acre in the village of East Hampton, with expanses of lawn and many shade trees and flowering shrubs in a pleasant surround of woods. They have—or had at the time she took sick—a small brown dog and a large white cat. Her illness, she told me in a talk we had at her home, had its beginnings just after the Fourth of July weekend in the summer of 1982.

"It was July 7," she said, glancing at a sheaf of papers. "A

Wednesday—I keep a diary. It's not a 'Dear Diary' diary, it's just a kind of social record, but if something interesting happens I make a note of that, too. Well, I'd been feeling mean for a couple of days. Not actually sick, but just not feeling well. I blamed it on the big weekend of the Fourth. Summer weekends here are always a strain. Too many parties. So I dragged around, and then, all of a sudden, it struck. It started with a headache—a horrible, terrible headache. And I felt burning up. I took my temperature. It was 103.5 degrees. That was around midafternoon. At around seven, I took it again: 104 degrees. That was pretty frightening. I went to bed and spent a miserable night. I was still burning up the next morning. Jeffrey called Dr. Medler—Raymond Medler, in East Hampton. We were given his first available appointment: one o'clock. When I got there, I didn't have to say much. Dr. Medler took my temperature. It was 104.6 degrees. He said he didn't know what was wrong with me, but with a fever like that the only place for me was the hospital. And right away. I said can't I even go home and get my toothbrush? He said no, that I should go straight to the hospital—Southampton Hospital. So I went.

"I was fortunate enough to get a private room. My headache was horrible. They got me into bed, and Dr. Medler arrived and started me on aspirin. It was amazing. In a couple of hours, my fever was down to 99.9 degrees, and my headache had practically vanished. I felt well enough to ask for some paper and a pen; and I started making these notes. Then the aspirin wore off and my headache came back and my fever went up to 102.5 degrees. They seemed to go together. First the fever, then the headache. Or maybe it was the other way around. My memory is a little hazy about those next few days. I made my notes, but the details are a little dim. Dr. Medler and the nurses kept asking me if I remembered having a tick bite. I said no—not as far as I knew. It was somewhere around that time that they began to speak of Lyme disease. I don't think I'd ever even heard of Lyme disease. I wrote that down in my notes, only I wrote 'l-i-m-e,' and 'arthritis.' That's what they called it then. They asked me about a rash. I hadn't noticed any kind of rash, and I didn't have a rash then. Not at that moment. That was Thursday and Friday and Saturday. My fever went up and down, and so did the headache. Then, on Sunday, there it was—a rash. It was a circular red rash, about three or four inches in diameter. It was on my leg, my thigh. Then I saw two more circles, also on my leg. One of the nurses found another one, on my back. So now it was established. I had 'lime arthritis.'

"On Monday morning, July 12, a man walked into my room carrying a camera and a black bag. He stopped at the foot of the bed, and said, 'I can tell you one thing that's the matter.' I stared at him, and he said, 'Your kimono is not tied the right way. I've been to Japan, and the Japanese are very careful about whether the fold is to the left or the right.' My way—I forget which way it was—meant I was moribund or something. Well, that was my introduction to Dr. Berger—Bernard W. Berger, a dermatologist and, I later learned, an expert on Lyme disease. Dr. Medler had called him in as a consultant. Anyway, he made me laugh with his kimono joke. We talked, and I told him my sad story, and he opened up his camera and took some photographs of my rash. He told me what there was to tell about Lyme disease, and put me on a course of penicillin. Maybe I was already on it; I don't remember. One thing he told me was that in four weeks or so I might have a strange reaction. We talked about the tick bite. He said it could have been a few days before I got sick, or it could have been two weeks. There were certainly plenty of opportunities for me to pick up a tick—the garden, the shrubbery, the woods, the cat. I felt much better after talking with Dr. Berger. I asked Jeffrey to bring me some drawing paper, and I did a little drawing—a self-portrait of me sitting up in bed. There's a color print of it on the wall over there."

The picture was almost a miniature. It was not much bigger than a copy of *TV Guide*. I had to go over to get a good look. It showed a dark-haired woman (more hair than face) sitting up in a raised hospital bed. She was wearing a red kimono—a red garment, anyway. A blue-and-white blanket (in July!) was pulled up in pleasing disarray to her waist. There was a suggestion of yellow flowers in a vase in the background. Just beyond her head was the grim shepherd's-crook shaft of an I.V. machine.

"Don't ask me about that I.V.," Miss Bowden said. "I must have had one at some point, but I simply don't remember. The blanket? Well, this is a picture. It's not a scientific illustration. Maybe I was having a chill. That doesn't seem likely, though, does it? I mean, I was discharged the next day. That was Tuesday, July 13—my sixth hospital day. I wasn't exactly well. I was still on oral penicillin. My temperature when I got home in the early afternoon was 98.8 degrees. But at seven o'clock that night it was up to 102.2 degrees. Lyme disease is very strange. I woke up around midnight, and took my temperature again. It was down again—100.2 degrees. The next day, it was normal all day. And so was I. I was well."

Miss Bowden squared her litter of notes and turned a page. "Now," she said, "it's August 27. A Friday. I'd been feeling pretty mean off and on for several weeks. I took a nap every afternoon, so I could go out at night and join the fun. But then on that Friday I began running a little fever, and I didn't seem to have any strength. All my joints and muscles were stiff. The next day, I felt even meaner, and my temperature was 99.6 degrees. By Sunday, I knew I was sick, and I spent a horrible night. It hurt to lie down. There seemed to be weights on my head and chest. I telephoned Dr. Medler on Monday afternoon, and he put me on aspirin. That may have helped some. But my stiffness turned to pain. Particularly in my back, neck, and shoulders. I've never been poked with an electric cattle prod, but I'm sure it would feel exactly like that: a stab of pain, then nothing for a moment or two, then another stab, in a different place. It jumped all over—back, neck, chest, arms, head. Then, maybe because of those jumping pains, I began to feel nauseated. And I developed diarrhea. I called Dr. Medler again the next day, but he was away—his father had died. I talked to an assistant, who suggested that I switch from aspirin to Tylenol. Which I did. I couldn't see that it made much difference. The shooting pains kept shooting, and in between prods I had that pins-and-needles feeling running up and down my hands and feet. I also had a very funny feeling here and there on my midriff and on my arms and cheeks—as if I were being splashed with cold water. About the only misery missing was fever. My temperature stayed at normal. It was very peculiar.

"Dr. Berger had told me that an attack of Lyme disease often had a sequel of some sort. It could be an arthritis. It could involve the heart. It could be neurological. I supposed that I was experiencing one of those sequels, and that mine was taking a neurological form. I don't remember when I began to realize that. I'm confused about a lot of things that aren't set down in my notes. I was too sick to talk to the doctor. Jeffrey did all that for me. At some point, I was put back on penicillin. Around September 1, I began to be really nauseated. I couldn't keep anything on my stomach. I even had trouble with pills. I was vomiting, or trying to, every ten or fifteen minutes. But some of the pains were easing up a bit. I gather from my notes that the next couple of days were much the same. The days weren't as bad as the nights—trying to get comfortable enough to sleep. Then, on Friday, September 3, something new and awful happened. The left side of my mouth began to feel stiff, paralyzed. I remember trying to wash out my mouth with mouthwash, and my mouth wouldn't quite close. The mouthwash just dribbled out. I woke up the next morning, the beginning of the big Labor Day weekend, with the right side of my mouth partly paralyzed. It was almost impossible to eat or drink. I was drooling. And at some point I looked in the bathroom mirror, and I had Chinese eyes. The muscles around the corners of my eyes had gone limp, like my mouth. I learned later that there's a name for all that. I had Bell's palsy. That was the worst— that was the peak. Then, little by little, I began to improve. I had a consultation with an otolaryngologist in Southampton. I had a phone call from Dr. Jorge Benach, an important Lyme-disease investigator at Stony Brook. The days went by. By the middle of the month, I was almost back to normal. But even then, when I really felt well again, I wasn't. It was two months before I could play the flute properly. For a long time, all I could manage was the middle register. It was just a horrible experience. It changed my life. I don't know where I got my tick, but I never go into the woods anymore, I'm careful about petting any animal, and I've given up gardening. I'm going to exercise class, something I never dreamed of doing, but I have to get my strength back. And I never go out in the sun without dark glasses. It seems I have photophobia."

Jeffrey Potter had come into the room while we were talking. He stood just inside a pair of French doors opening on a terrace. He cleared his throat. "I suppose this sounds a bit silly," he said. "But there was a time, when Priscilla was at her worst, when I had an awful thought: We had never talked about death as something that might happen to one of us. I realized that I didn't know if she wanted a burial or cremation. I didn't know if she wanted a funeral, or even a church service. Or a memorial service. Or whatever. I know this sounds morbid and silly. But I thought I'd just mention it. It gives you some idea of how sick I thought she was."

Driving home from the Potters', I thought back to my earlier conversation with Dr. Grunwaldt. The telephone had rung three or four times in the course of our talk. The calls were all from patients, and I gathered from Dr. Grunwaldt's comments that they all had to do with ticks. "Yes," he said, as I was leaving, as he walked me to the door. "There's a lot of concern around here. Not everybody who finds a tick has Lyme disease. But they have the possibility on their minds. So do I. Not everybody I see who has a high fever, with the general symptoms of the flu, has Lyme disease. But when I see those symptoms here in the tick season—in the summer and early fall—I don't think flu. Lyme disease is essentially easily cured if it is properly treated early enough. Even then, of course, it can be very unpleasant. But those of us who practice here on the East End have an advantage over doctors outside the endemic areas. We have a high level of awareness. Of course, all doctors everywhere are more alert to Lyme disease than they once were. But if you were to come down with Lyme disease, I would hope it happened here. The unlucky ones are the ones who come out here for a weekend or a week or so and then go back home to the city and get sick."

[*1988*]

* * *

AUTHOR'S NOTE: A report published in the June 14, 1990, issue of *The New England Journal of Medicine* suggests that the avoidance of ticks or tick country may not be enough to avoid exposure to Lyme disease. Dr. Steven W. Luger of Lyme, Connecticut, the home place of Lyme disease, offers both laboratory and anecdotal clinical evidence that biting flies—deerflies and horseflies—may carry the *Borrelia burgdorferi* spirochete and transmit it to human beings. He does, however, offer a word of consolation: "In contrast to the painless bite of *I. dammini,* the bite of flies is painful and not likely to be overlooked."

Index